Springer Monographs in Mathematics

More information about this series at
http://www.springer.com/series/3733

Askold Khovanskii

Topological Galois Theory

Solvability and Unsolvability of Equations in Finite Terms

Springer

Askold Khovanskii
Department of Mathematics
University of Toronto
Toronto, Ontario
Canada

Translators:
V. Timorin and V. Kirichenko: Chapters 1–7.
Lucy Kadets: Appendices A and B.

Appendices C and D were written jointly with Yura Burda.

Based on Russian edition entitled "Topologicheskaya Teoriya Galua, Razreshimost i nerazreshimost uravnenii v konechnom vide", published by MCCME, Moscow, Russia, 2008.

ISSN 1439-7382 ISSN 2196-9922 (electronic)
Springer Monographs in Mathematics
ISBN 978-3-642-38870-5 ISBN 978-3-642-38871-2 (eBook)
DOI 10.1007/978-3-642-38871-2
Springer Heidelberg New York Dordrecht London

Library of Congress Control Number: 2014952224

Mathematics Subject Classification (2010): 55-02, 34M15, 32Q55, 12F10, 30F10

Printed on acid-free paper

Springer is part of Springer Science+Business Media (www.springer.com)

To the memory of Vladimir Igorevich Arnold

Preface

Numerous unsuccessful attempts to solve certain algebraic and differential equations "in finite terms" (i.e., "explicitly") led mathematicians to the belief that explicit solutions of such equations simply do not exist. This book is devoted to the question of the unsolvability of equations in finite terms, and in particular to the topological obstructions to solvability. This question has a rich history.

The first proofs of the unsolvability of algebraic equations by radicals were given by Abel and Galois. While thinking about the problem of explicit indefinite integration of an algebraic differential form, Abel laid the foundations for the theory of algebraic curves. Liouville continued Abel's work and proved that indefinite integrals of many algebraic and elementary differential forms are not elementary functions. Liouville was also the first to prove the unsolvability by quadratures of many linear differential equations.

It was Galois who first saw that the question of solvability by radicals is related to the properties of a certain finite group (now called the Galois group of an algebraic equation). Indeed, the notion of a finite group as introduced by Galois was motivated exactly by this question. Sophus Lie introduced the notion of a continuous transformation group while trying to solve differential equations explicitly by reducing them to a simpler form. To each linear differential equation, Picard associated its Galois group, which is a Lie group (and moreover, a linear algebraic group). Picard and Vessiot then showed that this particular group is responsible for the solvability of equations by quadratures. Next, Kolchin elaborated the theory of algebraic groups, completed the development of Picard–Vessiot theory, and generalized it to the case of holonomic systems of linear partial differential equations.

Vladimir Igorevich Arnold discovered that many classical questions in mathematics are unsolvable for topological reasons. In particular, he showed that a generic algebraic equation of degree 5 or higher is unsolvable by radicals precisely for topological reasons. Developing Arnold's approach, I constructed in the early 1970s a one-dimensional version of topological Galois theory. According to this theory, the way the Riemann surface of an analytic function covers the plane of complex numbers can obstruct the representability of this function by explicit formulas. The strongest known results on the unexpressibility of functions by explicit formulas have been obtained in this way. I had always been under the impression that a full-fledged multidimensional version of this theory was impossible. Then in spring 1999, I suddenly realized that, in fact, one can generalize the one-dimensional version of topological Galois theory to the multivariable case.

This book covers topological Galois theory. First, a complete and detailed exposition of the one-dimensional version is given, followed by a more schematic exposition of the multidimensional version. The topological theory is closely related to usual (algebraic) Galois theory as well as to differential Galois theory.

Algebraic Galois theory is simple, and its main ideas are connected with topological Galois theory. In the "permissive" part of topological Galois theory, not only is linear algebra used, but also results from Galois theory. In this book, Galois theory and its applications to the solvability of algebraic equations by radicals are presented with complete proofs. Apart from the problem of solvability by radicals, other closely related problems are also considered, including the problem of solvability of an equation with the help of radicals and auxiliary equations of degree at most k.

The main theorems of Picard–Vessiot theory are stated without proof, and the similarity with Galois theory is emphasized. We shall explain why, at least in principle, Picard–Vessiot theory answers the questions of solvability of linear differential equations in explicit form. The "permissive" part of topological Galois theory (which proves, in particular, that linear Fuchsian equations with solvable monodromy group are solvable by quadratures) uses only the simple, linear-algebraic, part of Picard–Vessiot theory. This linear algebra is covered in the book. The "prohibitive" part of topological Galois theory (which says, in particular, that linear differential equations with unsolvable monodromy group are not solvable by quadratures) will be explained in full detail. It is stronger than the "prohibitive" part of Picard–Vessiot theory.

This book also discusses beautiful constructions, due to Liouville, of the class of elementary functions, the class of functions expressible by quadratures, and so on, and his theory of elementary functions, which had a strong impact on all subsequent work in this area.

We will discuss three versions of Galois theory—algebraic, differential, and topological. These versions are unified by the same group-theoretic approach to the problems of solvability and unsolvability of equations. However, it is not true that all results on solvability and unsolvability are related to group theory. A number of brilliant results based on a different approach are contained in the theory of Liouville. To give a flavor of Liouville's theory, we provide a complete proof of his theorem stating that certain indefinite integrals are not elementary functions (this includes indefinite integrals of nonzero holomorphic differential forms on algebraic curves of higher genus).

We do not always follow the historical sequence of events. For example, the Picard–Vessiot theorem on the solvability of linear differential equations by quadratures was proved before the main theorem of differential Galois theory.

However, the Picard–Vessiot theorem is a direct corollary of this fundamental theorem, and it is presented here in this way.

A few words about the bibliography: The first modern book on integration in finite terms was written by Ritt [86]. Bronstein's book [16] contains a modern treatment of the subject together with many algorithms and includes much of what is in Sects. 1.6–1.9. Algebraic Galois theory is explained well in many textbooks; see, for example, [24, 25]. A clear and concise exposition of differential Galois theory is contained in Kaplansky's book [43]. For a more detailed and modern treatment, see the book [96] by van den Put and Singer. Kolchin's theory is covered in [64–67]. An interesting survey of work on the solvability and unsolvability of equations together with an extensive bibliography can be found in [93, 94].

My first results in topological Galois theory appeared in the early 1970s, when I was Arnold's student, to whom I am greatly indebted. Unfortunately, I did not publish my results in a timely manner: At first, I was unable to reconstruct the complicated history of the subject, and then I became interested in a totally different kind of mathematics. Much later, Andrei Bolibrukh convinced me to revisit the subject. My wife, Tatiana Belokrintskaya, prepared the Russian edition of this book for publication.

In this English-language edition, extra material has been added (Appendices A–D), the last two of which were written jointly with Yuri Burda. Vladlen Timorin and Valentina Kirichenko translated the Russian text into English. Michael Singer read the book and made many useful remarks and suggestions. David Kramer performed a careful editing of the book. I am grateful to all of them.

Toronto, Canada Askold Khovanskii

Contents

Chapter 1
Construction of Liouvillian Classes of Functions and Liouville's Theory

Some algebraic and differential equations are explicitly solvable. What does this mean? If an explicit solution is presented, the question answers itself. However, in most cases, every attempt to solve an equation explicitly is doomed to failure. We are then tempted to prove that certain equations have no explicit solutions. It is now necessary to define exactly what we mean by explicit solutions (otherwise, it is unclear what we are trying to prove).

From the modern viewpoint, the classical works on the subject lack rigorous definitions and statements of theorems. Nonetheless, it is clear that Liouville understood exactly what he was proving. He not only stated the problems on solvability of equations by elementary functions and by quadratures, but he also algebraized them. His work made it possible to define all such notions over an arbitrary differential field. But the standards of mathematical rigor were different in the time of Liouville. Indeed, according to Kolchin [64], even Picard failed to give accurate, unambiguous definitions. Kolchin's work satisfies modern standards, but his definitions are given for abstract differential fields from the very beginning.

However, the indefinite integral of an elementary function and the solution of a linear differential equation are functions rather than elements of an abstract differential field. In function spaces, for example, apart from differentiation and algebraic operations, an absolutely nonalgebraic operation is defined, namely composition. Anyhow, function spaces provide greater means for writing "explicit formulas" than abstract differential fields. Moreover, we should take into account that functions can be multivalued, can have singularities, and so on.

In function spaces, it is not hard to formalize the problem of unsolvability of equations in explicit form, and in this book, we are interested in this particular problem. One can proceed as follows: fix a class of functions and say that an equation is solvable explicitly if its solution belongs to this class. Different classes of functions correspond to different notions of solvability.

A. Khovanskii, *Topological Galois Theory*, Springer Monographs in Mathematics,
DOI 10.1007/978-3-642-38871-2_1

1.1 Defining Classes of Functions by Lists of Basic Functions and Admissible Operations

A class of functions can be introduced by specifying a list of *basic functions* and a list of *admissible operations*. Given these two lists, the class of *functions* is defined as the set of all functions that can be obtained from the basic functions by repeated application of admissible operations. In Sect. 1.2, we define Liouvillian classes of functions in exactly this way.

Liouvillian classes of functions, which appear in problems of solvability in finite terms, contain multivalued functions. Thus the basic terminology should be made clear. In this section, we work with multivalued functions "globally," which leads to a more general understanding of classes of functions defined by lists of basic functions and admissible operations. In this global version, a multivalued function is regarded as a single entity, and we can define *operations on multivalued functions*.

The result of such an operation is a set of multivalued functions; every element of this set is referred to as a function obtained from the given functions by the given operation. The *class of functions* is defined as the set of all (multivalued) functions that can be obtained from the basic functions by repeated application of admissible operations.[1]

Let us define, for example, the sum of two multivalued functions of one variable.

Definition 1.1 Take an arbitrary point a on the complex line, a germ f_a of an analytic function f at the point a, and a germ g_a of an analytic function g at the same point a. We say that the multivalued function φ generated by the germ $\varphi_a = f_a + g_a$ is *representable as the sum of the functions f and g.*

For example, it is easy to see that exactly two functions are representable in the form $\sqrt{x} + \sqrt{x}$, namely, $f_1 = 2\sqrt{x}$ and $f_2 \equiv 0$. Other operations on multivalued functions are defined in exactly the same way. *For a class of multivalued functions, being stable under addition means that together with any pair of its functions, this class contains all functions representable as their sum.* The same applies to all other operations on multivalued functions understood in the same sense as above.

In the definition given above, it is not only the operation of addition that plays a key role but also the operation of analytic continuation hidden in the notion

[1] If f and g are multivalued functions and $\wedge$ is, say, a binary operation, then $f \wedge g$ is a set of multivalued functions. The class defined by a list $\{f_1, \dots, f_n\}$ of basic functions and a list $\{\wedge_1, \dots, \wedge_m\}$ of admissible binary operations is, by definition, the minimal set $\mathscr{C}$ of functions such that all $f_i \in \mathscr{C}$ and $f \wedge_j g \subseteq \mathscr{C}$ whenever $f, g \in \mathscr{C}$. An obvious modification can be made to include infinite sets of basic functions and admissible functions, such as unary, ternary, etc., operations.

of multivalued function. Indeed, consider the following example. Let f_1 be an analytic function defined on an open subset U of the complex line $\mathbb{C}^1$ and admitting no analytic continuation outside of U, and let f_2 be an analytic function on U given by the formula $f_2 = -f_1$. According to our definition, the zero function is representable in the form $f_1 + f_2$ *on the entire complex line*. From the commonly accepted viewpoint, the equality $f_1 + f_2 = 0$ holds inside the region U but not outside.

In working with multivalued functions globally, we do not insist on the existence of a *common region* where all necessary operations would be performed on single-valued branches of multivalued functions. A first operation can be performed in a first region, then a second operation can be performed in a second, different, region on analytic continuations of functions obtained in the first step. In essence, this more general understanding of operations is equivalent to including analytic continuation in the list of admissible operations on analytic germs. For functions of a single variable, it is possible to obtain topological obstructions even with this more general understanding of operations on multivalued analytic functions.

In the sequel, in considering topological obstructions to the membership of an analytic function of a single variable in a certain class, we will always mean this global definition of the function class via lists of basic functions and admissible operations.

For functions of several variables, things do not work in this general setting, and we are forced to adopt a more restrictive formulation (see Sect. 7.1.1) dealing with germs of functions. It is, however, no less natural, and perhaps even more so. The only place in the book where we use this more restrictive formulation is Chap. 7, in which we deal with multivariable functions.

1.2 Liouvillian Classes of Functions of a Single Variable

In this section, we define Liouvillian classes of functions of a single variable (for the multivariable case, the corresponding definitions are given in Chap. 7). We will describe these classes by lists of basic functions and admissible operations.

1.2.1 Functions of One Variable Representable by Radicals

List of basic functions:

- All complex constants
- An independent variable x

List of admissible operations:

- Arithmetic operations
- The operation of taking the nth root $\sqrt[n]{f}$, $n = 2, 3, \ldots$, of a given function f

The function $g(x) = \sqrt[3]{5x + 2\sqrt[2]{x}} + \sqrt[7]{x^3 + 3}$ is an example of a function representable by radicals.

The famous problem of solvability of equations by radicals is related to this class. Consider the algebraic equation

$$y^n + r_1 y^{n-1} + \cdots + r_n = 0,$$

in which the r_i are rational functions of one variable. A complete answer to the question of solvability of such equations by radicals is given by Galois theory (see Chap. 2).

To define other classes, we will need the list of *basic elementary functions*. In essence, this list contains functions that are studied in high-school and college precalculus courses. They are the functions frequently found on pocket calculators.

List of basic elementary functions:

1. All complex constants and an independent variable x.
2. The exponential, the logarithm, and the power x^α, where α is any complex constant.
3. The trigonometric functions sine, cosine, tangent, cotangent.
4. The inverse trigonometric functions arcsine, arccosine, arctangent, arccotangent.

Let us now proceed with the list of *classical operations* on functions. We begin the list here. It will be continued in the following section.

List of classical operations:

1. The operation *composition* takes functions f, g to the function $f \circ g$.
2. The *arithmetic operations* take functions f and g to the functions $f + g$, $f - g$, fg, and f/g.
3. The operation *differentiation* takes a function f to the function f'.
4. The operation *integration* takes a function f to its indefinite integral y (i.e., to any function y such that $y' = f$; the function y is determined by the function f up to an additive constant).
5. The operation *solving an algebraic equation* takes functions $f_1, \ldots, f_n$ to the function y such that $y^n + f_1 y^{n-1} + \cdots + f_n = 0$ (the function y is not quite uniquely determined by the functions $f_1, \ldots, f_n$, since an algebraic equation of degree n can have n solutions).

We can now return to the definition of Liouvillian classes of functions of a single variable.

1.2.2 Elementary Functions of One Variable

List of basic functions:

- Basic elementary functions.

List of admissible operations:

- Compositions
- Arithmetic operations
- Differentiation

All elementary functions are given by formulas such as the following:

$$f(x) = \arctan(\exp(\sin x) + \cos x).$$

1.2.3 Functions of One Variable Representable by Quadrature

List of basic functions

- Basic elementary functions

List of admissible operations:

- Composition
- Arithmetic operations
- Differentiation
- Integration

For example, the elliptic integral

$$f(x) = \int_{x_0}^{x} \frac{dt}{\sqrt{P(t)}},$$

where P is a cubic polynomial, is representable by quadratures. However, Liouville showed that if the polynomial P has no multiple roots, then the function f is not elementary.

Generalized elementary functions of one variable This class of functions is defined in the same way as the class of elementary functions. We only need to add the operation of solving algebraic equations to the list of admissible operations.

Functions of one variable representable by generalized quadratures This class of functions is defined in the same way as the class of functions representable by quadratures. We only need to add the operation of solving algebraic equations to the list of admissible operations. Let us now define two more classes of functions similar to Liouvillian classes.

Functions of one variable representable by k-radicals This class of functions is defined in the same way as the class of functions representable by radicals. We only need to add the operation of solving algebraic equations of degree $\leq k$ to the list of admissible operations.

Functions of one variable representable by k-quadratures This class of functions is defined in the same way as the class of functions representable by quadratures. We only need to add the operation of solving algebraic equations of degree at most k to the list of admissible operations.

1.3 A Bit of History

The first rigorous proofs of unsolvability of some equations by quadratures and by elementary functions were obtained in the middle of the nineteenth century by Liouville (see [75–77, 86]). Later work of Chebyshev, Mordukhai-Boltovski, Ostrovskii, Ritt, Risch, Rosenlicht, Davenport, Singer, and Bronstein have elaborated on Liouville's results. A bibliography on this subject can be found in [93].

According to Liouville's theory of elementary functions, "sufficiently simple" equations have either "sufficiently simple" solutions or no explicit solutions at all. In some cases, the results go all the way to algorithms that either provide a proof of unsolvability of an equation in explicit form or construct an explicit solution.

Liouville's theory answers questions such as the following:

1. Under what conditions is an indefinite integral of an elementary function also an elementary function?
2. Under what conditions are all solutions of a linear differential equation all of whose coefficients are rational functions representable by generalized quadratures?

To demonstrate Liouville's method, we will give a proof of his theorem about integrals (see Sect. 1.6) and consider several applications of this theorem. Let $\alpha = R(z, u)\, dz$ be a 1-form, where R is a rational function of two variables, z is a complex variable, and u is a function of z. In Sect. 1.7, we consider the case that u is the natural logarithm of a rational function f of z, that is, $u = \log f(z)$. A procedure will be explained that allows us either to find an indefinite integral of α explicitly or to prove that it is not a generalized elementary function. In Sect. 1.8, a similar result is described in the case that u is the exponential of a rational function f of z, $u = \exp f(z)$. The case of an abelian 1-form α, where u is an algebraic function of z, is considered in Sect. 1.9. Necessary and sufficient conditions for the elementarity of an abelian integral are described. These conditions are hard to verify. In this sense, the algebraic case is more complicated than the logarithmic and the exponential cases. Sections 1.6–1.9 are not necessary for understanding the remainder of the book and can be omitted. To avoid references to these sections, we repeat, in Chap. 3, simple and short computations related to adjoining an integral, an exponential of an integral, and a root of an algebraic equation to a differential field.

In Sect. 1.4, we give significantly simpler definitions of Liouvillian classes of functions, due to Liouville (for example, that of the class of elementary functions). We explain how exactly Liouville succeeded in algebraizing the questions of solvability of equations by elementary functions or by other Liouvillian classes of functions. Liouville extensions of functional differential fields are constructed in Sect. 1.5.

In Sect. 1.10, we state some results from Liouville's theory concerning questions of solvability of linear differential equations. A more complete answer to this question is given by differential Galois theory (see Chap. 3).

1.4 New Definitions of Liouvillian Classes of Functions

Liouville algebraized the problem of solvability by elementary functions and by quadratures. The main obstacle in the algebraization is the absolutely nonalgebraic operation of composition. Liouville circumvented this obstacle in the following way: He associated to every function g from the list of basic functions the operation of postcomposition with this function. This operation takes a function f to the function $g \circ f$. Liouville noted that all basic elementary functions can be reduced to the logarithm and the exponential (see Lemma 1.2 below). The compositions $y = \exp f$ and $z = \log f$ can be regarded as solutions of the equations $y' = f'y$ and $z' = f'/f$. Thus, within Liouvillian classes of functions, it suffices to consider operations of solving some simple differential equations. After that, the solvability problem for Liouvillian classes of functions becomes differential-algebraic, and carries over to abstract differential fields. Let us proceed with the realization of this plan.

We will now continue the list of classical operations (the beginning of the list is given in the previous section).

List of classical operations (continued):

6. The operation *exponentiation* takes a function f to the function $\exp f$.
7. The operation of taking the logarithm, which we shall call *logarithmation*, takes a function f to the function $\log f$.

We will now give new definitions for transcendental Liouvillian classes of functions.

1.4.1 Elementary Functions of One Variable

List of basic functions:

- All complex constants
- An independent variable x

List of admissible operations:

- Exponentiation
- Logarithmation
- Arithmetic operations
- Differentiation

1.4.2 Functions of One Variable Representable by Quadratures

List of basic functions

- All complex constants

List of admissible operations:

- Exponentiation
- Arithmetic operations
- Differentiation
- Integration

1.4.3 Generalized Elementary Functions of One Variable and Functions of One Variable Representable by Generalized Quadratures and k-Quadratures

These functions are defined in the same way as the corresponding nongeneralized classes of functions; we have only to add the operation of solving algebraic equations or the operation of solving algebraic equations of degree $\leq k$ to the list of admissible operations.

Lemma 1.2 *Basic elementary functions can be expressed through exponentials and logarithms with the help of complex constants, arithmetic operations, and compositions.*

Proof For a power function x^α, the required expression is given by the equality $x^\alpha = \exp(\alpha \log x)$. For the trigonometric functions, the required expressions follow from Euler's formula $e^{a+bi} = e^a(\cos b + i \sin b)$. For real values of x, we have

$$\sin x = \frac{1}{2i}\left(e^{ix} - e^{-ix}\right) \quad \text{and} \quad \cos x = \frac{1}{2}\left(e^{ix} + e^{-ix}\right).$$

By analyticity, the same formulas remain true for all complex values of x. The tangent and the cotangent functions are expressed through the sine and the cosine. Let us now show that for all real x, the equality

$$\arctan x = \frac{1}{2i} \log z$$

holds, where

$$z = \frac{1 + ix}{1 - ix}.$$

Obviously,

$$|z| = 1, \quad \arg z = 2 \arg(1 + ix), \quad \tan(\arg(1 + ix)) = x,$$

which proves the desired equality. By analyticity, the same equality also holds for all complex values of x. The remaining inverse trigonometric functions can be expressed through the arctangent. Namely,

$$\operatorname{arccot} x = \frac{\pi}{2} - \arctan x, \quad \arcsin x = \arctan \frac{x}{\sqrt{1 - x^2}}, \quad \arccos = \frac{\pi}{2} - \arcsin x.$$

The square root that appears in the expression for the function arcsin can be expressed through the exponential and the logarithm: $x^{1/2} = \exp\left(\frac{1}{2} \log x\right)$. The lemma is proved. □

Theorem 1.3 *For every transcendental Liouvillian class of functions, the definitions in this section and those in Sect. 1.2 are equivalent.*

Proof In one direction, the theorem is obvious: it is clear that every function belonging to some Liouvillian class of functions in the sense of the new definition belongs to the same class in the sense of the old definition.

Let us prove the converse. By Lemma 1.2, the basic elementary functions lie in the class of elementary functions and in the class of generalized elementary functions in the sense of the new definition. It follows from the same lemma that the classes of functions representable by quadratures, generalized quadratures, and k-quadratures in the sense of the new definition also contain the basic elementary functions. Indeed, the independent variable x belongs to these classes, since it can be obtained as the integral of the constant function 1, since $x' = 1$. Instead of taking the logarithm, which is not among the admissible operations in these classes, one can use integration, since $(\log f)' = f'/f$.

It remains to show that the Liouvillian classes of functions in the sense of the new definition are stable under composition. The reason that they are is the following: composition commutes with all other operations that appear in the new definition of function classes, except for differentiation and integration. Thus, for example, the result of the operation exp applied to the composition $g \circ f$ coincides with the composition of the functions $\exp g$ and f, i.e., $\exp(g \circ f) = (\exp g) \circ f$. Similarly,

$$\log(g \circ f) = (\log g) \circ f,$$

$$(g_1 \pm g_2) \circ f = (g_1 \circ f) \pm (g_2 \circ f),$$
$$(g_1 g_2) \circ f = (g_1 \circ f)(g_2 \circ f),$$
$$(g_1/g_2) \circ f = (g_1 \circ f)/(g_2 \circ f).$$

If a function y satisfies an equation of the form $y^n + g_1 y^{n-1} + \cdots + g_n = 0$, then the function $(y \circ f)$ satisfies the equation $(y \circ f)^n + (g_1 \circ f)(y \circ f)^{n-1} + \cdots + (g_n \circ f) = 0$.

For differentiation and integration, we have the following simple commutation relations with the operation of composition: $(g)' \circ f = (g \circ f)'(f')^{-1}$ (if a function f is constant, then the function $(g)' \circ f$ is also constant), and if y is an indefinite integral of a function g, then $y \circ f$ is an indefinite integral of the function $(g \circ f)f'$ (in other words, composing the integral of a function g with a function f corresponds to the integration of the function $g \circ f$ multiplied by the function f').

This implies that the Liouvillian classes in the sense of the new definition are stable under composition. Indeed, if a function g is obtained from constants (or from constants and the independent variable) by operations discussed above, then the function $g \circ f$ is obtained by applying the same operations, or almost the same as in the case of integration and differentiation, to the function f. The theorem is proved. □

Remark 1.4 It is easy to see that differentiation can also be excluded from the lists of admissible operations for the Liouvillian classes of functions. To prove this, it suffices to use the explicit computation for the derivatives of the exponential and the logarithmic functions and the rules for differentiating formulas containing compositions and arithmetic operations. However, the exclusion of differentiation does not help in the problem of solvability of equations in finite terms (sometimes, the exclusion of differentiation makes it possible to state a result in a more invariant form; see the second formulation of Liouville's theorem on abelian integrals from Sect. 1.9).

1.5 Liouville Extensions of Abstract and Functional Differential Fields

A field K is said to be a *differential field* if an additive map $a \mapsto a'$ is defined that satisfies the Leibniz rule $(ab)' = a'b + ab'$. Such a map $a \mapsto a'$ is called a *derivation*. If a particular derivation is fixed, the element a' is sometimes called the *derivative* of a. The operation of taking derivatives is called *differentiation*.

An element y of a differential field K is called a *constant* if $y' = 0$. All constants in a differential field form a subfield, which is called the *field of constants*. In all cases that are of interest to us, the field of constants is the field of complex numbers. We shall always assume in the sequel that the differential field has characteristic zero and an algebraically closed field of constants.

An element y of a differential field is said to be

- An *exponential* of an element a if $y' = a'y$
- An *exponential of integral* of an element a if $y' = ay$ (we use "exponential of integral" as an indivisible term)
- A *logarithm* of an element a if $y' = a'/a$
- An *integral* of an element a if $y' = a$

In each of these cases, y is defined only up to an additive or multiplicative constant.

Suppose that a differential field K and a set M lie in some differential field F. The *adjunction* of the set M to the differential field K is the minimal differential field $K\langle M\rangle$ containing both the field K and the set M. We will refer to the transition from K to $K\langle M\rangle$ as *adjoining* the set M to the field K.

A differential field F containing a differential field K and having the same field of constants is said to be an *elementary extension* of the field K if there exists a chain of differential fields $K = F_1 \subseteq \cdots \subseteq F_n = F$ such that for every $i = 1, \ldots, n-1$, the field $F_{i+1} = F_i\langle x_i\rangle$ is obtained by adjoining an element x_i to the field F_i, and x_i is an exponential or a logarithm of some element a_i from the field F_i. An element $a \in F$ is said to be *elementary* over K, $K \subset F$, if it is contained in some elementary extension of the field K.

A *generalized elementary extension*, a *Liouville extension*, a *generalized Liouville extension*, and a *k-Liouville extension* of a field K are defined in a similar way. In the construction of generalized elementary extensions, one is allowed to adjoin exponentials and logarithms and to take algebraic extensions. In the construction of Liouville extensions, one is allowed to adjoin integrals and exponentials of integrals. In generalized Liouville extensions and k-Liouville extensions, one is also allowed to take algebraic extensions and to adjoin solutions of algebraic equations of degree at most k. An element $a \in F$ is said to be *generalized elementary* (*representable by quadratures, by generalized quadratures, by k-quadratures*) over K, $K \subset F$, if a is contained in some generalized elementary extension (Liouville extension, generalized Liouville extension, k-Liouville extension) of the field K.

Remark 1.5 The equation for an exponential of integral is simpler than the equation for an exponential. That is why in the definition of Liouville extensions, etc., we adjoin exponentials of integrals. Instead, we could adjoin exponentials and integrals separately.

Let us now turn to functional differential fields. We will be dealing with this particular type of field in this book (although some results can be easily extended to abstract differential fields).

Let K be a subfield in the field of all meromorphic functions on a connected domain U of the Riemann sphere. Suppose that K contains all complex constants and is stable under differentiation (i.e., if $f \in K$, then $f' \in K$). Then K provides an example of a functional differential field. Let us now give a general definition. Let V, v be a pair consisting of a connected Riemann surface V and a meromorphic vector field v defined on it. The Lie derivative L_v along the vector field v acts on the field F of all meromorphic functions on the surface V and defines the derivation

$f' = L_v f$ in this field. A *functional differential field* is any differential subfield of F containing all complex constants.

It is sometimes more convenient to use another definition of the derivation in a field of functions, which uses a meromorphic 1-form α instead of a meromorphic vector field. The derivative f' of a function f can be defined by the following formula: $f' = df/\alpha$ (the ratio of two meromorphic 1-forms is a well-defined meromorphic function). The derivation we just described is the Lie derivative L_v along the vector field v connected with the form α as follows: the value of α on v is identically equal to one.

The following construction helps to *extend* functional differential fields. Let K be a subfield in the field of meromorphic functions on a connected Riemann surface V equipped with a meromorphic form α, and suppose that the subfield is invariant under the derivation $f' = df/\alpha$, (i.e., if $f \in K$, then $f' \in K$). Consider any connected Riemann surface W together with a nonconstant analytic map $\pi : W \to V$. Fix the form $\beta = \pi^*\alpha$ on W. The differential field F of all meromorphic functions on W with the differentiation $\varphi' = d\varphi/\beta$ contains the differential subfield $\pi^* K$ consisting of functions of the form $\pi^* f$, where $f \in K$. The differential field $\pi^* K$ is isomorphic to the differential field K, and it lies in the differential field F. For a suitable choice of the surface W, an extension of the field $\pi^* K$ that is isomorphic to K can be done within the field F.

Suppose that we need to extend the field K, say by an integral y of some function $f \in K$. This can be done in the following way. Consider the covering of the Riemann surface V by the Riemann surface W of an indefinite integral y of the form $f\alpha$. By the very definition of the Riemann surface W, there exists a natural projection $\pi : W \to V$, and the function y is a single-valued meromorphic function on the surface W. The differential field F of meromorphic functions on W with the differentiation $\varphi' = d\varphi/\pi^*\alpha$ contains the element y as well as the field $\pi^* K$ isomorphic to K. That is why the extension $\pi^* K\langle y\rangle$ is well defined as a subfield of the differential field F. We mean this particular construction of the extension whenever we talk about extensions of functional differential fields.

The same construction allows us to adjoin a logarithm, an exponential, an integral, or an exponential of integral of any function f from a functional differential field K to K. Similarly, for any functions $f_1, \dots, f_n \in K$, one can adjoin a solution y of an algebraic equation $y^n + f_1 y^{n-1} + \dots + f_n = 0$ or all the solutions $y_1, \dots, y_n$ of this equation to K (the adjunction of all the solutions $y_1, \dots, y_n$ can be implemented on the Riemann surface of the vector function $\mathbf{y} = y_1, \dots, y_n$). In the same way, for any functions $f_1, \dots, f_{n+1} \in K$, one can adjoin the n-dimensional $\mathbb{C}$-affine space of all solutions of the linear differential equation $y^{(n)} + f_1 y^{(n-1)} + \dots + f_{n}y + f_{n+1} = 0$ to K. (Recall that a germ of any solution of this linear differential equation admits an analytic continuation along a path on the surface V not passing through the poles of the functions $f_1, \dots, f_{n+1}$.)

Thus, all the above-mentioned extensions of functional differential fields can be implemented without leaving the class of functional differential fields. When we talk about extensions of functional differential fields, we always mean this particular procedure.

The differential field of all complex constants and the differential field of all rational functions of one variable can be regarded as differential fields of functions defined on the Riemann sphere.

Let us restate Theorem 1.3 using definitions from abstract differential algebra and the construction of extensions of functional differential fields.

Theorem 1.6 *A function of one complex variable (possibly multivalued) belongs to:*

1. *The class of elementary functions if and only if it belongs to some elementary extension of the field of all rational functions of one variable;*
2. *The class of generalized elementary functions if and only if it belongs to some generalized elementary extension of the field of rational functions;*
3. *The class of functions representable by quadratures if and only if it belongs to some Liouville extension of the field of all complex constants;*
4. *The class of functions representable by k-quadratures if and only if it belongs to some k-Liouville extension of the field of all complex constants;*
5. *The class of functions representable by generalized quadratures if and only if it belongs to a generalized Liouville extension of the field of all complex constants.*

1.6 Integration of Elementary Functions

Elementary functions are easy to differentiate but hard to integrate. As Liouville proved, an indefinite integral of an elementary function is usually not an elementary function.

Theorem 1.7 (Liouville's theorem on integrals) *An indefinite integral y of a function f lying in a functional differential field K belongs to some generalized elementary extension of that field if and only if the integral is representable in the form*

$$y(x) = \int_{x_0}^{x} f(t)\,dt = A_0(x) + \sum_{i=1}^{n} \lambda_i \log A_i(x), \qquad (1.1)$$

where A_i are functions in the field K for $i = 0, \ldots, n$.

Theorem 1.7 implies the following corollary.

Corollary 1.8 *An indefinite integral y of a generalized elementary function f is a generalized elementary function if and only if it is representable in the form*

$$y(x) = A_0(x) + \sum_{i=1}^{n} \lambda_i \log A_i(x),$$

where A_i are rational functions of f and of its derivatives with complex coefficients for $i = 0, \dots, n$.

To deduce Corollary 1.8 from Theorem 1.7, we set K to be the minimal functional differential field containing all complex constants and the function f. Then every element of K is a rational function of f and its derivatives with complex coefficients.

A priori, the integral of an elementary function f could be a very complicated elementary function. Liouville's theorem shows that nothing like that can happen. The integral of an elementary function is either nonelementary or of the simple form described in the corollary. Liouville's theorem is a prominent result on solvability and unsolvability of equations in explicit form not connected with group theory. In Sect. 1.6, we give a complete proof of this theorem.

In differential terms, condition (1.1) in Liouville's theorem means that the element $f \in K$ is representable in the form

$$f = A_0' + \sum_{i=1}^{n} \lambda_i \frac{A_i'}{A_i}, \tag{1.2}$$

where the A_i are elements of the field K for $i = 0, \dots, n$. There is an analogue of Liouville's theorem in abstract differential algebra [87]. In the statement of the abstract theorem, one takes an abstract differential field for K and uses condition (1.1) in its differential version (1.2).

Liouville's argument (see [86]) was significantly simplified in [16,83,87–89] (see also [82]). With minor modifications, the proof extends to abstract differential fields. However, since we are not interested in abstract differential fields, we do not discuss the existence of certain extensions needed in the proof such as the existence of a differential field containing an integral of a certain element of a given differential field. For functional differential fields, questions of this sort are obvious and have already been considered above (see Sect. 1.5).

For many elementary differential forms, Liouville's theorem allows us either to find an explicit integral or to prove that the integral is not an elementary function. For example, consider a form $\alpha = R(z, u)\, dz$, where R is a rational function of two variables, and u is a function of z. The question whether the integral of α is elementary for the case $u = \log f$, where f is a rational function of z, is discussed in Sect. 1.7. The case $u = \exp f$ and f is a rational function of z is worked out in Sect. 1.8.

Is an algebraic function integrable in explicit form? For a special class of algebraic functions, this question was studied in pioneering work of Abel that laid the foundations of the theory of algebraic curves and abelian integrals. Applying results of topology and of the theory of algebraic curves allows us to give a rather complete explanation of the reasons behind the nonelementarity of abelian integrals (see Sect. 1.9). However, the verification of necessary and sufficient conditions for the elementarity of the abelian integrals from Sect. 1.9 is itself a complicated problem (cf. [27]), one that we do not consider in this book.

A few remarks on terminology: From Sect. 1.6.4 till the end of Sect. 1.8, the term "rational function" may mean one of several things. In order to avoid confusion, we provide here some explanation. We will be dealing with a field F generated over a field K by a single element t that is transcendental over K. It is natural to identify F with the field $K\langle x\rangle$ of rational functions over the field K: each element $g \in F$ is the value at t of a unique rational function $G \in K\langle x\rangle$. We will identify the element $g \in F$ with the function G as well as with its value $G(t)$ at t. We will refer to elements of F as rational functions, and we will perform such operations as decomposing them into partial fractions. The operation of differentiation in F gives rise to a derivation $G \mapsto DG$ on rational functions. The operation D depends on the equation satisfied by the element t: it has the form $DG = (a'/a)G'$ if t is a logarithm of some element $a \in K$, and the form $DG(t) = a'tG'(t)$ if t is an exponential of $a \in K$. In Sects. 1.7 and 1.8, we also encounter the field K of all rational functions of a complex variable. To avoid confusion, in talking about elements of this field, we will emphasize that they are rational functions *of a complex variable* z, and this field K will be denoted by $\mathbb{C}\langle z\rangle$.

1.6.1 Liouville's Theorem: Outline of a Proof

We need to prove that if the derivative of a generalized elementary function[2] y over a field K lies in K, i.e., if $y' \in K$, then the function y is representable in the form (1.1). We can introduce the notion of complexity for generalized elementary functions over a functional differential field K. We say that a function y is a *generalized elementary function of complexity less than or equal to* k *over* K if there exists a chain $K = F_0 \subset F_1 \subset \cdots \subset F_k$ of functional differential fields F_i in which F_i is an extension of the field F_{i-1} for $i = 1, \ldots, k$ obtained by adjoining an exponential, a logarithm, or an algebraic function over the field F_{i-1}.

Theorem 1.7 can be proved by induction. The induction hypothesis $I(m)$ reads: *Liouville's theorem is true for every generalized elementary integral* y *of complexity at most* m *over an arbitrary functional differential field* K. The statement $I(0)$ is obviously true: if the integral y lies in K, then it is representable in the form $y = A_0$, where $A_0 \in K$.

Let y be a generalized elementary integral over K of complexity at most k, i.e., $y' \in K$, $y \in F_k$, and $K = F_0 \subset F_1 \subset \cdots \subset F_k$ is a chain of fields in which every field F_i is obtained from the preceding field F_{i-1} by adjoining an exponential, a logarithm, or an algebraic function over the field F_{i-1}. Since $y' \in F_1$, we can assume by the induction hypothesis $I(k-1)$ that

[2] A generalized elementary function over a functional differential field K is, by definition, an element of a generalized elementary extension of K.

$$y = \sum_{i=1}^{q} \lambda_i \log R_i + R_0, \tag{1.3}$$

where $R_1, \ldots, R_q, R_0 \in F_1$.

The field F_1 is obtained from the field K by adjoining an algebraic element, a logarithm, or an exponential over the field K. These three cases are considered separately below (see Sect. 1.6.6). We will show that if F_1 is an algebraic extension of the field K, then the element y can be expressed through the elements of the field K by a formula similar to (1.3) containing the same number of logarithmic terms. If F_1 is obtained from K by a logarithmic extension, then the function R_0 can have an additive logarithmic term. However, the functions $R_1, \ldots, R_q$ cannot contain a logarithm. Therefore, an expression for y in terms of elements of K has the same form as (1.3) but can contain $q+1$ logarithmic terms. Finally, if F_1 is obtained from K by an exponential extension, then the exponential cannot appear in the function R_0, and it can appear in the functions R_i for $i > 0$ only as a factor. Therefore, the exponential disappears after logarithmation, and the corresponding factors become some additive terms, which are then added to R_0.

We will begin this inductive proof in Sect. 1.6.6. Before that, we will sharpen the statement of Liouville's theorem (Sect. 1.6.2) and discuss general properties of differential field extensions of transcendence degree 0 (Sect. 1.6.3) and 1 (Sect. 1.6.4); among these, we distinguish the adjunctions of an integral and those of an exponential of integral (Sect. 1.6.5).

1.6.2 *Refinement of Liouville's Theorem*

We now prove that in the formulation of Liouville's theorem, one can require additionally that the coefficients $\lambda_1, \ldots, \lambda_n$ be linearly independent over the field of rational numbers (see the "toric lemma," Lemma 1.10, below). We begin with an obvious lemma:

Lemma 1.9 *If $g = f_1^{k_1} \cdots f_n^{k_n}$, where the k_i are integers and $f_1, \ldots, f_n$ are nonzero elements of a certain differential field, then*

$$\frac{g}{'}g = \sum k_i \frac{f_i'}{f_i}.$$

Proof This follows from the Leibniz rule (or from the fact that the logarithm of a product is equal to the sum of the logarithms). □

Consider a collection $A_1, \ldots, A_n$ of nonzero elements of a differential field K and a linear combination

$$S = \lambda_1 \frac{A_1'}{A_1} + \cdots + \lambda_n \frac{A_n'}{A_n}$$

of their logarithmic derivatives with constant coefficients $\lambda_1, \ldots, \lambda_n \in \mathbb{C}$.

Lemma 1.10 *If the numbers $\lambda_1, \ldots, \lambda_n$ are linearly dependent over $\mathbb{Q}$, then in the multiplicative group G generated by the elements $A_1, \ldots, A_n$, one can choose fewer than n elements whose logarithmic derivatives taken with some constant coefficients sum to S.*

Proof We can assume that no A_i is constant, for otherwise, we have $A_i'/A_i = 0$, and the number of summands in S can be reduced. We can assume that G is free and contains no constants different from 1. Indeed, a nontrivial identity $A_1^{k_1} \cdots A_n^{k_n} = c$, where c is a constant, implies a nontrivial linear relation $\sum k_i A_i'/A_i = 0$ by Lemma 1.9, which helps to reduce the number of summands in S. If the group G is free, then one can choose a different set of generators $B_1, \ldots, B_n$ in G such that

$$A_1 = B_1^{m_{1,1}} \cdots B_n^{m_{1,n}}, \quad \ldots, \quad A_n = B_1^{m_{n,1}} \cdots B_n^{m_{n,n}},$$

where $M = \{m_{i,j}\}$ is an arbitrary integer $n \times n$ matrix with determinant 1. By Lemma 1.9, we have

$$\frac{A_i'}{A_i} = m_{i,1} \frac{B_1'}{B_1} + \cdots + m_{i,n} \frac{B_n'}{B_n}.$$

Hence S is a linear combination of logarithmic derivatives of B_i.

The logarithmic derivative of the function B_1 enters this linear combination with coefficient $\lambda_1 m_{1,1} + \cdots + \lambda_n m_{n,1}$. Let $p_1\lambda_1 + \cdots + p_n\lambda_n = 0$ be the relation on the coefficients $\lambda_1, \ldots, \lambda_n$, where $p_1, \ldots, p_n$ are relatively prime integers. Choose an integer matrix with determinant 1 such that $m_{1,1} = p_1, \ldots, m_{n,1} = p_n$. This can be done because the primitive integer vector $p = (p_1, \ldots, p_n)$ can be included in a basis of the integer lattice $\mathbb{Z}^n$. This choice of matrix corresponds to a certain set of generators $B_1, \ldots, B_n$. The element S is a linear combination of the logarithmic derivatives of $B_2, \ldots, B_n$, and the lemma is proved. □

We have shown that Liouville's theorem implies the refined statement of it given at the beginning of this subsection.

1.6.3 Algebraic Extensions of Differential Fields

Let $K \subset F$ be functional differential fields, and $P \in K[x]$ an irreducible polynomial of degree n over K. Suppose that the field F contains all n roots $x_1, \ldots, x_n$ of the polynomial P. For $i = 1, \ldots, n$, let K_i denote the field obtained from K by the algebraic adjunction of x_i. The fields K_i are isomorphic to each other: for every

$i = 1, \dots, n$, the field K_i is isomorphic to the quotient $K[x]/(P)$ of the polynomial ring $K[x]$ by the ideal (P) generated by P.

Lemma 1.11 *For every $i = 1, \dots, n$, the field K_i is stable under differentiation. For every pair of indices $1 \le i, j \le n$, the map that fixes all elements of K and takes x_i to x_j extends to a differential field isomorphism between K_i and K_j.*

Proof For a polynomial $P(x) = x^n + a_1x^{n-1} + \dots + a_n$ over the field K, we let $\frac{\partial P}{\partial x}$ and $\frac{\partial P}{\partial t}$ denote the polynomials $nx^{n-1} + \dots + a_{n-1}$ and $a_1'x^n + \dots + a_n'$, respectively. Since the polynomial P is irreducible over K, the polynomial $\frac{\partial P}{\partial x}$ has no common roots with P and is different from zero in the field $K[x]/(P)$. Let M denote a polynomial of degree less than n that satisfies the congruence

$$M\frac{\partial P}{\partial x} \equiv -\frac{\partial P}{\partial t} \pmod P.$$

For each root x_i, by differentiating the identity $P(x_i) = 0$ in the field F, we obtain

$$\frac{\partial P}{\partial x}(x_i)x_i' + \frac{\partial P}{\partial t}(x_i) = 0,$$

which implies $x_i' = M(x_i)$. Thus the derivative of the element x_i coincides with the value at x_i of a polynomial M that does not depend on the choice of x_i. Both claims of the lemma follow from this. □

1.6.4 Extensions of Transcendence Degree One

In this subsection, we perform simple computations on which the proof of Liouville's theorem (Sect. 1.6.6) as well as criteria for the elementarity of logarithmic (Sect. 1.7) and exponential (Sect. 1.8) integrals are based.

Let $K \subset F$ be a nested pair of differential fields, where F is generated as a field (not only as a differential field) over K by a single element $t \in F$, and the element t is transcendental over K. In this case, the field F can be regarded as the field of rational functions over K equipped with a new operation of differentiation. Indeed, every element of the field F can be represented as the value of a unique rational function $G(x)$ over K at the element $x = t$: two different rational functions cannot have the same value at t, because that element is transcendental over K. In particular, the derivative t' of the element t is equal to $G(t)$, where G is some rational function over the field K. In this situation, the differential field F is isomorphic to the field of rational functions over K with the new operation D of taking derivatives defined by the formula $D\varphi = G\varphi'$, where φ' is the usual derivative in the field of rational functions over the differential field K.

In what follows, we confine ourselves to the case in which the function G defining the operation of taking derivatives is a polynomial P over the field K.

We identify the field F with the field of rational functions over K equipped with the differentiation $D\varphi = P\varphi'$. This differentiation takes the polynomial ring over K to itself. A rational function over an arbitrary field K admits both additive and multiplicative representations. We now recall the properties of these representations.

Multiplicative representation Every rational function R can be represented as a product

$$R = AP_1^{k_1} \cdots P_l^{k_l},$$

where P_j is a polynomial with leading coefficient 1 irreducible over K, the exponent k_j is an integer, and A is an element of the field K. Such a representation is unique up to a permutation of the factors.

Additive representation Every rational function can be expanded into partial fractions, i.e., represented as a sum

$$R = Q + \sum_{j,m} \frac{Q_{m,j}}{L_j^m},$$

where Q is a polynomial, L_j is a polynomial with leading coefficient 1 irreducible over K, and $Q_{m,j}$ is a polynomial of degree strictly less than the degree of L_j. This representation is unique up to a permutation of summands. We will call the polynomial Q the *polynomial part* of R. The difference $R - Q$ will be called the *polar part* of the function R; the sum $\sum_m Q_{m,j}/L_j^m$ will be called the L_j*-polar part* of the function R; the term $Q_{m,j}/L_j^m$ in the L_j-polar part corresponding to the maximal exponent m in the denominator will be called the *leading term* of the L_j-polar part, and the number m will be called the *order* of the L_j-polar part.

The following two propositions are obvious.

Proposition 1.12 *Suppose that polynomials L_j and DL_j are relatively prime. Then for every rational function R, the L_j-polar part of the derivative depends only on the L_j-polar part of the function. If Q_m/L_j^m is the leading term in the L_j-polar part of R, and Q is the remainder in the division of the polynomial $(-m)Q_mDL_j$ by the polynomial L_j, then Q/L_j^{m+1} is the leading term of the L_j-polar part of the derivative DR.*

Proposition 1.13 *Let P_k be irreducible polynomials with leading coefficient* 1, *and let μ_k be complex numbers. Then the polar part of the function*

$$\sum_{k=1}^{p} \mu_k \frac{DP_k}{P_k}$$

is equal to $\sum_{k=1}^{p} \mu_k Q_k/P_k$, where Q_k is the remainder in the division of the polynomial DP_k by the polynomial P_k (if the polynomials DP_k and P_k have a *common divisor,*[3] *then the remainder* Q_k *is equal to zero and does not affect the sum displayed above).*

We say that an element $g \in F$ regarded as the value of a rational function G over the field K at the element t has a Liouville representation if the function G is representable in the form

$$G = \sum_{i=1}^{q} \lambda_i \frac{DR_i}{R_i} + DR_0.$$

In this representation, each of the rational functions R_i, for $i = 1, \dots, q$, can be written in the multiplicative form

$$R_i = A_i P_{i_1}^{k_{i_1}} \cdots P_{i_l}^{k_{i_l}},$$

where P_{i_j} are irreducible polynomials over K with leading coefficient 1, k_{i_j} are integers, and A_i is an element of the field K. With the help of Lemma 1.9, we can represent the linear combination $\sum_{i=1}^{q} \lambda_i DR_i/R_i$ of logarithmic derivatives of the functions R_i as a sum of linear combinations $\sum_{i=1}^{q} \lambda_i DA_i/A_i$ of logarithmic derivatives of the elements A_i of the field K and linear combinations $\sum_{k=1}^{p} \mu_k DP_k/P_k$ of logarithmic derivatives of the polynomials P_k. Thus a function G admitting a Liouville representation can be written in the form

$$G = \sum \lambda_i \frac{DA_i}{A_i} + \sum \mu_k \frac{DP_k}{P_k} + DR_0.$$

Our next objective is to figure out how the L_j-polar parts of the function G are related to those of R_0 and DP_k/P_k. Let us compute the L_j-polar parts in the case that the polynomials L_j and DL_j are relatively prime. In this case, it follows from Propositions 1.12 and 1.13 that if the L_j-polar part of the function R_0 is not equal to zero, then the order of the L_j-polar part of its derivative DR_0 is greater than 1, whereas the order of the L_j-polar part of the logarithmic derivative DL_j/L_j is equal to 1. Using this remark, it is easy to deduce the following corollaries from Propositions 1.12 and 1.13.

Corollary 1.14 *Assume that the polynomials* L_j *and* DL_j *are relatively prime. Then the* L_j*-polar part of the function* G *has order greater than* 1 *if and only if the* L_j*-polar part of the function* R_0 *is not equal to zero. In this case, the leading term in the* L_j*-polar part of* G *is equal to the leading term in the* L_j*-polar part of the function* DR_0.

Corollary 1.15 *Assume that the polynomials* L_j *and* DL_j *are relatively prime. Then the* L_j*-polar part of the function* G *has order* 1 *if and only if the* L_j*-polar*

[3]Since P_k is irreducible by our assumption, this simply means that DP_k is divisible by P_k.

part of the function R_0 is equal to zero and the linear combination of the logarithmic derivatives includes the term $\mu_k DP_k/P_k$, in which $P_k = L_j$ and $\mu_k \neq 0$. In this case, the leading term in the L_j-polar part of G is equal to the leading term in the L_j-polar part of the function $\mu_k DL_j/L_j$.

1.6.5 Adjunction of an Integral and an Exponential of Integral

In this subsection, we continue the computations that were begun in Sect. 1.6.4 for the case in which the field F is obtained from the field K by adjoining an integral over K or an exponential of integral over K.

Lemma 1.16 *Let $t \in F$ be a transcendental element over the field K, and $L \in K[x]$ an irreducible polynomial over the field K. Then:*

1. *If t is an integral over K, i.e., $t' = f$, $f \in K$, then the polynomials DL and L do not have nontrivial common divisors (equivalently, DL is not divisible by L).*
2. *If t is an exponential of integral over the field K, i.e., $t' = ft$, $f \in K$, then the polynomials L and DL have a nontrivial common divisor if and only if $L \equiv ax$, where a is some element of K.*

Proof First note that the existence or nonexistence of a nontrivial common factor for L and DL does not change with the multiplication of L by an element $a \in K \setminus \{0\}$. This can be seen from the Leibniz rule $D(aL) = a'L + aD(L)$. If t is an integral over K, $t' = f$, then, possibly after multiplication of L by an element of the field K, we may assume that the leading coefficient of the polynomial L is equal to 1: $L(x) = x^n + a_1 x^{n-1} + \cdots + a_n$. In this case, the polynomial $DL = (nf + a_1)x^{n-1} + \cdots$ has smaller degree than the polynomial L and cannot be divisible by the irreducible polynomial L.

If t is an exponential of integral over K, $t' = ft$, and the irreducible polynomial L does not coincide with the polynomial $L = x$ or a multiple of it, then, possibly after multiplication of L by an element of the field K, we may assume that the constant term of L is equal to 1: $L(x) = a_n x^n + \cdots + 1$ (an irreducible polynomial has vanishing constant term only if $L(x) = ax$). In this case, the polynomial $DL = (a_n' + na_n f)x^n + \cdots$ has the same degree as the polynomial L, but the constant term of the polynomial DL is equal to zero. Therefore, this polynomial cannot be divisible by L. □

Remark 1.17 Using Rolle's theorem and computations from Lemma 1.16, it is easy to show that every real-valued Liouville function (see [62]) has only a finite number of real roots; moreover, the number of roots admits an explicit upper bound (in particular, the sine function is not a real-valued Liouville function). The theory of "fewnomials" (see [49]) contains far-reaching multidimensional generalizations of this kind of estimate.

Lemma 1.18 *Let $t \in F$ be a transcendental element over the field K that is an integral over K, i.e., $t' = f$, where $f \in K$, and let $Q \in K[x]$ be a polynomial of degree n. The derivative of the element $Q(t)$ is the value at the element t of the polynomial $DQ = fQ'$. If the leading coefficient of the polynomial Q is not a constant, then the degree of DQ is equal to n. In the opposite case, the degree of the polynomial DQ is equal to $n-1$. In particular, $Q(t)$ is an integral over K if and only if $Q = cx + b$, where c is a constant, and $b \in K$.*

Proof Let $Q = a_n x^n + a_{n-1} x^{n-1} + \cdots$. Then

$$DQ = a_n' x^n + (a_{n-1}' + n a_n f) x^{n-1} + \cdots .$$

The degree of the polynomial DQ is less than n if and only if the element a_n is a constant, i.e., $a_n = c_1 \in \mathbb{C}$. This polynomial cannot have degree less than $n-1$. Indeed, if $a_{n-1}' + n a_n f = 0$, then $(-a_{n-1}/nc_1)' = f$. Since $t' = f$, we have

$$t = \frac{-a_{n-1}}{nc_1} + c_2,$$

where $c_2 \in \mathbb{C}$, whence $t \in K$. The inclusion $t \in K$ contradicts our assumption that t is transcendental over K. □

Lemma 1.19 *Let $t \in F$ be a transcendental element over the field K that is an exponential of integral over K, i.e., $t' = ft$, where $f \in K$, and let $Q(x) = \sum_{m \le k \le n} a_k x^k$ be a Laurent polynomial over K. The derivative of the element $Q(t)$ is equal to $DQ(t)$, where*

$$DQ(x) = \sum_{m \le k \le n} (a_k' + k a_k f) x^k$$

is a Laurent polynomial in which the coefficient $a_k' + k a_k f$ is not equal to zero if $k \neq 0$ and $a_k \neq 0$. In particular, the element $Q(t)$ is an integral over K if and only if the Laurent polynomial Q coincides with the constant term a_0.

Proof Let us show that if $k \neq 0$ and $a_k \neq 0$, then the coefficient $a_k' + k a_k f$ cannot vanish. Indeed, otherwise, the elements a_k and t^{-k} satisfy the same differential equation: $a_k' = -kfa_k$ and $(t^{-k})' = -kt^{-k-1}ft = -kft^{-k}$, and therefore $t^{-k} = ca_k$, where c is a constant. It follows that the element t is algebraic over K, which contradicts the assumption that it is transcendental. □

1.6.6 Proof of Liouville's Theorem

Let us return to the proof of Liouville's theorem.

The case of an algebraic extension Suppose that a field $F_1 = K\langle x_1\rangle$ is obtained from K by adjoining the root x_1 of some polynomial P of degree n irreducible over K. Every element of F_1 is the value at x_1 of some polynomial over the field K whose degree is less than n. By the induction hypothesis, there exist polynomials $M_1, \dots, M_q$ and M_0 of degree less than n such that

$$f = \sum_{i=1}^{q} \lambda_i \frac{(M_i(x_1))'}{M_i(x_1)} + (M_0(x_1))'.$$

Let F be a differential field obtained from K by adjoining all roots $x_1, \dots, x_n$ of the polynomial P, and let $K\langle x_i\rangle$ be a subfield in F obtained by adjoining the element x_i to K. Since the fields $K\langle x_1\rangle, \dots, K\langle x_n\rangle$ are isomorphic (see Lemma 1.11), we have

$$f = \sum_{i=1}^{q} \lambda_i \frac{(M_i(x_k))'}{M_i(x_k)} + (M_0(x_k))'$$

for every $k = 1, \dots, n$. Now take the arithmetic mean of the n obtained equalities in the field F. By Lemma 1.9, for every i, we have

$$\sum_{k=1}^{n} \frac{(M_i(x_k))'}{M_i(x_k)} = \frac{Q_i'}{Q_i},$$

where $Q_i = M_i(x_1) \cdots M_i(x_n)$. The elements Q_i and $Q_0 = \frac{1}{n}(M_0(x_1) + \cdots + M_0(x_n))$ depend symmetrically on the roots of the polynomial P and therefore belong to the field K. Thus

$$f = \sum_{i=1}^{q} \frac{\lambda_i}{n} \frac{Q_i'}{Q_i} + Q_0',$$

where $Q_1, \dots, Q_q, Q_0 \in K$. This proves the inductive step in the case of an algebraic extension $K \subset F_1$.

The case of adjoining a logarithm Suppose that the field F_1 is obtained from K by adjoining a transcendental element t over K that is a logarithm over K (i.e., $t' = a'/a$, where $a \in K$). A logarithm over the field K is, in particular, an integral over K: its derivative a'/a lies in the field K. We regard F_1 as the field of rational functions over the field K with the operation of differentiation $D\varphi = (a'/a)\varphi'$. By Lemma 1.16, every irreducible polynomial L is coprime to its derivative DL.

By the induction hypothesis, the element f admits a Liouville representation over the field F_1, i.e., the element f can be written in the form

$$f = \sum \lambda_i \frac{A_i'}{A_i} + \sum \mu_k \frac{DP_k}{P_k} + DR_0,$$

where the A_i are elements of the field K, the polynomials P_k are irreducible over K and have leading coefficient 1, and R_0 is a rational function over the field K (see Sect. 1.6.4).

Now apply Corollaries 1.14 and 1.15 to the function $G = f$, which does not depend on t. Since all L_j-polar parts of the function f are equal to zero, so are all L_j-polar parts of the function R_0 and all logarithmic terms $\mu_k DP_k/P_k$, i.e., $f = \sum \lambda_i A_i'/A_i + DQ$, where Q is the polynomial part of the function R_0. The derivative of the polynomial Q must lie in the field K. By Lemma 1.18, we have $Q(t) = ct + A$, where c is a complex constant and $A \in K$. By our assumption, $t' = a'/a$, whence

$$f = \sum \lambda_i \frac{A_i'}{A_i} + c\frac{a'}{a} + A'.$$

This proves the inductive step in the case of a logarithmic extension $K \subset F_1$.

The case of adjoining an exponential Suppose that the field F_1 is obtained from K by adjoining a transcendental element t over K that is an exponential over K (i.e., $t' = a't$, where $a \in K$). An exponential over the field K is a particular case of an exponential of integral over K. We regard F_1 as the field of rational functions over the field K with the operation of differentiation $D\varphi = (a't)\varphi'$. By Lemma 1.16, every irreducible polynomial L_j that is not of the form $L_j(x) = ax$, $a \in K$, is coprime to its derivative.

By the induction hypothesis, the element f admits a Liouville representation over the field F_1, i.e. (see Sect. 1.6.4), the element f can be written in the form

$$f = \sum \lambda_i \frac{A_i'}{A_i} + \sum \mu_k \frac{DP_k}{P_k} + DR_0,$$

where the A_i are elements of the field K, the polynomials P_k are irreducible over K and have leading coefficient 1, and R_0 is a rational function over the field K. Apply Corollaries 1.14 and 1.15 to the function $G = f$, which does not depend on t. Since all L_j-polar parts of the function f are equal to zero, so are all L_j-polar parts of the function R_0 (provided that L_j is not a multiple of x over K) and all logarithmic terms $\mu_k DP_k/P_k$, provided that P_k is not a multiple of x, i.e.,

$$f = \sum \lambda_i \frac{A_i'}{A_i} + \sum D\left(\frac{a_m}{x^m}\right) + \mu\frac{Dx}{x} + DQ,$$

where Q is the polynomial part of the function R_0 (on the right-hand side, we must keep the derivative $\sum (a_m/x^m)$ of the x-polar part of the function R_0 and the logarithmic term $\mu Dx/x$). By definition, the value of the rational function $\mu Dx/x$ at the element t is equal to $\mu a' \in K$. Hence the derivative $D\left(Q + \sum a_m/x^m\right)$

lies in the field K. By Lemma 1.19, this is possible only if the Laurent polynomial $Q + \sum a_m/t^m$ coincides with its constant term A. We have

$$f = \sum \lambda_m \frac{A'_m}{A_m} + (\mu a + A)'.$$

This completes the induction step in the case of an exponential extension $K \subset F_1$, and with it, the proof of Liouville's theorem on integrals.

1.7 Integration of Functions Containing the Logarithm

In this section, we give a criterion for the elementarity of the antiderivatives of 1-forms $R(z,u)\,dz$, where R is a rational function of two variables, z is a complex variable, and $u = \log a$ for some rational function a of the complex variable z. In other words, we give a criterion for the elementarity of integrals of functions lying in a logarithmic extension F of the differential field K of rational functions of the complex variable z. That is, $K = \mathbb{C}\langle z\rangle$, $F = K\langle t\rangle$, $t' = a'/a$, $a \in K$. We will regard the field F as the field of rational functions over the field K with the operation of differentiation D, where $D\varphi = (a'/a)\varphi'$ (see the introduction to Sects. 1.6 and 1.6.4). By Liouville's theorem, the function $G(t) \in F$ has an elementary integral if and only if it is representable in the form

$$G = \sum \lambda_i \frac{A'_i}{A_i} + \sum \mu_k \frac{DP_k}{P_k} + DR_0,$$

where the A_i are elements of the field K, the polynomials P_k are irreducible over K and have leading coefficient 1, and R_0 is a rational function over the field K.

We now give several definitions. The *multiplicity* of a nonzero L_j-polar part of a rational function R is defined as the number $q - 1$, where q is the order of this part. We say that an L_j-polar part is *multiplicity-free* if its multiplicity is equal to zero. We say that a rational function R has a *multiplicity-free polar part* if all its L_j-polar parts are multiplicity-free.

1.7.1 The Polar Part of an Integral

A function Ψ will be called the *polar part of the integral* of a function G if the polynomial part of the function Ψ is equal to zero, and the polar part of the function $G - D\Psi$ is multiplicity-free.[4] For every function G, there exists at most one polar

[4]We use "polar part of the integral" as a single piece of terminology.

part of the integral. Indeed, different functions Ψ_1 and Ψ_2 without polynomial parts must have different L_j-polar parts for a suitable polynomial L_j. For this polynomial, the polar part of the function $D\Psi_1 - D\Psi_2$ has positive multiplicity. Therefore, the functions Ψ_1 and Ψ_2 cannot simultaneously be polar parts of the integral of the function G.

Proposition 1.20 *There exists a polar part of the integral for every function G. Moreover, it can be explicitly computed from the collection of L_j-polar parts of the function G having positive multiplicity.*

Remark 1.21 The computational method described in Propositions 1.20 and 1.27 is known as *Hermite reduction*. It is exploited in algorithms to find elementary integrals (see, for example, the book [16]).

Proof (of Proposition 1.20) We will describe an iterative construction of the polar part of the integral. At each step, the problem reduces to a similar problem for a new rational function that has smaller total multiplicity of polar parts.

Suppose that for some polynomial L_j of degree p, the leading term in the L_j-polar part of the function G is equal to Q/L_j^{m+1}, where Q is a polynomial of degree less than p, and $m > 0$. Choose a polynomial T of degree less than p such that the leading term in the L_j-polar part of the function $D(T/L_j^m)$ is equal to Q/L_j^{m+1}, i.e., $(-m)T\,DL_j \equiv Q \pmod{L_j}$. Indeed, let l_j denote a polynomial for which the following congruence holds: $(DL_j)l_j \equiv 1 \pmod{L_j}$. The polynomial l_j can be constructed explicitly from the polynomial DL_j with the help of the Euclidean algorithm (recall that the polynomials DL_j and L_j are relatively prime). We can now choose T as the remainder in the division of $Ql_j/(-m)$ by the polynomial L_j.

The function $G_1 = G - D(T/L_j^m)$ has a smaller total polar multiplicity than the function G. Therefore, we can assume that the polar part Ψ_1 of the integral of the function G_1 has already been found. By construction, the polar part Ψ of the integral of the function G is equal to $\Psi_1 + T/L_j^m$. □

Proposition 1.20 reduces the problem of integration of rational functions to the problem of integration of rational functions with a multiplicity-free polar part.

1.7.2 The Logarithmic Derivative Part

Let G be a rational function with a multiplicity-free polar part. The function $\Phi = \sum \mu_k DP_k/P_k$, where the P_k are irreducible polynomials with leading coefficient 1 and the μ_k are complex numbers, is called the *logarithmic derivative part* of the function G if the function $G - \Phi$ is a polynomial.

Consider the additive representation of a function G that has a multiplicity-free polar part: $G = \sum_{0 \le j \le n} Q_j/L_j + Q$, where the L_j are irreducible polynomials with leading coefficient 1, and Q and Q_j are polynomials such that the degree of the polynomial Q_j is less than the degree of the polynomial L_j.

Proposition 1.22 *Suppose that the function G defined above is representable in Liouville form. Then for every j, $0 \le j \le n$, we have $Q_j = \mu_j DL_j$, where μ_j is a complex number. Under these assumptions, the function $\Phi = \sum Q_j/L_j$ is equal to the derivative of the function $\sum \mu_j \log L_j$ and coincides with the logarithmic derivative part of the function G.*

Proof We shall deal with a logarithmic extension of the field K. The derivative DL_j of the polynomial L_j has smaller degree than the polynomial L_j, since the leading coefficient of the polynomial L_j is equal to 1. Therefore, the leading term of the L_j-polar part of the function DL_j/L_j is equal to DL_j/L_j. This computation reduces Proposition 1.22 to Corollaries 1.14 and 1.15. □

Corollary 1.23 *A function G whose polynomial part is equal to zero and whose polar part is multiplicity-free has an elementary integral if and only if it satisfies the conditions of Proposition 1.22.*

For most rational functions having a multiplicity-free polar part, the conditions of Proposition 1.22 do not hold. Hence in most cases, such functions have nonelementary integrals.

Example 1.24 Let f and g be rational functions of the variable z, and suppose that the function f is not a constant. Then the integral $\int g\,dz/\log f$ is a generalized elementary function if and only if the function gf/f' is identically equal to a constant. In particular, the integral $\int dz/\log z$ is not elementary.

1.7.3 Integration of a Polynomial of a Logarithm

Now let G be a polynomial, $G(t) = a_n t^n + \cdots + a_0$, $t' = a'/a$, and $a, a_0, \ldots, a_n \in K = \mathbb{C}\langle z\rangle$. The binomial $\Delta_n = ct^{n+1} + b_n t^n$ will be called the nth polynomial component of the integral of G if the polynomial $G - D\Delta_n$ has degree less than n. We will use the fact that the element a taking part in the definition of the logarithmic extension $F = K\langle t\rangle$, $t' = a'/a$, and the coefficients a_k of the polynomial G are rational functions of the complex variable z. Consider the following two 1-forms of the complex variable z: $(a'/a)\,dz$ and $a_n\,dz$. We will regard the residues $\operatorname{res}_q(a'/a)\,dz$ and $\operatorname{res}_q a_n\,dz$ of these forms at a point q as functions of $q \in \mathbb{C}$ (these functions in the complex plane vanish everywhere except at finitely many points).

Proposition 1.25 *If a polynomial $G = a_n t^n + \cdots$ of degree $n > 0$ is representable in Liouville form, then for some complex number μ and every point $q \in \mathbb{C}$, the following identity holds: $\operatorname{res}_q(a'/a)\,dz \equiv \mu \operatorname{res}_q a_n\,dz$. Under this condition, there exists a binomial $\Delta_n = ct^{n+1} + b_n t^n$ in which the coefficient c is equal to $\mu/(n+1)$, and the coefficient b_n is a rational function of the complex variable z defined up to an additive constant by the equation $b_n' = a_n - \mu a'/a$. This binomial Δ_n is the nth polynomial component of the integral of G.*

Proof Suppose that G is representable in Liouville form (see Sect. 1.6.4). It can be seen from Corollaries 1.14 and 1.15 that the polynomial G must be equal to the derivative of some polynomial G_0, i.e., $G = DG_0$. By Lemma 1.18, the highest-degree monomials of the polynomial G_0 have the form $G_0 = ct^{n+1} + b_n t^n + \cdots$, where c is a complex constant (possibly equal to zero). Differentiating, we obtain

$$DG_0(t) = ((n+1)c(a'/a) + b_n')t^n + \cdots .$$

The rational function b_n of the complex variable z must satisfy the equation $b_n' = a_n - (n+1)c(a'/a)$. This equation has a rational solution if and only if all residues of the form $(a_n - (n+1)c(a'/a))dz$ vanish; the statement of the proposition follows. □

For most polynomials of positive degree n, the assumptions of Proposition 1.25 do not hold, and hence polynomials of logarithms usually have nonelementary integrals.

Example 1.26 Let f, g be rational functions of the variable z, and suppose that f is not a constant. Then the integral $\int g \log f \, dz$ is a generalized elementary function if and only if the function g is representable in the form $cf'/f + \varphi'$, where c is a constant and φ is a rational function. In particular, the integral

$$\int \frac{\log z \, dz}{z-1}$$

is not elementary.

1.7.4 Integration of Functions Lying in a Logarithmic Extension of the Field $\mathbb{C}\langle z \rangle$

We now describe a procedure that makes it possible either to find the integral of a function G or to prove that the integral cannot be expressed in terms of generalized elementary functions.

Step 1. If the rational function G has a multiple polar part, then using Proposition 1.20, one can find the polar part Ψ of the integral of the function G and pass to the function $G_s = G - D\Psi$, whose polar part is multiplicity-free.

Step 2. For the rational function G_s with a multiplicity-free polar part, one needs to verify the conditions of Proposition 1.22. If these conditions are not satisfied, then the integral of the function G is not expressible by generalized elementary functions. If the conditions of Proposition 1.22 hold, then one can find the logarithmic derivative part Φ of the function G_s. By construction, the integral of the function Φ is a linear combination of logarithms, and the function $G_s - \Phi$ is a polynomial G_n of some degree n.

Step 3_n. For the polynomial G_n, one needs to check the conditions of Proposition 1.25. If they fail to be satisfied, then the integral of the function G cannot be expressed by generalized elementary functions. If the conditions are satisfied, then one can find a binomial Δ_n that is the nth polynomial component of the integral of the polynomial G_n. The function $G_n - D\Delta_n$ is a polynomial G_{n-1} of degree $n - 1$.

Steps $3_{n-1}, \ldots, 3_1$. Repeating the procedure of Step 3_n, we will either pass to polynomials of smaller and smaller degree or, at some step, prove the nonelementarity of the integral.

Step 3_0. If we reach a polynomial G_0 of degree 0, then the original integral is elementary. Indeed, a polynomial of degree zero is a rational function of the complex variable z, and the integral of it is always expressible by elementary functions.

1.8 Integration of Functions Containing an Exponential

In this section, we give a criterion for the elementarity of the antiderivatives of 1-forms $R(z, u)\,dz$, where R is a rational function of two variables, z is a complex variable, and $u = \exp a$ for some rational function a of the complex variable z. In other words, we give a criterion for the elementarity of functions lying in an exponential extension F of the differential field K of rational functions of the complex variable z. That is, $K = \mathbb{C}\langle z\rangle$, $F = K\langle t\rangle$, $t' = a't$, $a \in K$. We will regard the field F as the field of rational functions over the field K with the operation of differentiation D, where $D\varphi(t) = a't\varphi'(t)$ (see the introduction to Sects. 1.6 and 1.6.4).

By Liouville's theorem, a function $G(t) \in F$ has an elementary integral if and only if it is representable in the form

$$G = \sum \lambda_i \frac{A_i'}{A_i} + \sum \mu_k \frac{DP_k}{P_k} + DR_0,$$

where the A_i are elements of the field K, the polynomials P_k are irreducible over K and have leading coefficient 1, and R_0 is a rational function over the field K (see Sect. 1.6.4). We will now modify the definitions from Sect. 1.7 for the exponential case. The irreducible polynomial x plays a special role in exponential extensions: it is the unique (up to a factor that is an element of K) polynomial L that is a divisor of its derivative DL (see Lemma 1.16).

1.8.1 *Principal Polar Part of the Integral*

We define the *principal polar part* of a rational function R as the sum of its L_j-polar parts over all monic (i.e., with leading coefficient 1) irreducible polynomials L_j except the polynomial $L = x$.

Consider the polynomial that is the polynomial part of the rational function R. The sum of all monomials in this polynomial except the constant term will be called the *principal polynomial part* of the function R.

Define the *Laurent part* of the function R as the sum of its polynomial part and its x-polar part.

We say that a function Ψ is the *principal polar part of the integral* of the function G if its Laurent part is equal to zero and the principal polar part of the function $G - D\Psi$ is multiplicity-free.

Proposition 1.27 *For every function G, there exists a principal polar part of the integral. Moreover, it can be explicitly computed if the collection of all L_j-polar parts entering the principal polar part of G and having positive multiplicity is known.*

Remark 1.28 As we noted in Remark 1.21, the computational method described in the above proposition is known as *Hermite reduction*. It is exploited in algorithms to find elementary integrals (see, for example, the book [16]).

We will not prove Proposition 1.27 here. The proof is almost word for word the same as the proof of Proposition 1.20. The only difference is that in the process of iteratively constructing the principal polar part of the integral of the function G, one disregards the x-polar part of the function.

Proposition 1.27 reduces the problem of integration of rational functions to the problem of integration of rational functions with multiplicity-free principal polar part.

1.8.2 Principal Logarithmic Derivative Part

Let G be a rational function with a multiplicity-free principal polar part. The function $\Phi = \sum \mu_k DP_k/P_k$, where P_k is a monic irreducible polynomial different from x, and μ_k is a complex number, is called the *principal logarithmic derivative part of the function* G if the function $G - \Phi$ is a Laurent polynomial.

Consider the additive representation of the function G that has a multiplicity-free principal polar part,

$$G = \sum_{0 \le j \le n} \frac{Q_j}{L_j} + Q,$$

where L_j is a monic irreducible polynomial different from x, Q_j is a polynomial whose degree is less than the degree of the polynomial L_j, and Q is a Laurent polynomial. Let $[DL_j]$ denote the remainder on division of the polynomial DL_j by the polynomial L_j.

Proposition 1.29 *Suppose that the function G defined above is representable in Liouville form. Then for every j, $0 \leq j \leq n$, the following identity holds: $Q_j = \mu_j[DL_j]$, where μ_j is a complex number. Under these conditions, the function*

$$\Phi = \sum \mu_j \frac{DL_j}{L_j} = \sum \mu_j (\log L_j)'$$

is the principal logarithmic part of the integral of the function G, and the difference

$$\Phi - \sum_{0 \leq j \leq n} \mu_j \frac{[DL_j]}{L_j}$$

lies in the field $K = \mathbb{C}\langle z \rangle$.

Proof We shall deal with an exponential extension of the field K. The derivative DL_j of the polynomial L_j has the same degree as the polynomial L_j. Therefore, the difference

$$\frac{DL_j}{L_j} - \frac{[DL_j]}{L_j}$$

lies in the field $K = \mathbb{C}\langle z \rangle$. This computation reduces Proposition 1.29 to Corollaries 1.14 and 1.15. □

Corollary 1.30 *A function G whose principal polar part is multiplicity-free and whose Laurent part is equal to zero has an elementary integral if and only if the conditions of Proposition 1.29 hold for this function.*

Proof If a function G satisfies the assumptions of Proposition 1.29, then it has a principal logarithmic derivative part Φ that has an elementary integral. The difference $G - \Phi$ is a rational function of the complex variable z. Its integral is also elementary. □

For most rational functions having a multiplicity-free principal polar part, the assumptions of Proposition 1.29 do not hold. Therefore, most such functions have nonelementary integrals.

Example 1.31 Let f, g, h be rational functions of the complex variable z; suppose that the function f is not a constant and that the function g is nonzero. Consider the integral

$$\int \frac{g\,dz}{\exp f + h}.$$

For the elementarity of this integral, it is necessary and sufficient that the function $g/(h' - f'h)$ be constant (indeed, in this example, $L = t + h$, $DL = f't + h'$, $[DL] = h' - f'h$). In particular, the integral

$$\int \frac{g\,dz}{\exp z + 1}$$

is elementary if and only if the rational function g is constant.

1.8.3 Integration of Laurent Polynomials of the Exponential

Let now $G(t) = \sum_{m \le k \le n} a_k t^k$ be a Laurent polynomial over K, i.e., $a_m, \dots, a_k \in K$, with vanishing constant term a_0, and let t be a solution of the equation $t' = a't$ for some $a \in K$.

Proposition 1.32 1. *Suppose that the Laurent polynomial G is representable in Liouville form. Then there exists a Laurent polynomial Δ with vanishing constant term such that $D\Delta = G$.*

2. *For the existence of the Laurent polynomial Δ, it is necessary and sufficient that for every k such that $m \le k \le n$ and $k \ne 0$, the linear differential equation $b_k' + k b_k a' = a_k$ have a solution in the field K, $b_k \in K$. In that case, $\Delta = \sum_{m \le k \le n} b_k t^k$.*

Proof From Corollaries 1.14 and 1.15, it follows that the Liouville form of a Laurent polynomial has zero principal polar part and zero principal logarithmic part. Hence it is a Laurent polynomial. By Lemma 1.19, a Laurent polynomial Δ satisfies the equality $D\Delta = G$ if and only if it satisfies part 2 of Proposition 1.32. □

Most differential equations over the field $K = \mathbb{C}\langle z \rangle$ that appear in Proposition 1.32 do not have solutions in rational functions of the complex variable z. Therefore, most Laurent polynomials of the function $u = \exp(a(z))$ over the field $\mathbb{C}\langle z \rangle$ have nonelementary integrals. Below, we discuss a criterion for the solvability by rational functions of the differential equations just encountered.

1.8.4 Solvability of First-Order Linear Differential Equations

We turn to the question of solvability (by rational functions of the complex variable z) of linear differential equations of the form $y' + f'y = g$, where f and g are rational functions of z. This question can be resolved in the following way. If a rational solution y exists, then the coefficients of f and g determine the poles of y and the orders of these poles (see Corollary 1.34). Thus we can a priori specify a finite-dimensional vector space of functions in which a rational solution of the equation must lie, provided that it exists. After that, the indeterminate coefficients method makes it possible either to find a rational solution of the equation explicitly or to prove that it does not exist (in fact, the nonexistence of a rational solution is often easily seen without any computation).

For every nonzero rational function φ, let $\operatorname{ord}_a(\varphi)$ denote the order of the function φ at a point a of the Riemann sphere. If $a \neq \infty$, then the orders of a function and its derivative satisfy the following relations: $\operatorname{ord}_a(\varphi) \neq 0$ implies that $\operatorname{ord}_a(\varphi') = \operatorname{ord}_a(\varphi) - 1$. If $\operatorname{ord}_a(\varphi) = 0$ and the function φ is not constant, then $\operatorname{ord}_a(\varphi') \geq 0$. In particular, the order of the derivative at a finite point is never equal to -1. At the point ∞, the relations take the following form: $\operatorname{ord}_\infty(\varphi) \neq 0$ implies $\operatorname{ord}_\infty(\varphi') = \operatorname{ord}_\infty(\varphi) + 1$. If $\operatorname{ord}_\infty(\varphi) = 0$ and the function φ is not a constant, then $\operatorname{ord}_\infty(\varphi') \geq 2$. In particular, the order of the derivative at the point ∞ is never equal to 1.

Lemma 1.33 *Suppose that a rational function y has a pole at a point a and that a rational function f is not equal to a constant. Then:*

1. *If $a \in \mathbb{C}$, then the order of the function $y' + f'y$ at the point a is equal to the minimum of the numbers* $\operatorname{ord}_a(y) - 1$ *and* $\operatorname{ord}_a(f') + \operatorname{ord}_a(y)$.
2. *If $a = \infty$, then the order of the function $y' + f'y$ at ∞ is equal to the minimum of the numbers* $\operatorname{ord}_\infty(y) + 1$ *and* $\operatorname{ord}_\infty(f') + \operatorname{ord}_\infty(y)$.

Proof Under the assumptions we have made, the functions y' and $f'y$ have different orders at the point a. Therefore, the order of the sum of these two functions is equal to the minimum of their orders. □

Suppose that the equation $y' + f'y = g$ has a rational solution. The following corollary describes the set of poles of the solution and their orders.

Corollary 1.34 *A point $a \in \mathbb{C}$ is a pole of the function y in the following two cases:*

1. $\operatorname{ord}_a(f) \geq 0$, $\operatorname{ord}_a(g) < -1$. *Then* $\operatorname{ord}_a(y) = \operatorname{ord}_a(g) + 1$.
2. $\operatorname{ord}_a(f) < 0$, $\operatorname{ord}_a(g) < \operatorname{ord}_a(f) - 1$. *Then* $\operatorname{ord}_a(y) = \operatorname{ord}_a(g) + 1 - \operatorname{ord}_a(f)$.

The point ∞ is a pole of the function y in the following two cases:

1. $\operatorname{ord}_\infty(g) \leq 0$, $\operatorname{ord}_\infty(f) \geq 0$. *Then* $\operatorname{ord}_\infty(y) = \operatorname{ord}_\infty(g) - 1$.
2. $\operatorname{ord}_\infty(f) < 0$, $\operatorname{ord}_\infty(g) < 1 + \operatorname{ord}_\infty(f)$. *Then* $\operatorname{ord}_\infty(y) = \operatorname{ord}_\infty(g) - 1 - \operatorname{ord}_\infty(f)$.

Suppose that finite poles $a \in A \subset \mathbb{C}$ of the function y have orders $k_a = -\operatorname{ord}_a y$ and that the point ∞ is a pole of y of order $m = -\operatorname{ord}_\infty y$. Then y belongs to the finite-dimensional space of functions l of the form

$$l = \sum_{\substack{a \in A \\ 0 < i \leq -k_a}} \frac{c_{i,a}}{(z-a)^i} + c_0 + \sum_{0 < p \leq m} d_p z^p.$$

Substituting the function l with indeterminate coefficients $c_{i,a}, c_0, d_p$ and with the poles found from Corollary 1.34 for y into the equation $y' + f'y = g$, we obtain a system of linear equations in the indeterminate coefficients. If this system has no solutions, then the equation has no solutions by rational functions. If the system has a solution, then this solution gives rise to a rational solution y.

Example 1.35 Let f, g be polynomials, and suppose that the degree $\deg(g)$ of the polynomial g is less than $\deg(f)-1$. Then the equation $y'+f'y = g$ has no rational solutions. Indeed, because of the inequality on the degrees of the polynomials, the equation clearly has no constant solutions. The set of poles of a solution is empty by Corollary 1.34. Indeed, at the point ∞, we have the inequality $\mathrm{ord}_\infty(f) < 0$, but the inequality $\mathrm{ord}_\infty(g) < 1 + \mathrm{ord}_\infty(f)$ does not hold. A nonconstant rational function must have poles. Therefore, the equation does not have rational solutions.

Example 1.36 If f, g are polynomials from Example 1.35, then the integral

$$\int g(z) \exp f(z)\, dz$$

cannot be computed by generalized elementary functions. In particular, the integral $\int \exp(z^2) dz$ is not elementary.

Example 1.37 Suppose that a function g has a simple pole at some point $a \in \mathbb{C}$ and that a function f is regular at the point a. Then the equation $y' + f'y = g$ has no rational solutions. Indeed, suppose that a rational solution exists. By Corollary 1.34, it cannot have a pole at the point a. Hence the function $y' + f'y$ is regular at the point a and cannot have a pole at that point.

Example 1.38 The integral $\int \exp z\, dz/z$ cannot be expressed by generalized elementary functions. Indeed, the integral is associated with the extension of the field $K = \mathbb{C}\langle z\rangle$ by an element t such that $t' = t$ and with the polynomial $G(t) = (1/z)t$. The integral is not elementary, since the equation $y' + y = 1/z$ has no rational solutions (see Example 1.37).

Example 1.39 The integral $\int \sin z\, dz/z$ is not expressible by generalized elementary functions. Indeed, the integral is associated with the extension of the field $K = \mathbb{C}\langle z\rangle$ by an element t such that $t' = it$ and with the Laurent polynomial

$$G(t) = \frac{1}{2iz}t - \frac{1}{2iz}t^{-1}.$$

The integral is not elementary, since the equations

$$y' + iy = \frac{1}{2iz} \quad \text{and} \quad y' - iy = -\frac{1}{2iz}$$

have no rational solutions (see Example 1.37).

1.8.5 *Integration of Functions Lying in an Exponential Extension of the Field* $\mathbb{C}\langle z\rangle$

We now describe a procedure that makes it possible either to find the integral of a function G or to prove that the integral cannot be expressed by generalized elementary functions.

Step 1. If the rational function G has a multiple principal polar part, then using Proposition 1.27, one can find the principal polar part Ψ of the integral of the function G and pass to the function $G_s = G - D\Psi$, whose principal polar part is multiplicity-free.

Step 2. For the rational function G_s with a multiplicity-free principal polar part, one needs to verify the conditions of Proposition 1.29. If these conditions are not satisfied, then the integral of the function G is not expressible by generalized elementary functions. If the conditions of Proposition 1.29 hold, then one can find the principal logarithmic derivative part Φ of the function G_s. By construction, the integral of the function Φ is a linear combination of logarithms, and the function $G_s - \Phi$ is a Laurent polynomial G_L. The polynomial G_L is the sum of the constant term $a_0 \in \mathbb{C}\langle z\rangle$ and a Laurent polynomial $G_{L,0}$ with vanishing constant term. The rational function a_0 of the complex variable z has an elementary integral.

Step 3. For the Laurent polynomial $G_{L,0}$, one needs to check the conditions of Proposition 1.32. To this end, we need to know whether the differential equations given in Proposition 1.32 are solvable by rational functions. The question of solvability of these equations was worked out in Sect. 1.8.4. As a result, we either find the integral of the function G or prove that this integral cannot be found among generalized elementary functions.

1.9 Integration of Algebraic Functions

If the Riemann surface of an algebraic function has genus zero, then its integral is always representable by generalized elementary functions. If the genus of the Riemann surface is positive, then the integral is usually nonelementary and is representable by generalized elementary functions in exceptional cases only. In this section, we discuss these exceptional cases. (A qualitative discussion of these cases can be found in the recent paper [59].)

Theorem 1.40 (Liouville's theorem on abelian integrals) *An indefinite integral y of an algebraic function A of a complex variable x is representable by generalized elementary functions if and only if it is representable in the form*

$$y(x) = \int_{x_0}^{x} A(x)\,dx = A_0(x) + \sum_{i=1}^{k} \lambda_i \log A_i(x),$$

where A_i, $i = 0, 1, \dots, k$ are algebraic functions that are single-valued on the Riemann surface W of the integrand A.

Proof The theorem follows from Liouville's theorem on integrals of elementary functions applied to the field F of all meromorphic functions on the surface W with the following differentiation: $f' = df/\alpha$, where $\alpha = \pi^* dx$ and $\pi : W \to \overline{\mathbb{C}}$ is the natural projection of the Riemann surface of the function A onto the Riemann sphere $\overline{\mathbb{C}}$ of the complex variable x. □

Define the class of *generalized elementary functions on a Riemann surface W* as the class of multivalued functions that are obtained from the meromorphic functions on W by arithmetic operations, solving algebraic equations, and compositions with the functions log and exp. Let $\pi : W \to \overline{\mathbb{C}}$ be any nonconstant holomorphic map of the surface W to the Riemann sphere $\overline{\mathbb{C}}$ of the variable x, i.e., a meromorphic function on W. As can be easily deduced from the definitions, a generalized elementary function on the Riemann surface W is a function of the form $\pi^* f$, where f is a generalized elementary function of the variable x. Let us restate Theorem 1.40 in a more invariant form.

Theorem 1.41 (Liouville's theorem on abelian integrals) *An indefinite integral of a meromorphic form α on a compact Riemann surface W is a generalized elementary function on W if and only if the form α is representable in the form $\alpha = \beta + \gamma$, where $\beta = dA_0$, $\gamma = \sum_{i=1}^{k} \lambda_i dA_i/A_i$, the functions $A_0, \dots, A_k$ are meromorphic, and $\lambda_1, \dots, \lambda_k$ are complex numbers.*

In connection with Liouville's theorem on abelian integrals, it is natural to discuss problems 1.42 and 1.46 stated below (see Sects. 1.9.1 and 1.9.2), the first being related to the Riemann–Roch theorem, and the second to Abel's theorem.

1.9.1 The Rational Part of an Abelian Integral

The problem of extracting the rational part from the integral of an algebraic function can be stated as follows.

Problem 1.42 Represent a meromorphic form α on the surface W as $\alpha = \beta + \alpha_1$, where $\beta = dA_0$ is an exact meromorphic form, and the form α_1 has poles of order at most 1.

Lemma 1.43 *If a meromorphic form β on the surface W has poles of order at most* 1 *and defines the trivial cohomology class of $W \setminus P$, where P is a finite set containing the poles of β, then the form β is identically equal to zero.*

Proof All residues of the form β are equal to zero; otherwise, the form cannot be exact. It follows that β has no poles at all, and its integral A_0 is a holomorphic function on W. A holomorphic function on a compact surface is constant. Therefore, $\beta = dA_0 = 0$. □

Corollary 1.44 *If Problem 1.42 is solvable for a form α, then it has exactly one solution.*

Let $O \subset W$ be the set of poles of the form α. In a neighborhood of every pole $x \in O$, fix a local coordinate z such that $z(x) = 0$. Near the point x, the form α can be represented uniquely in the form

$$\alpha = \left(\frac{c_k}{z^k} + \cdots + \frac{c_2}{z^2} + \frac{c_1}{z} + \varphi\right) dz,$$

where φ is the germ of the holomorphic function at the point x. The germ

$$\left(\frac{c_k}{z^k} + \cdots + \frac{c_2}{z^2} + \frac{c_1}{z}\right) dz$$

is called the *principal part* of the form α at the point x (the principal part depends on the choice of a local coordinate z).

The germ

$$A_{0x} = \frac{c_k}{(-k+1)z^{k-1}} + \cdots + \frac{c_2}{(-z)}$$

of a meromorphic function at a pole x of the form α is called the *principal part in the meromorphic component of the integral* of the form α at the point x. The germ A_{0x} has the following property: the form $\alpha - dA_{0x}$ has a pole of order at most 1 at the point x. This property defines the germ A_{0x} uniquely up to adding the germ of a holomorphic function.

Proposition 1.45 (Condition for the solvability of Problem 1.42) *Problem 1.42 is solvable for a form α if and only if for the principal parts A_{0x} in the meromorphic components of the integral of the form α at its poles $x \in O$ and for every holomorphic form ω on W, the following relation holds: $\sum_{x \in O} \operatorname{res}_x(A_{0x}\omega) = 0$.*

Proof Set $\alpha = \beta + \alpha_1$ and $\beta = dA_0$. For every holomorphic form ω, the sum of residues of the form $A_0\omega$ is equal to zero, since W is a compact Riemann surface. This implies the desired equality. Conversely, if $\sum \operatorname{res}_x(A_{0x}\omega) = 0$ for every holomorphic form ω, then by the Riemann–Roch theorem, there exists a meromorphic function A_0 having poles at points $x \in O$ only and such that for every point $x \in O$, the germ of $A_0 - A_{0x}$ is holomorphic. Obviously, the form $\alpha - dA_0$ has poles of order at most 1. □

There are no nonzero holomorphic forms on an algebraic curve of genus zero, and Problem 1.42 is always solvable. Problem 1.42 is usually not solvable on a curve of positive genus.

1.9.2 Logarithmic Part of an Abelian Integral

The problem of extracting the logarithmic part from the integral of an algebraic function can be stated as follows.

Problem 1.46 1. For a given meromorphic form α on a surface W, find a form γ with the same residues as α that is a linear combination of differentials of the logarithms of meromorphic functions.
2. If a desired form γ exists, represent it as a sum $\gamma = \sum_{i=1}^{n} \lambda_i dA_i/A_i$, where the A_i are meromorphic functions on W, containing the smallest possible number n of terms.

Let P be the set of points x at which the residue $\mathrm{res}_x\, \alpha$ of the form α is not equal to zero. On the set P, a function $\mathrm{res}\, \alpha : P \to \mathbb{C}^*$ is defined that assigns the residue $\mathrm{res}_x\, \alpha$ to every point $x \in P$. We associate to the function $\mathrm{res}\, \alpha$, the vector subspace $V(\mathrm{res}\, \alpha) \subset \mathbb{C}$ over the field $\mathbb{Q}$ spanned by the values of the function $\mathrm{res}\, \alpha$.

Lemma 1.47 *Suppose that the sum $\gamma = \sum_{i=1}^{n} \lambda_i dA_i/A_i$ is a solution of part 2 of Problem 1.46 for the form α. Then:*

1. *The numbers $\lambda_1, \ldots, \lambda_n$ belong to the vector space $V(\mathrm{res}\, \alpha)$ and form a basis of it.*
2. *The supports of the divisors (A_i) of the functions A_i, $i = 1, \ldots, n$, lie in the set P.*

Proof By Lemma 1.10, if the numbers $\lambda_1, \ldots, \lambda_n$ are linearly dependent over $\mathbb{Q}$, then the number of terms in the representation of the form γ can be reduced (by choosing a different collection of meromorphic functions), which contradicts the assumptions of the lemma.

If x is a zero or a pole of one of the functions $A_1, \ldots, A_n$, then the residue of the form $\sum \lambda_i dA_i/A_i$ at the point x is not equal to zero, since it is a nontrivial linear combination of the numbers $\lambda_1, \ldots, \lambda_n$. Therefore, $x \in P$.

Consider integer-valued functions φ_i on the set P that assign to a point $x \in P$ the order of the function A_i at that point: $\varphi_i = \mathrm{res}_x\, dA_i/A_i$. We now prove that the functions φ_i are linearly independent over the set P. Indeed, the existence of a nontrivial linear relation $\sum \mu_i \varphi_i = 0$ implies that the form $\omega = \sum \mu_i dA_i/A_i$ is holomorphic on W. We now show that the form ω is equal to zero. Represent ω as a linear combination of the least possible number of logarithmic differentials of meromorphic functions. As we have just shown, the supports of the divisors of these functions must be contained in the set of poles of the form ω, which means that all these divisors are equal to zero, whence $\omega = 0$. It follows that the forms dA_i/A_i are linearly dependent, and the numbers of terms in the representation of the form γ can be reduced, which gives a contradiction. We have thus shown that the functions φ_i are linearly independent.

We now show that the numbers $\lambda_1, \ldots, \lambda_n$ lie in the vector space $V(\mathrm{res}\, \alpha)$. Indeed, the form $\sum_{i=1}^{n} \lambda_i dA_i/A_i$ has the same residues as the form α, i.e., $\sum \lambda_i \varphi_i(x) = \text{res}_x \alpha$ for $x \in P$. Since φ_i are independent integer-valued functions, the numbers $\lambda_1, \dots, \lambda_n$ are linear combinations with rational coefficients of the values of the function $\text{res}\,\alpha$, i.e., they lie in the vector space $V(\text{res}\,\alpha)$. The numbers $\lambda_1, \dots, \lambda_n$ are linearly independent over $\mathbb{Q}$. They generate the space $V(\text{res}\,\alpha)$, since $\sum \lambda_i \varphi_i(x) = \text{res}_x \alpha$. □

Corollary 1.48 *If Problem 1.46 is solvable for a form α, then there exists a unique form γ satisfying part 1 of that problem.*

Proof If there are two forms satisfying part 1 of Problem 1.46, then all residues of the difference of these forms are equal to zero. Repeating the argument from the proof of Lemma 1.47, we obtain that the difference is equal to zero. □

Corollary 1.49 *If Problem 1.46 is solvable for a form α, then the number of summands in a solution of part 2 of that problem for the form α is equal to the dimension of the space $V(\text{res}\,\alpha)$ over the field $\mathbb{Q}$.*

We say that a divisor $D = \sum r_i x_i$, $x_i \in W$, with rational coefficients $r_i = p_i/q_i$ of degree $\sum r_i = 0$ is *almost principal* if there exists a positive integer N such that the divisor ND is principal, i.e., $ND = (A)$, where (A) is the divisor of a meromorphic function A. Let us restate this definition. Let k be the least common multiple of the denominators q_i of the coefficients r_i appearing in D. A divisor $D = \sum r_i x_i$ with rational coefficients is almost principal if the divisor kD with integer coefficients has finite order in the Jacobian variety of the curve W.

Proposition 1.50 *The following properties hold:*

1. *The sum of almost principal divisors is an almost principal divisor.*
2. *The product of an almost principal divisor and a rational number is an almost principal divisor.*

Proof 1. If $N_1 D_1 = (A_1)$ and $N_2 D_2 = (A_2)$, then $(N_1 N_2)(D_1 + D_2) = (A_1^{N_2} A_2^{N_1})$.
2. If $ND = (A)$ and $r = p/q$, then $Nq(rD) = (A^p)$. □

For a finite subset P of a compact Riemann surface, let $J_0(P)$ denote the set of functions $\psi : P \to \mathbb{Q}$ taking rational values and such that the divisor $D_\psi = \sum_{x \in P} \psi(x) x$ is almost principal. By the proposition that we have just proved, the sets $J_0(P)$ are vector spaces over the field $\mathbb{Q}$. The space $J_0(P)$ contains the lattice $\bar{J}_0(P)$ of functions corresponding to the principal divisors. The space $J_0(P)$ is spanned by the lattice $\bar{J}_0(P)$ over the rational numbers.

We now state a necessary and sufficient condition for the solvability of Problem 1.46. Let P be the set of all points x at which the residue $\text{res}_x \alpha$ of the form α is not equal to zero. On the set P is the function $\text{res}\,\alpha : P \to \mathbb{C}^*$ assigning the residue $\text{res}_x \alpha$ to each point $x \in P$. To the function $\text{res}\,\alpha$, we have previously associated the vector space $V(\text{res}\,\alpha)$ over the field $\mathbb{Q}$ spanned by the values of the function $\text{res}\,\alpha$. We now define yet another space, $F(\text{res}\,\alpha)$. Let $\lambda_1, \dots, \lambda_n$ be a basis of the space $V(\text{res}\,\alpha)$ over the field $\mathbb{Q}$. Consider the coordinate functions $\varphi_i : P \to \mathbb{Q}$, $i = 1, \dots, n$, defined by the identity $\text{res}_x \alpha = \sum \varphi_i(x) \lambda_i$.

The space $F(\operatorname{res}\alpha)$ is the vector space over the field $\mathbb{Q}$ spanned by the functions $\varphi_1, \dots, \varphi_n$. The space $F(\operatorname{res}\alpha)$ is well defined. Indeed, let $u_1, \dots, u_n$ be a different basis of the space $V(\operatorname{res}\alpha)$ and $\lambda_i = \sum_j a_{i,j} u_j$, where $\{a_{i,j}\}$ is an invertible $n \times n$ matrix with rational entries. Then $\operatorname{res}\alpha = \sum \rho_j u_j$, where $\rho_j = \sum_i a_{i,j}\varphi_i$. Therefore, the vector space over the field $\mathbb{Q}$ spanned by the functions ρ_j is contained in the vector space over $\mathbb{Q}$ spanned by the functions φ_i. The opposite inclusion is proved in the same way.

Proposition 1.51 *For every function $\varphi : P \to \mathbb{C}$ from the space $F(\operatorname{res}\alpha)$, the following equality holds: $\sum_{x \in P} \varphi(x) = 0$.*

Proof The sum of all residues of a form α is equal to zero. The residue $\operatorname{res}_x \alpha$ can be regarded as a vector in the space $V(\operatorname{res}\alpha)$. If the sum of vectors is equal to zero, then no matter how a basis $\lambda_1, \dots, \lambda_n$ is chosen, the sum of the ith coordinates of these vectors is also equal to zero for every i, $1 \le i \le n$. □

Proposition 1.52 (Condition for solvability of Problem 1.46) *Problem 1.46 for a form α is solvable if and only if the space $F(\operatorname{res}\alpha)$ is contained in the space $J_0(P)$.*

Proof Suppose that Problem 1.46 is solvable, and a sum $\sum \lambda_i dA_i/A_i$ with the same residues as α has the minimal possible number of terms. Consider integer-valued functions on the set P assigning to a point $x \in P$ the order of the function A_i at that point: $\varphi_i(x) = \operatorname{res}_x dA_i/A_i$. The form $\gamma = \sum \lambda_i dA_i/A_i$ has the same residues as α, i.e., $\sum \lambda_i \varphi_i(x) = \operatorname{res}_x \alpha$. By Lemma 1.10, the numbers $\lambda_1, \dots, \lambda_n$ form a basis of the vector space $V(\operatorname{res}\alpha)$. Therefore, the space $F(\operatorname{res}\alpha)$ is generated by the functions $\varphi_1, \dots, \varphi_n$. These functions lie in the space $J_0(P)$, since the divisors $D_i = \sum \varphi_i(x)x$ are the divisors of the functions A_i.

Suppose that the space $F(\operatorname{res}\alpha)$ is contained in the space $J_0(P)$. Choose a basis $\mu_1, \dots, \mu_n$ in the space $V(\operatorname{res}\alpha)$. By our assumption, the function $\operatorname{res}\alpha$ is representable in the form $\operatorname{res}\alpha = \sum \mu_i \rho_i$, where the functions ρ_i lie in the space $J_0(P)$. This means that for every i, there exist a positive integer N_i and a meromorphic function B_i such that the value of the function $N_i \rho_i$ at a point $x \in P$ is equal to the residue at this point of the form dB_i/B_i. It follows that the form

$$\sum \frac{\mu_i}{N_i} \frac{dB_i}{B_i}$$

has the same residues as the form α. □

Thus, according to the condition just proved, Problem 1.46 is solvable if and only if every divisor from a finite number of explicitly constructed divisors of degree zero becomes principal after multiplication by a sufficiently large integer. Abel's theorem gives a description of principal divisors. Therefore, the question of solvability of Problem 1.46 reduces in principle to Abel's theorem.

Remark 1.53 Is a given (explicitly constructed) divisor on an algebraic curve principal or almost principal? This question may turn out to be hard, since Abel's theorem is not constructive. Abel faced this problem in his work on the integrability by elementary functions of pseudo-hyperelliptic integrals. This work of Abel was completed by Zolotarev [26], and this problem was solved in [83]; see also [11,27]. An effective version of Liouville's theorem is discussed in the paper [15].

Example 1.54 Let W be a curve of genus 1. Fix a structure of an algebraic abelian group on W by specifying a point $* \in W$ that plays the role of the neutral element in the group. Consider the dense subset F of W consisting of all elements of W of finite order. Problem 1.46 is solvable for every form α for which the set P is contained in the set F.[5]

Let us now discuss the opposite situation. We say that a finite set F on a curve of positive genus is a *generic set* if there is no nonzero principal divisor D whose support is contained in the set F. As can be seen from Abel's theorem, among all the sets in a curve of positive genus consisting of the same number of points, the generic sets have full measure.

Example 1.55 Let W be a curve of positive genus, and F a generic subset of it. Problem 1.46 is unsolvable for every form α for which the set P is nonempty and is contained in F. Indeed, in this case, $J_0(P)$ is trivial, but $F(\operatorname{res} \alpha)$ is nontrivial.

1.9.3 Elementarity and Nonelementarity of Abelian Integrals

The question whether the integral of an algebraic function is elementary reduces to Problems 1.42 and 1.46 (see Sects. 1.9.1 and 1.9.2). Recall that an *antiderivative* (or a *primitive*) of a one-form α is a function whose differential is equal to α.

Theorem 1.56 *An antiderivative of a meromorphic form α is generalized elementary if and only if Problems 1.42 and 1.46 are solvable for the form α, and the form α is equal to the sum of the solutions of these problems, i.e., $\alpha = \beta + \gamma$, where β is the solution of Problem 1.42 for the form α, and γ is the solution of Problem 1.46 for the form α.*

Proof If an antiderivative of the form α is a generalized elementary function, then by Liouville's theorem, Problems 1.42 and 1.46 are solvable for the form α, and the form α is the sum of the solutions of these problems. The converse is obvious. □

Corollary 1.57 *Suppose that for a meromorphic form α on an algebraic curve of positive genus, the set of points P at which the residue of α is different from zero is a generic subset of the curve. Then an antiderivative of the form α cannot be a generalized elementary function.*

---

[5]It suffices to prove that every function $\varphi : P \to \mathbb{Q}$ belongs to $J_0(P)$. Indeed, a function $\varphi : P \to \mathbb{Q}$ belongs to $J_0(P)$ if and only if the point $\sum_{a \in P}(k\varphi(a))a$ has finite order in W, where k is the least common multiple of all the values of φ.

Let P be a finite subset of a compact algebraic curve W. Let $\overline{\Omega}_P$ denote the space of all meromorphic 1-forms on the curve W whose residues at every point of the set $W \setminus P$ are equal to zero. Every form $\alpha \in \overline{\Omega}_P$ defines the following one-dimensional de Rham cohomology class $[\alpha]$ of the space $W \setminus P$: the value $[\alpha](\gamma)$ of the class $[\alpha]$ on a 1-cycle γ is equal to $\int_{\tilde{\gamma}} \alpha$, where $\tilde{\gamma}$ is a 1-cycle avoiding the poles of the form α and homologous to γ in the region $W \setminus P$. The value of the integral does not depend on the choice of 1-cycle $\tilde{\gamma}$, since by our assumption, the residue of the form α at every pole lying in $W \setminus P$ is equal to zero.

Let $D = (f)$ be a principal divisor on the curve W whose support is contained in the set P. Associated with the divisor D, there is an integer cohomology class $[D]$ of the space $W \setminus P$ given by the 1-form

$$\frac{1}{2\pi i}\frac{df}{f}$$

(the function f is determined by the divisor D up to a multiplicative constant, whose choice does not affect the 1-form). Let $L(P)$ denote the complex vector subspace in the first cohomology space of $W \setminus P$ spanned by the integer classes $[D]$ of the principal divisors D whose supports lie in the set P (such divisors correspond to points of the lattice $\overline{J}_0(P)$).

Theorem 1.58 *An antiderivative of a form α on a curve W that belongs to the space $\overline{\Omega}_P$ is a generalized elementary function if and only if the cohomology class $[\alpha] \in H^1(W \setminus P, \mathbb{C})$ lies in the space $L(P)$.*

Proof If the class $[\alpha]$ belongs to the space $L(P)$, then the form α on the set $W \setminus P$ defines the same cohomology class as a certain form $\gamma = \sum \lambda_i dA_i/A_i$, where the A_i are meromorphic functions whose divisors have support in P. The form $\beta = \alpha - \gamma$ defines the zero cohomology class of $W \setminus P$. Therefore, the antiderivative of the form β is a single-valued function on $W \setminus P$. This antiderivative grows at most polynomially near the poles of the form β and hence is a meromorphic function on the curve W. The converse follows from Liouville's theorem. □

We now give several corollaries of the theorems proved above. First of all, observe the following topological obstruction to the elementarity of an antiderivative of an algebraic 1-form.

Corollary 1.59 *If an antiderivative of a meromorphic form α belonging to the space $\overline{\Omega}_P$ is a generalized elementary function, then $1/2\pi i$ times the value of the cohomology class $[\alpha]$ at any 1-cycle $\gamma \in H_1(W \setminus P, \mathbb{Z})$ belongs to the space $V(\operatorname{res} \alpha)$.*

Proof Indeed, if an antiderivative of the form α is elementary, then

$$\alpha = dA_0 + \sum \lambda_j \frac{dA_j}{A_j},$$

and the numbers λ_j belong to the space $V(\operatorname{res}\alpha)$. The periods of the form dA_0 are equal to zero, and the periods of the forms

$$\frac{1}{2\pi i}\frac{dA_j}{A_j}$$

are integers. □

Corollary 1.60 ([26]) *If all residues of a meromorphic form α on a compact Riemann surface W are equal to zero, then an antiderivative of α is elementary if and only if it is a single-valued function on W.*

Proof An antiderivative of a meromorphic form grows at most polynomially near the poles of the form. Therefore, if an antiderivative is single-valued, then it is a meromorphic function. The converse follows from the preceding corollary. □

Corollary 1.61 ([26]) *An antiderivative of a nonzero holomorphic form on a compact Riemann surface is never an elementary function. For example, an elliptic integral $\int_{x_0}^{x} dt/\sqrt{P(t)}$, where P is a cubic polynomial with no multiple roots, is not elementary.*

Indeed, the 1-form

$$\frac{dx}{y} = \frac{dx}{\sqrt{P(x)}}$$

on the projective algebraic curve E given in affine coordinates (x, y) by the equation $y^2 = P(x)$ is holomorphic. This follows from the identity

$$\frac{dx}{y} = \frac{2dy}{P'(x)},$$

which holds on E.

Proof An antiderivative of a holomorphic form is single-valued if and only if the form is equal to zero. □

Proposition 1.62 *Suppose that a form α has at most simple poles, and all its residues are rational. Then an antiderivative of the form α is a generalized elementary function if and only if all periods of the form $\frac{1}{2\pi i}\alpha$ are rational.*

Proof The necessity of the requirement that all periods be rational follows from Corollary 1.59. We now verify the sufficiency. By our assumption, for some positive integer N, all periods of the form $N\alpha/2\pi i$ are integers. Therefore, the function F defined by the equality $F(x) = \exp\int_{x_0}^{x} N\alpha$ is a single-valued function on W. The function F is meromorphic, since the form α has simple poles only. The equality

$$\int_{x_0}^{x}\alpha = \frac{1}{N}\log F(x) + c$$

proves the desired statement. □

Corollary 1.63 *Suppose that all residues of a meromorphic form α are rational. Then an antiderivative of the form α is a generalized elementary function if and only if Problem 1.42 is solvable for the form α, and all periods of the form $\frac{1}{2\pi i}\alpha$ are rational.*

Proof Since Problem 1.42 is solvable for the form α, there exists a meromorphic function A_0 such that the form $(\alpha - dA_0)$ has only simple poles. The form $(\alpha - dA_0)$ has the same periods as the form α, and Proposition 1.62 is applicable to it. □

1.10 The Liouville–Mordukhai-Boltovski Criterion

The first results on the nonsolvability of linear differential equations in explicit form are due to Liouville (see [77, 86]).

Theorem 1.64 *A differential equation $y'' + py' + qy = 0$ with coefficients from a functional differential field K all of whose elements are representable by generalized quadratures can be solved by generalized quadratures if and only if it has a solution of the form $y_1(x) = \exp \int_{x_0}^{x} f(t)\,dt$, where f is a function that satisfies an algebraic equation with coefficients in the field K.*

In one direction, the theorem is obvious. If one solution y_1 of a linear second-order differential equation is known, then the equation can be solved by reducing its order. The proof in the other direction is rather involved.

It took more than half a century to generalize this theorem of Liouville to equations of order n. Using Liouville's method, Mordukhai-Boltovski proved in 1910 the following criterion, which makes it possible to reduce the question of solvability of an equation to the question of solvability of another equation whose order is lower.

Theorem 1.65 (The Liouville–Mordukhai-Boltovski criterion) *A linear differential equation*

$$y^{(n)} + p_1 y^{(n-1)} + \cdots + p_{ny} = 0$$

of order n with coefficients in a functional differential field K all of whose elements are representable by generalized quadratures is solvable by generalized quadratures if and only if (1) it has a solution of the form $y_1(x) = \exp \int_{x_0}^{x} f(t)\,dt$, where f is a function that belongs to a certain algebraic extension K_1 of the field K, and (2) the differential equation of order $n-1$ in the function

$$z = y' - \frac{y_1'}{y_1} y$$

with coefficients in the field K_1 obtained from the initial equation by order reduction (see Sect. 2.1.2) is solvable by generalized quadratures over the field K_1.

In the same year, 1910, the Picard–Vessiot theorem appeared, in which the question of solvability of linear differential equations was answered in a fundamentally different manner, namely, from the viewpoint of differential Galois theory.

In the third chapter of this book, we discuss the foundations of this theory. The Liouville–Mordukhai-Boltovski criterion is essentially equivalent to the Picard–Vessiot theorem. Picard–Vessiot theory not only explains this criterion but also gives the possibility of making it into an explicit algorithm that would make it possible to determine for a differential equation over the field of rational functions (i.e., with rational coefficients) whether it is solvable by generalized quadratures (see [92] and Sect. 3.7).

Chapter 2
Solvability of Algebraic Equations by Radicals and Galois Theory

Is a given algebraic equation solvable by radicals? Can one solve a given algebraic equation of degree n using solutions of auxiliary algebraic equations of smaller degree and radicals? In this chapter, we discuss how Galois theory answers these questions (at least in principle).

The questions we have posed are purely algebraic by nature and can be stated over any field K. We will assume in this chapter that the field K has characteristic zero. This case is slightly simpler than the case of general characteristic, and we are mostly interested in functional differential fields, which contain all complex constants. Other interesting examples of fields to which the results of this chapter are applicable are subfields of the field of complex numbers (in particular, the field $\mathbb{Q}$ of all rational numbers).

The "permissive" part of Galois theory (see Sect. 2.1) that allows us to solve equations by radicals is very simple. It depends neither on the fundamental theorem of Galois theory nor on the theory of fields, and it essentially belongs to linear algebra. Only these linear-algebraic considerations are used in the topological version of Galois theory in relationship to the question of representability of algebraic functions by radicals. However, a sufficient condition for solvability of an equation by solving auxiliary equations of smaller degree and taking radicals depends not only on linear algebra, but also on the fundamental theorem of Galois theory. This is one of the reasons that we give a complete proof of that theorem.

The well-known properties of solvable groups and the symmetric groups $S(k)$ are used without proof. In Sect. 2.9.1, we prove a considerably less well known characteristic property of the subgroups of $S(k)$. These facts from group theory are applied in usual Galois theory as well as in its differential and topological versions.

We often need to extend the field under consideration by adjoining one or several roots of an algebraic equation. For functional differential fields, this construction is simple and was already described in Sect. 1.5. For subfields of the field $\mathbb{C}$ of complex numbers, the construction of such extensions is obvious. Since we are

A. Khovanskii, *Topological Galois Theory*, Springer Monographs in Mathematics,
DOI 10.1007/978-3-642-38871-2_2

mostly interested in fields of these two types, we will use such extensions below without giving details on how to construct them.

Several words are in order on the organization of the material. In Sects. 2.1–2.4, we consider a field P on which a finite group G acts by field automorphisms. Elements of the field P fixed under the action of G form a subfield $K \subseteq P$ called the *invariant subfield of* G.

In Sect. 2.1, we show that if the group G is solvable, then the elements of the field P are representable by radicals through the elements of the invariant subfield K of G. (Here an additional assumption is needed that the field K contains all roots of unity of degree equal to the cardinality of G.) When P is the field of rational functions of n variables, G is the symmetric group acting by permutations of the variables, and K is the subfield of symmetric functions of n variables, this result provides an explanation for the fact that algebraic equations of degrees 2, 3, and 4 in one variable are solvable by radicals.

In Sect. 2.2, we show that for every subgroup G_0 of the group G, there exists an element $x \in P$ whose stabilizer is equal to G_0. The results of Sects. 2.1 and 2.2 are based on simple considerations from group theory; they use the explicit formula for the Lagrange interpolating polynomial.

In Sect. 2.3, we show that every element of the field P is algebraic over the field K. We prove that if the stabilizer of a point $z \in P$ contains the stabilizer of a point $y \in P$, then z is the value at y of some polynomial over the field K. This proof is also based on the study of the Lagrange interpolating polynomial (see Sect. 2.3.3).

In Sect. 2.4, we introduce the class of k-solvable groups. We show that if a group G is k-solvable, then the elements of the field P are representable in k-radicals through the elements of the field K (that is, they can be obtained by taking radicals and solving auxiliary algebraic equations of degree k or less from the elements of the field K). Here we need to assume additionally that the field K contains all roots of unity of order the cardinality of G.

Consider now a different situation. Suppose that a field P is obtained from a field K by adjoining all roots of a polynomial equation over K with no multiple roots. In this case, there exists a finite group G of automorphisms of the field P whose invariant subfield coincides with K. To construct the group G, the initial equation needs to be replaced with an equivalent Galois equation, i.e., an equation each of whose roots can be expressed through any other root (see Sect. 2.5). The group G of automorphisms is constructed in Sect. 2.6.

Thus Sects. 2.2, 2.3, 2.5, and 2.6 contain proofs of the central theorems of Galois theory. In Sect. 2.7, we summarize the results obtained so far and then state and prove the fundamental theorem of Galois theory.

An algebraic equation over some field is solvable by radicals if and only if its Galois group is solvable (Sect. 2.8), and it is solvable by k-radicals if and only if its Galois group is k-solvable (Sect. 2.9). In Sect. 2.10, we discuss the question of solvability of algebraic equations with higher complexity by solving equations with lower complexity. We give a necessary condition for such solvability in terms of the Galois group of the equation.

A major focus in this chapter is the applications of Galois theory to problems of solvability of algebraic equations in explicit form. However, the exposition of

Galois theory does not refer to these applications. The fundamental principles of Galois theory are covered in Sects. 2.2, 2.3, and 2.5–2.7, and those sections can be read independently of the rest of the chapter.

A recipe for solving algebraic equations by radicals (including solutions of general equations of degree 2, 3, and 4) is given in Sect. 2.1; it is independent of the remainder of the text.

2.1 Action of a Solvable Group and Representability by Radicals

In this section, we prove that if a finite solvable group G acts on a field P by field automorphisms, then (under certain additional assumptions on the field P) all elements of P can be expressed through the elements of the invariant subfield K of G by radicals and arithmetic operations.

A construction of a representation by radicals is based on linear algebra (see Sect. 2.1.1). In Sect. 2.1.2, we use this result to prove solvability of equations of low degree. To obtain explicit solutions, the linear-algebraic construction needs to be done explicitly. In Sect. 2.1.3, we introduce the technique of Lagrange resolvents, which allows us to perform an explicit diagonalization of an abelian linear group. In Sect. 2.1.4, we explain how Lagrange resolvents can help us in writing down explicit formulas with radicals for the solutions of equations of degree 2, 3, and 4.

The results of this section are applicable in the general situation considered in Galois theory. If a field P is obtained from the field K by adjoining all roots of an algebraic equation without multiple roots, then there exists a group G of automorphisms of the field P whose invariant subfield is the field K (see Sect. 2.7.1). This group is called the *Galois group of the equation.*

It follows from the results of this section (the sufficient condition for solvability by radicals from Theorem 2.50) that an equation whose Galois group is solvable can be solved by radicals. The existence of the Galois group is by no means obvious; it is one of the central results of Galois theory. In this section, we do not prove this theorem (a proof is given in Sect. 2.7.1); we assume from the very beginning that the group G exists.

In a variety of important cases, the group G is given a priori. This is the case, for example, if K is the field of rational functions of a single variable, P is the field obtained by adjoining all solutions of an algebraic equation to K, and G is the monodromy group of the algebraic function defined by this equation (see Chap. 4).

2.1.1 A Sufficient Condition for Solvability by Radicals

The fact that we shall be dealing with fields is barely used in the construction of representation by radicals. To emphasize this, we describe this construction in a general setup whereby a field is replaced with an algebra V, which may even be noncommutative. (In fact, we do not even need to multiply different elements of the algebra. We will use only the operation of taking an integer power k of an element and the fact that this operation is homogeneous of degree k under multiplication by elements of the base field: $(\lambda a)^k = \lambda^k a^k$ for all $a \in V, \lambda \in K$.)

Let V be an algebra over the field K containing all roots of unity. A finite abelian group of linear transformations of a finite-dimensional vector space over the field K can be diagonalized in a suitable basis (see Sect. 2.1.3).

Proposition 2.1 *Let G be a finite abelian group of order n acting by automorphisms of the algebra V. Suppose that K contains all roots of unity of degree n. Then every element of the algebra V is representable as a sum of $k \leq n$ elements $x_i \in V$, $i = 1, \ldots, k$, such that x_i^n lies in the invariant subalgebra V_0 of G, i.e., in the fixed-point set of the group G.*

Proof Consider a finite-dimensional vector subspace L in the algebra V spanned by the G-orbit of an element x. The space L splits into a direct sum $L = L_1 \oplus \cdots \oplus L_k$ of eigenspaces for all operators from G (see Sect. 2.1.3). Therefore, the vector x can be represented in the form $x = x_1 + \cdots + x_k$, where $x_1, \ldots, x_k$ are eigenvectors for all the operators from the group. The corresponding eigenvalues are nth roots of unity. Therefore, the elements $x_1^n, \ldots, x_k^n$ belong to the invariant subalgebra V_0. □

Definition 2.2 We say that an element x of the algebra V is an *nth root of an element a* if $x^n = a$.

We can now restate Proposition 2.1 as follows: every element x of the algebra V is representable as a sum of nth roots of some elements of the invariant subalgebra.

Theorem 2.3 *Let G be a finite solvable group of automorphisms of the algebra V of order n. Suppose that the field K contains all roots of unity of degree n. Then every element x of the algebra V can be obtained from the elements of the invariant subalgebra V_0 by root extractions and summations.*

We first prove the following simple statement about an action of a group on a set. Suppose that a group G acts on a set X, that H is a normal subgroup of G, and that X_0 is the subset of X consisting of all points fixed under the action of G.

Proposition 2.4 *The subset X_H of the set X consisting of the fixed points under the action of the normal subgroup H is invariant under the action of G. There is a natural action of the quotient group G/H on the set X_H with the fixed-point set X_0.*

Proof Suppose that $g \in G$, $h \in H$. Then the element $g^{-1}hg$ belongs to the normal subgroup H. Let $x \in X_H$. Then $g^{-1}hg(x) = x$, or $h(g(x)) = g(x)$, which means that the element $g(x) \in X$ is fixed under the action of the normal subgroup H. Thus the set X_H is invariant under the action of the group G. Under the action of G on X_H, all elements of H correspond to the identity transformation. Hence the action of G on X_H reduces to an action of the quotient group G/H. □

We now proceed with the proof of Theorem 2.3.

Proof (of Theorem 2.3) Since the group G is solvable, it has a chain of nested subgroups $G = G_0 \supset \cdots \supset G_m = e$ in which the group G_m consists of the identity element e only, and every group G_i is a normal subgroup of the group G_{i-1}. Moreover, the quotient group G_{i-1}/G_i is abelian.

Let $V_0 \subset \cdots \subset V_m = V$ denote the chain of invariant subalgebras of the algebra V with respect to the action of the groups $G_0, \ldots, G_m$. By Proposition 2.4, the abelian group G_{i-1}/G_i acts naturally on the invariant subalgebra V_i, leaving the subalgebra V_{i-1} pointwise fixed. The order m_i of the quotient group G_{i-1}/G_i divides the order of the group G. Therefore, Proposition 2.1 is applicable to this action. We conclude that every element of the algebra V_i can be expressed with the help of summation and root extraction through the elements of the algebra V_{i-1}. Repeating the same argument, we will be able to express every element of the algebra V through the elements of the algebra V_0 by a chain of summations and root extractions. □

2.1.2 The Permutation Group of the Variables and Equations of Degree 2, 3, and 4

Theorem 2.3 explains why equations of low degree are solvable by radicals. Suppose that the algebra V is the polynomial ring in the variables $x_1, \ldots, x_n$ over the field K. The symmetric group $S(n)$ consisting of all permutations of n elements acts on this ring, permuting the variables $x_1, \ldots, x_n$ in polynomials from this ring. The invariant subalgebra of this action consists of all symmetric polynomials. Every symmetric polynomial can be represented explicitly as a polynomial in the elementary symmetric functions $\sigma_1, \ldots, \sigma_n$, where

$$\sigma_1 = x_1 + \cdots + x_n, \quad \sigma_2 = \sum_{i<j} x_i x_j, \quad \ldots, \quad \sigma_n = x_1 \cdots x_n.$$

Consider the general algebraic equation $x^n + a_1 x^{n-1} + \cdots + a_n = 0$ of degree n. According to Viète's formulas, the coefficients of this equation are equal, up to sign, to the elementary symmetric functions of its roots $x_1, \ldots, x_n$. Namely, $\sigma_1 = -a_1, \ldots, \sigma_n = (-1)^n a_n$.

For $n = 2, 3, 4$, the group $S(n)$ is solvable. Suppose that the field K contains all roots of unity of degree less than or equal to 4. Applying Theorem 2.3, we obtain that every polynomial in $x_1, \ldots, x_n$ can be expressed through the elementary symmetric polynomials $\sigma_1, \ldots, \sigma_n$ using root extraction, summation, and multiplication by rational numbers. Therefore, Theorem 2.3 for $n = 2$, 3, and 4 establishes the representability of the roots of a degree-n algebraic equation through the coefficients of this equation using root extraction, summation, and multiplication by rational numbers.

To obtain explicit formulas for the roots, we need to repeat all these arguments, performing all necessary constructions explicitly. We will do this in Sects. 2.1.3 and 2.1.4.

2.1.3 Lagrange Polynomials and Abelian Linear-Algebraic Groups

Let T be a monic polynomial of degree n over an arbitrary field K. Suppose that the polynomial T has exactly n distinct roots $\lambda_1, \dots, \lambda_n$. To every root λ_i, we associate the polynomial

$$T_i(t) = \frac{T(t)}{T'(\lambda_i)(t - \lambda_i)},$$

which is the unique polynomial of degree at most $n - 1$ that is equal to 1 at the root λ_i and to zero at all other roots of the polynomial T. Let $c_1, \dots, c_n$ be any collection of elements of the field K. The polynomial $L(t) = \sum c_i T_i(t)$ is called the *Lagrange interpolating polynomial* with interpolation points $\lambda_1, \dots, \lambda_n$ and interpolation data $c_1, \dots, c_n$. This is the unique polynomial of degree less than n that takes the value c_i at every point λ_i, $i = 1, \dots, n$.

Consider a vector space V (possibly infinite-dimensional) over the field K and a linear operator $A : V \to V$. Suppose that the operator A satisfies a linear equation

$$T(A) = A^n + a_1 A^{n-1} + \dots + a_{n-1} A + a_n E = 0,$$

where $a_i \in K$ and E is the identity operator. Assume that the polynomial $T(t) = x^n + a_1 x^{n-1} + \dots + a_n$ has n distinct roots $\lambda_1, \dots, \lambda_n$ in the field K. The operator $L_i = T_i(A)$, where

$$T_i(t) = \frac{T(t)}{T'(\lambda_i)(t - \lambda_i)},$$

will be called the *generalized Lagrange resolvent* of the operator A corresponding to the root λ_i. For every vector $x \in V$, the vector $x_i = L_i x$ will be called the *generalized Lagrange resolvent (corresponding to the root λ_i)* of the vector x.

Proposition 2.5 *The following statements hold:*

1. *The generalized Lagrange resolvents L_i of the operator A satisfy the following relations: $L_1 + \dots + L_n = E$, $L_i L_j = 0$ for $i \neq j$, $L_i^2 = L_i$, $AL_i = \lambda_i L_i$.*
2. *Every vector $x \in V$ is representable as the sum of its generalized Lagrange resolvents, i.e., $x = x_1 + \dots + x_n$. Moreover, the nonzero resolvents x_i of the vector x are linearly independent and are equal to eigenvectors of the operator A with corresponding eigenvalues λ_i.*

Proof 1. Let $\Lambda = \{\lambda_i\}$ be the set of all roots of the polynomial T. By definition, the polynomial T_i is equal to 1 at the point λ_i and is equal to zero at all other points of this set. It is obvious that the following polynomials vanish on the set Λ: $T_1 + \cdots + T_n - 1$, $T_i T_j$ for $i \neq j$, $T_i^2 - T_i$, $tT_i - \lambda_i T_i$. Therefore, each of the polynomials indicated above is divisible by the polynomial T, which has simple roots at the points of the set Λ. Since the polynomial T annihilates the operator A, i.e., $T(A) = 0$, this implies the relations $L_1 + \cdots + L_n = E$, $L_i L_j = 0$ for $i \neq j$, $L_i^2 = L_i$, $AL_i = \lambda_i L_i$.

2. The second part of the statement is a formal consequence of the first. Indeed, since $E = L_1 + \cdots + L_n$, every vector x satisfies $x = L_1 x + \cdots + L_n x = x_1 + \cdots + x_n$. Assume that the vector x is nonzero and that some linear combination $\sum \mu_j x_j$ of the vectors $x_1, \dots, x_n$ vanishes. Then

$$0 = L_i \sum \mu_j L_j x = \sum L_i L_j \mu_j x = \mu_i x_i,$$

i.e., every nonzero vector x_i enters this linear combination with coefficient zero: $\mu_i = 0$. The identity $AL_i = \lambda_i L_i$ implies that $AL_i x = \lambda_i L_i x$, i.e., either the vector $x_i = L_i x$ is an eigenvector of L_i with the eigenvalue λ_i, or $x_i = 0$.

□

The explicit construction for the decomposition of x into eigenvectors of the operator A carries over automatically to the case of several commuting operators. Let us discuss the case of two commuting operators in more detail. Suppose that along with the linear operator A on the space V, we are given another linear operator $B : V \to V$ that commutes with A and satisfies a polynomial relation of the form $Q(B) = B^k + b_1 B^{k-1} + \cdots b_k E = 0$, where $b_i \in K$. Assume that the polynomial $Q(t) = t^k + b_1 t^{k-1} + \cdots b_k$ has k distinct roots $\mu_1, \dots, \mu_k$ in the field K. To a root μ_j, we associate the polynomial

$$Q_j(t) = \frac{Q(t)}{Q'(\mu_j)(t - \mu_j)}$$

and the operator $Q_j(B)$, i.e., the generalized Lagrange resolvent of the operator B corresponding to the root μ_j. We call the operator $L_{i,j} = T_i(A) Q_j(B)$ the *generalized Lagrange resolvent of the operators A and B corresponding to the pair of roots* λ_i, μ_j. The vector $x_{i,j} = L_{i,j} x$ will be called the *generalized Lagrange resolvent of the vector* $x \in V$ *(corresponding to the pair of roots* λ_i *and* μ_j*) with respect to the operators A and B*.

Proposition 2.6 *The following statements hold:*

1. *The generalized Lagrange resolvents* $L_{i,j}$ *of commuting operators A and B satisfy the following relations:*

$$\sum L_{i,j} = E, \quad L_{i_1,j_1} L_{i_2,j_2} = 0 \text{ for } (i_1, j_1) \neq (i_2, j_2),$$

$$L_{i,j}^2 = L_{i,j}, \quad AL_{i,j} = \lambda_i L_{i,j}, \quad BL_{i,j} = \mu_j L_{i,j}.$$

2. *Every vector $x \in V$ is representable as the sum of its generalized Lagrange resolvents, i.e., $x = \sum x_{i,j}$. Moreover, the nonzero resolvents $x_{i,j}$ of the vector x are linearly independent and are equal to eigenvectors of the operators A and B with the eigenvalues λ_i and μ_j, respectively.*

To prove the first part of the proposition, it suffices to multiply the corresponding identities for the generalized resolvents of the operators A and B. The second part of the proposition is a formal consequence of the first part.

We can now apply the propositions just proved to an operator A of finite order: $A^n = E$. Generalized Lagrange resolvents for such operators are particularly important for solving equations by radicals. These are the resolvents that Lagrange discovered, and we call them the *Lagrange resolvents* (omitting the word "generalized"). Suppose that the field K contains n roots of unity $\xi_1, \dots, \xi_n$ of degree n, $\xi^n = 1$. By our assumption, $T(A) = 0$, where $T(t) = t^n - 1$. Let us now compute the Lagrange resolvent corresponding to the root $\xi_i = \xi$. We have

$$T_i(t) = \frac{t^n - \xi^n}{n\xi^{n-1}(t - \xi)} = \frac{1}{n\xi^{n-1}}\left(t^{n-1} + \cdots + \xi^{n-1}\right) = \frac{1}{n}\left(\left(\xi^{-1}t\right)^{n-1} + \cdots + 1\right).$$

The Lagrange resolvent $T_i(A)$ of the operator A corresponding to a root $\xi_i = \xi$ will be denoted by $R_\xi(A)$. We obtain

$$R_\xi(A) = \frac{1}{n} \sum_{0 \le k < n} \xi^{-k} A^k.$$

Corollary 2.7 *Consider a vector space V over a field K containing all roots of unity of degree n. Suppose that an operator A satisfies the relation $A^n = E$. Then for every vector $x \in V$, either the Lagrange resolvent $R_\xi(A)(x)$ is zero, or it is equal to an eigenvector of the operator A with the eigenvalue ξ. The vector x is the sum of all its Lagrange resolvents.*

Remark 2.8 Corollary 2.7 can be verified directly, without reference to any of the preceding results.

Let G be a finite abelian group of linear operators on a vector space V over the field K. Let n denote the order of the group G. Suppose that the field K contains all roots of unity of degree n. Then the space V is a direct sum of subspaces that are simultaneously eigenspaces for all operators from the group G. Let us make this statement more precise. Suppose that the group G is the direct sum of k cyclic groups of orders $m_1, \dots, m_k$. Suppose that the operators $A_i \in G, \dots, A_k \in G$ generate these cyclic subgroups. In particular, $A_1^{m_1} = E, \dots, A_k^{m_k} = E$. For every collection $\lambda = \lambda_1, \dots, \lambda_k$ of roots of unity of degree $m_1, \dots, m_k$, consider the *joint Lagrange resolvent* $L_\lambda = L_{\lambda_1}(A_1) \cdots L_{\lambda_k}(A_k)$ of all generators $A_1, \dots, A_k$ of the group G.

Corollary 2.9 *Every vector $x \in V$ is representable in the form $x = \sum L_\lambda x$. Each of the vectors $L_\lambda x$ is either zero or a common eigenvector of the operators $A_1, \dots, A_k$ with the respective eigenvalues $\lambda_1, \dots, \lambda_k$.*

2.1.4 Solving Equations of Degrees 2, 3, and 4 by Radicals

In this subsection, we revisit equations of low degree (see Sect. 2.1.2). We will use the technique of Lagrange resolvents and explain how the solution scheme for the equations from Sect. 2.1.2 can be promoted to explicit formulas. The formulas themselves will not be written down. We use notation from Sects. 2.1.2 and 2.1.3. Lagrange resolvents of operators will be labeled by the eigenvalues of these operators. Joint Lagrange resolvents of pairs of operators will be labeled by pairs of the corresponding eigenvalues.

Equations of degree 2 The polynomial ring $K[x_1, x_2]$ carries a linear action of the permutation group $S(2) = \mathbb{Z}_2$ of two elements. This group consists of the identity map and the operator of order 2 that permutes the variables x_1 and x_2. The element x_1 has two Lagrange resolvents with respect to the action of this operator:

$$R_1 = \frac{1}{2}(x_1 + x_2) = \frac{1}{2}\sigma_1, \quad R_{-1} = \frac{1}{2}(x_1 - x_2).$$

The square of the Lagrange resolvent R_{-1} is a symmetric polynomial. We have

$$R_{-1}^2 = \frac{1}{4}\left((x_1 + x_2)^2 - 4x_1x_2\right) = \frac{1}{4}\left(\sigma_1^2 - 4\sigma_2\right).$$

We obtain a representation of the polynomial x_1 through the elementary symmetric polynomials

$$x_1 = R_1 + R_{-1} = \frac{\sigma_1 \pm \sqrt{\sigma_1^2 - 4\sigma_2}}{2},$$

which gives the usual formula for the solutions of a quadratic equation.

Equations of degree 3 Suppose that the field K contains all three cube roots of unity.[1] There is an action on the polynomial ring $K[x_1, x_2, x_3] = V$ of the permutation group $S(3)$ of three elements. The alternating group $A(3)$, which is a cyclic group of order 3, is a normal subgroup of the group $S(3)$. The group $A(3)$ is generated by the operator B defining the permutation x_2, x_3, x_1 of the variables

[1]Since the field K has characteristic 0, it automatically contains the two square roots of unity 1 and -1.

x_1, x_2, x_3. The quotient group $S(3)/A(3)$ is a cyclic group of order 2. Let V_1 denote the invariant subalgebra of the group $A(3)$ (consisting of all polynomials that remain unchanged under all even permutations of the variables), and V_2 the algebra of symmetric polynomials. The element x_1 has three Lagrange resolvents with respect to the generator B of the group $A(3)$:

$$R_1 = \frac{1}{3}(x_1 + x_2 + x_3),$$

$$R_{\xi_1} = \frac{1}{3}\left(x_1 + \xi_2 x_2 + \xi_2^2 x_3\right),$$

$$R_{\xi_2} = \frac{1}{3}\left(x_1 + \xi_1 x_2 + \xi_1^2 x_3\right),$$

where

$$\xi_1, \xi_2 = \frac{-1 \pm \sqrt{-3}}{2}$$

are the cube roots of unity different from 1.

We have $x_1 = R_1 + R_{\xi_1} + R_{\xi_2}$, and R_1^3, $R_{\xi_1}^3$, $R_{\xi_2}^3$ lie in the algebra V_1. Moreover, the resolvent R_1 is a symmetric polynomial, and the polynomials $R_{\xi_1}^3$ and $R_{\xi_2}^3$ are interchanged by the action of the group $\mathbb{Z}_2 = S(3)/A(3)$ on the ring V_1. Applying the construction used for solving quadratic equations to the polynomials $R_{\xi_1}^3$ and $R_{\xi_2}^3$, we obtain that these polynomials can be expressed through the symmetric polynomials $R_{\xi_1}^3 + R_{\xi_2}^3$ and $(R_{\xi_1}^3 - R_{\xi_2}^3)^2$. We finally obtain that the polynomial x_1 can be expressed through the symmetric polynomials $R_1 \in V_2$, $R_{\xi_1}^3 + R_{\xi_2}^3 \in V_2$, and $(R_{\xi_1}^3 - R_{\xi_2}^3)^2 \in V_2$ with the help of square and cube root extractions and the arithmetic operations. To write down an explicit formula for the solution, it remains only to express these symmetric polynomials in terms of the elementary symmetric polynomials. A simple explicit solution of a degree-3 equation can be found in Appendix B.

Equations of degree 4 Equations of the fourth degree are solvable because the group $S(4)$ is solvable, and the group $S(4)$ is solvable because there exists a homomorphism $\pi : S(4) \to S(3)$ whose kernel is the abelian group $\mathrm{Kl} = \mathbb{Z}_2 \oplus \mathbb{Z}_2$. (This group is called the Klein four-group, whence the notation.) The homomorphism π can be described in the following way. There exist exactly three ways to split a four-element set into pairs of elements. Every permutation of the four elements gives rise to a permutation of these splittings. This correspondence defines the homomorphism π. The kernel Kl of this homomorphism is a normal subgroup of the group $S(4)$ consisting of four permutations: the identity permutation and the three permutations that are a product of two disjoint transpositions.

Suppose that the field K contains all three cube roots of unity. (As noted in footnote 1 on page 55, K also contains the two square roots of unity, ± 1.) The group

$S(4)$ acts on the polynomial ring $K[x_1, x_2, x_3, x_4] = V$. Let V_1 denote the invariant subalgebra of the normal subgroup Kl of the group $S(4)$. Thus the polynomial ring $V = K[x_1, x_2, x_3, x_4]$ carries an action of the abelian group Kl with the invariant subalgebra V_1. On the ring V_1, there is an action of the solvable group $S(3) = S(4)/\mathrm{Kl}$, and the invariant subalgebra with respect to this action is the ring V_2 of symmetric polynomials.

Let A and B be operators corresponding to the permutations x_2, x_1, x_4, x_3 and x_3, x_4, x_1, x_2 of the variables x_1, x_2, x_3, x_4. The operators A and B generate the group Kl. The following identities hold: $A^2 = B^2 = E$. The roots of the polynomial $T(t) = t^2 - 1$ annihilating the operators A and B are equal to $+1, -1$. The group Kl is the sum of two copies of the group with two elements, the first copy being generated by A, the second copy by B.

The element x_1 has four Lagrange resolvents with respect to the action of commuting operators A and B generating the group Kl:

$$R_{1,1} = \frac{1}{4}(x_1 + x_2 + x_3 + x_4), \quad R_{-1,1} = \frac{1}{4}(x_1 - x_2 + x_3 - x_4),$$

$$R_{1,-1} = \frac{1}{4}(x_1 + x_2 - x_3 - x_4), \quad R_{-1,-1} = \frac{1}{4}(x_1 - x_2 - x_3 + x_4).$$

The element x is equal to the sum of these resolvents: $x_1 = R_{1,1} + R_{-1,1} + R_{1,-1} + R_{-1,1}$, and the squares $R^2_{1,1}$, $R^2_{-1,1}$, $R^2_{1,-1}$, $R^2_{-1,1}$ of the Lagrange resolvents belong to the algebra V_1. Therefore, x_1 is expressible through the elements of the algebra V_1 with the help of the arithmetic operations and extraction of square roots. In turn, the elements of the algebra V_1 can be expressed through symmetric polynomials, since this algebra carries an action of the group $S(3)$ with the invariant subalgebra V_2 (see the solution of cubic equations above).

Let us show that this argument provides an explicit reduction of a fourth-degree equation to a cubic equation. Indeed, the resolvent $R_{1,1} = \frac{1}{4}\sigma_1$ is a symmetric polynomial, and the squares of the resolvents $R_{-1,1}$, $R_{1,-1}$, and $R_{-1,1}$ are permuted under the action of the group $S(4)$ (see the description of the homomorphism $\pi : S(4) \to S(3)$ above). Since the elements $R^2_{-1,1}$, $R^2_{1,-1}$, and $R^2_{-1,1}$ are only being permuted, the elementary symmetric polynomials in them are invariant under the action of the group $S(4)$ and hence belong to the ring V_2. Thus the polynomials

$$b_1 = R^2_{-1,1} + R^2_{1,-1} + R^2_{-1,1},$$

$$b_2 = R^2_{-1,1} R^2_{1,-1} + R^2_{1,-1} R^2_{-1,-1} + R^2_{-1,-1} R^2_{-1,1},$$

$$b_3 = R^2_{-1,1} R^2_{1,-1} R^2_{-1,-1}$$

are symmetric polynomials of x_1, x_2, x_3, and x_4. Therefore, b_1, b_2, and b_3 are expressible explicitly through the coefficients of the equation

$$x^4 + a_1 x^3 + a_2 x^2 + a_3 x + a_4 = 0, \tag{2.1}$$

whose roots are x_1, x_2, x_3, x_4. To solve (2.1), it suffices to solve the equation

$$r^3 - b_1 r^2 + b_2 r - b_3 = 0 \tag{2.2}$$

and set

$$x = \frac{1}{4}\left(-a_1 + \sqrt{r_1} + \sqrt{r_2} + \sqrt{r_3}\right),$$

where r_1, r_2, and r_3 are the roots of (2.2).

A beautiful explicit reduction of a fourth-degree equation to a third-degree equation based on consideration of a pencil of conics (published in [12]) can be found in Appendix B. We end this section with a brief description of it.

The coordinates of the intersection points of two conics $P = 0$ and $Q = 0$, where P and Q are given second-degree polynomials of x and y, can be found by solving one cubic and several quadratic equations. Indeed, every conic of the pencil $P + \lambda Q = 0$, where λ is an arbitrary parameter, passes through the points we are looking for. For some value λ_0 of the parameter λ, the conic $P + \lambda Q = 0$ splits into a pair of lines. This value satisfies the cubic equation $\det(\tilde{P} + \lambda \tilde{Q}) = 0$, where $\tilde{P}$ and $\tilde{Q}$ are 3×3 matrices of the quadratic forms corresponding to the equations of the conics in homogeneous coordinates.

The equation for each of the lines forming the degenerate conic $P + \lambda_0 Q = 0$ can be found by solving a quadratic equation. Indeed, the center of a degenerate conic given in affine coordinates by an equation $f(x, y) = 0$, i.e., the intersection point of the two lines forming the degenerate conic, can be found by solving the system $\partial f / \partial x = \partial f / \partial y = 0$. This is a linear system, and thus a solution can be expressed as a rational function of the coefficients. The intersection of a conic with any given line not passing through the center of the conic can be found by solving a quadratic equation. The two lines forming the degenerate conic are the lines connecting the center of the conic with the two intersection points. An equation of the line passing through two given points can be found with the help of arithmetic operations. If the equations of the lines into which the conic $P + \lambda_0 Q = 0$ splits are known, then to find the desired points, it remains only to solve the quadratic equations at the intersection points of the conic $P = 0$ and each of the two lines constituting the degenerate conic.

Therefore, the general equation of the fourth degree reduces to a cubic equation with the help of arithmetic operations and extraction of square roots. Indeed, the roots of the equation $a_0 x^4 + a_1 x^3 + a_2 x^2 + a_3 x + a_4 = 0$ are projections to the x-axis of the intersection points of the conics $y = x^2$ and $a_0 y^2 + a_1 xy + a_2 y + a_3 x + a_4 = 0$.

2.2 Fixed Points of Finite Group Actions

We prove here one of the central theorems of Galois theory, according to which distinct subgroups in a finite group of field automorphisms have distinct invariant subfields. The proof is based on a simple explicit construction using the Lagrange interpolating polynomial and on a geometrically obvious statement that a vector space cannot be covered by a finite number of proper vector subspaces.

We begin with the geometric statement. Let V be an affine space (possibly infinite-dimensional) over some field.

Proposition 2.10 *The space V cannot be represented as a union of a finite number of its proper affine subspaces.*

Proof We use induction on the number of affine subspaces. Suppose that the statement holds for every union of fewer than n proper affine subspaces. Suppose that the space V is representable as a union of n proper affine subspaces $V_1, \ldots, V_n$. Consider an arbitrary affine hyperplane $\hat{V}$ in the space V containing the first of these subspaces, V_1. The space V is the union of an infinite family of disjoint affine hyperplanes parallel to $\hat{V}$. (We always assume that the base field has characteristic zero, hence is infinite.) At most n hyperplanes from this family contain one of the subspaces $V_1, \ldots, V_n$. Take any other hyperplane from the family. The induction hypothesis applies to this hyperplane and its intersections with the affine subspaces $V_2, \ldots, V_n$, which concludes the proof. □

Corollary 2.11 *Suppose that a finite group of linear transformations acts on a vector space V. Then there exists a vector a such that the restriction of the action to the orbit of a is free.*

Proof The fixed-point set of a linear transformation is a vector subspace. If the linear transformation is different from the identity, then this subspace is proper. We can choose a to be any vector not belonging to the union of the fixed-point subspaces of nontrivial transformations from the group. □

The *stabilizer* $G_a \subset G$ of a vector $a \in V$ is defined as the subgroup consisting of all elements $g \in G$ that fix the vector a, i.e., such that $g(a) = a$.

In general, not every subgroup G_0 of a finite linear group G is the stabilizer of some vector a. As an example, consider the cyclic group of linear transformations of the complex line generated by multiplication by a primitive nth root of unity. If the number n is not prime, then this cyclic group has a nontrivial cyclic subgroup, but the stabilizers of all vectors are trivial subgroups (the identity subgroup for every element $a \neq 0$, and the entire group for $a = 0$). Thus the existence of a vector whose stabilizer coincides with G_0 is not obvious. Moreover, it is not true for all representations of the group G.

Lemma 2.12 *Let G_a and G_b be stabilizers of vectors a and b in some vector space V on which a finite group G acts by linear transformations. Then the subspace L spanned by the vectors a and b contains a vector c whose stabilizer G_c is equal to $G_a \cap G_b$.*

Proof The subgroup $G_a \cap G_b$ fixes all vectors of the space L. However, every element $g \notin G_a \cap G_b$ acts nontrivially either on the element a or on the element b. Vectors in L stable under the action of a fixed element $g \notin G_a \cap G_b$ form a proper subspace in L. By Proposition 2.10, such subspaces cannot cover the entire space L. □

Let G be a group of automorphisms of a field P. Fixed elements under the action of the group G form a subfield, which will be denoted by K. The field P can be viewed as a vector space over the field K.

The following theorem plays a major role in Galois theory.

Theorem 2.13 *Let G be a finite group of automorphisms of a field P. Then for every subgroup G_0 of the group G, there exists an element $x \in P$ whose stabilizer coincides with the subgroup G_0.*

Proof In proving this theorem, it will be convenient to use the space $P[t]$ of polynomials with coefficients in the field P. Every element f of the space $P[t]$ has the form $f = a_0 + \cdots + a_m t^m$, where $a_0, \dots, a_m \in P$. A polynomial $f \in P[t]$ defines a map $f : P \to P$ taking a point $x \in P$ to the point $f(x) = a_0 + \cdots + a_m x^m$. Every automorphism σ of the field P gives rise to the induced automorphism of the ring $P[t]$ mapping a polynomial $f = a_0 + \cdots + a_m t^m$ to the polynomial $f^\sigma = \sigma(a_0) + \cdots + \sigma(a_m)t^m$. For every element $x \in P$, the following identity holds: $f^\sigma(\sigma x) = \sigma(f(x))$. Thus the automorphism group of the field P acts on the ring $P[t]$. For every $k \geq 0$, the space $P_k[t]$ of polynomials of degree at most k is invariant under this action.

Lemma 2.14 *Suppose that a group G of automorphisms of the field P contains m elements. Then for every subgroup G_0 of the group G, there exists a polynomial f of degree less than m whose stabilizer coincides with the group G_0.*

Proof Indeed, by Corollary 2.11, there exists an element $a \in P$ on whose orbit O the action of the group G is free. In particular, the orbit O contains exactly m elements. Suppose that the subgroup G_0 contains k elements. Then the group G has $q = m/k$ right G_0-cosets. Under the action of the subgroup G_0, the set O splits into q orbits O_j, $j = 1, \dots, q$. Fix q distinct elements $b_1, \dots, b_q$ in the invariant field K, and consider the Lagrange polynomial of degree less than m that takes the value b_j at every element of the subset O_j, $j = 1, \dots, q$. This polynomial f satisfies the assumptions of the lemma.

Indeed, f is invariant under an automorphism σ if and only if for every element x of the field P, the equality $f(\sigma(x)) = \sigma(f(x))$ holds (recall that two polynomials of degree less than m coincide if their values coincide at more than m points). Since the polynomial f has degree less than m and the set O contains m elements, it suffices to verify the equality at all elements of the set O. By construction of f, the equality $f(\sigma(x)) = \sigma(f(x))$ holds if and only if $\sigma \in G_0$. □

We now proceed with the proof of the theorem. Consider a polynomial $f(x) = a_0 + \cdots + a_{m-1}x^{m-1}$ whose stabilizer is equal to G_0. The intersection of the stabilizers of the coefficients $a_0, \dots, a_{m-1}$ of this polynomial coincides with the subgroup G_0.

Consider the vector subspace L over the invariant subfield $K \subset P$ with respect to the action of the group G spanned by the coefficients $a_0, \dots, a_{m-1}$. By Lemma 2.12, there exists a vector $c \in L$ whose stabilizer is equal to G_0. □

2.3 Field Automorphisms and Relations Between Elements in a Field

In this section, we consider a finite group of field automorphisms. We prove the following two theorems from Galois theory. The first (Theorem 2.16) states that every element of a field is algebraic over the invariant subfield of the group action.

Suppose that y and z are two elements of a field. Under what conditions does there exist a polynomial T with coefficients from the invariant subfield such that $z = T(y)$? The second (Theorem 2.21) states that such a polynomial T exists if and only if the stabilizer of the element y lies in the stabilizer of the element z.

2.3.1 *Equations Without Multiple Roots*

Let $T(t)$ be a polynomial over the field K, $T'(t)$ its derivative, $D(t)$ the greatest common divisor of these polynomials, and $\tilde{T}$ the polynomial defined by the formula $\tilde{T} = T/D$.

Proposition 2.15 *The following statements hold:*

1. *A root of the polynomial T of multiplicity $k > 1$ is also a root of the polynomial D of multiplicity $k - 1$.*
2. *The polynomial $\tilde{T}$ has the same roots as the polynomial T. Moreover, all roots of the polynomial $\tilde{T}$ are simple.*

Proof Suppose that $T(t) = (t - x)^k Q(t)$ and $Q(x) \neq 0$. Then

$$T'(t) = k(t - x)^{k-1} Q(t) + (t - x)^k Q'(t).$$

Both statements of the proposition follow. □

2.3.2 *Algebraicity over an Invariant Subfield*

Let P be a commutative algebra with no zero divisors on which a group π acts by automorphisms, and let K be the invariant subalgebra of π. We do not assume that the group π is finite (although for Galois theory, it suffices to consider the actions of finite groups).

Theorem 2.16 *The following statements hold:*

1. *The stabilizer of an element $y \in P$ algebraic over K has finite index in the group π.*
2. *If the stabilizer of an element $y \in P$ has finite index n in the group π, then y satisfies some irreducible algebraic equation over K of degree n whose leading coefficient is equal to 1.*

Proof 1. Suppose that an element y satisfies an algebraic equation of the form

$$y^n + \cdots + p_n = 0 \tag{2.3}$$

with coefficients p_i from the invariant subalgebra K. Then every automorphism of the algebra P that fixes all elements of K maps the element y to one of the roots of (2.3). There are no more than n roots of this equation, and hence the index of the stabilizer of y in the group π does not exceed n.

2. Suppose that the stabilizer G of an element y has index n in the group π. Then the orbit of the element y under the action of the group π contains precisely n distinct elements. Let $y_1, \ldots, y_n$ denote the elements of this orbit. Consider the polynomial $Q(y) = (y - y_1) \cdots (y - y_n)$. The factors of the polynomial Q are permuted under every permutation of the points $y_1, \ldots, y_n$, but the polynomial itself does not change. Hence, the coefficients of Q belong to the invariant subalgebra K. The element y satisfies the algebraic equation $Q(y) = 0$ over the algebra K. This equation is irreducible (i.e., the polynomial T does not admit a factorization into two polynomials of positive degree whose coefficients lie in the algebra K). Indeed, if it were reducible, then y would satisfy an algebraic equation of smaller degree over the algebra K, and the orbit of y would contain fewer than n elements.

□

Remark 2.17 Viète's formulas allow one to find the coefficients of the polynomial Q. The elementary symmetric functions

$$\sigma_1 = y_1 + \cdots + y_n, \quad \sigma_2 = \sum_{i<j} y_i y_j, \quad \ldots, \quad \sigma_n = y_1 \cdots y_n$$

of the orbit points $y_1, \ldots, y_n$ belong to the invariant subalgebra K, and

$$Q(y) = y^n - \sigma_1 y^{n-1} + \sigma_2 y^{n-2} + \cdots + (-1)^n \sigma_n = 0.$$

2.3.3 Subalgebras Containing the Coefficients of a Lagrange Polynomial

In this subsection, we consider the Lagrange polynomial constructed by a special data set and estimate the subalgebra containing its coefficients. These results will be used in Sect. 2.3.4.

Let P be a commutative algebra without zero divisors, let $y_1, \ldots, y_n$ be distinct elements of the algebra P, and let $Q \in P[y]$ be a monic polynomial of degree n vanishing at the points $y_1, \ldots, y_n$, i.e., $Q(y) = (y - y_1) \cdots (y - y_n)$. Consider the following problem. For given elements $z_1, \ldots, z_n$ of the algebra P, find the Lagrange polynomial T taking the values $z_i Q'(y_i)$ at points y_i.

Let Q_i denote the polynomial $Q_i(y) = \prod_{j \neq i}(y - y_j)$. The following statement is obvious.

Proposition 2.18 *The desired Lagrange polynomial T is equal to $\sum_{i=1}^{n} z_i Q_i(y)$.*

Let us formulate a more general statement related to this problem, which we will not use and will not prove.

Proposition 2.19 *Suppose that a subalgebra K of the algebra P contains the coefficients of the polynomial Q and the elements $m_0, \dots, m_{n-1}$, where $m_k = \sum z_i y_i^k$. Then the coefficients of the polynomial T belong to the subalgebra K, i.e., $T \in K[y]$.*

We will need the following special case of this statement. Let π be the group of automorphisms of the algebra P, $K \subset P$ the algebra of invariants under the action of π, and Y the orbit of an element $y_1 \in P$ under the group action. Let $f : Y \to P$ be a map commuting with the action of the group π, that is, $f \circ g = g \circ f$ if $g \in \pi$. Set $z_1 = f(y_1), \dots, z_n = f(y_n)$.

Proposition 2.20 *The coefficients of the polynomial T defined in Proposition 2.18 belong to the algebra of invariants K.*

Proof The action of elements g of the group π permutes the summands of the polynomial T: if $g(y_i) = y_j$, then g maps the polynomial $z_i Q_i(y)$ to the polynomial $z_j Q_j(y)$. Hence, the polynomial T does not change under the action of the group π. That is, $T \in K[y]$. □

2.3.4 Representability of One Element Through Another Element over an Invariant Subfield

Let P be a field on which a group π of automorphisms acts, and let K be the corresponding invariant subfield. Suppose that x and y are elements of the field P algebraic over the field K, and G_y, G_z are their stabilizers. By Theorem 2.16, the element y (respectively the element z) is algebraic over the field K if and only if the group G_y (respectively the group G_z) has finite index in the group π. Under what conditions does z belong to the extension $K(y)$ of the field K obtained by adjoining the element y? The answer to this question is provided by the following theorem.

Theorem 2.21 *The element z belongs to the field $K(y)$ if and only if the stabilizer G_z of the element z includes the stabilizer G_y of the element y.*

Proof In one direction, the theorem is obvious: every element of the field $K(y)$ is fixed under the action of the group G_y. In other words, the stabilizer of every element of $K(y)$ contains the group G_y.

We will prove the converse statement in a stronger form. Let us relax the assumptions of Theorem 2.21. We will now assume that P is a commutative algebra with no zero divisors (not necessarily a field), π is a group of automorphisms of the algebra P, K is the invariant subalgebra, y and z are elements of the algebra P whose stabilizers G_y and G_z have finite index in the group π. Let Q denote an irreducible monic algebraic equation over the algebra K such that $Q(y) = 0$ (see part 2 of Theorem 2.16). The following result will conclude the proof. □

Proposition 2.22 *If $G_z \supseteq G_y$, then there exists a polynomial T with coefficients in the algebra K for which $zQ'(y) = T(y)$.*

Proof Let S denote the set of all right G_y-cosets in the group π. Suppose that the set S consists of n elements. Label the elements $s_1, \dots, s_n$ of this set in such a way that the coset of π containing the identity element is s_1. Let g_i be any representative of the coset s_i in the group π. The images $g_i(y)$, $g_i(z)$ of the elements y, z under the action of an automorphism g_i do not depend on the choice of a representative g_i in the class s_i. We write y_i and z_i, respectively, for these images. All elements $y_1, \dots, y_n$ are distinct by construction, whereas some of the elements $z_1, \dots, z_n$ may coincide. To conclude the proof, it remains to use Proposition 2.20.[2] □

2.4 Action of a k-Solvable Group and Representability by k-Radicals

In this section, we consider a field P with an action of a finite group G of automorphisms and the invariant subfield K. We will assume that the field P contains all roots of unity. The definition of a k-solvable group will be given. We will prove that if the group G is k-solvable, then every element of the field P can be expressed through the elements of the field K by radicals and solutions of auxiliary algebraic equations of degree at most k. The proof is based on the theorems given in preceding sections.

Definition 2.23 A group G is called k-solvable if it has a tower of subgroups

$$G = G_0 \supset G_1 \supset \cdots \supset G_n = e$$

such that for each i, $0 < i \leq n$, either the index of the subgroup G_i in the group G_{i-1} does not exceed k, or G_i is a normal divisor of G_{i-1} and the quotient group G_{i-1}/G_i is abelian.

Theorem 2.24 *Let G be a finite k-solvable group of automorphisms of a field P containing all roots of unity. Then every element x of the field P can be expressed through the elements of the invariant subfield K with the help of arithmetic operations, root extractions, and solving auxiliary algebraic equations of degree k or less.*

[2] If $z_i Q'(y_i) = T(y_i)$ for all i, then in particular, we obtain that $zQ(y) = T(y)$ by setting $i = 1$.

Proof Let $G = G_0 \supset G_1 \supset \cdots \supset G_n = e$ be a chain of nested subgroups satisfying the assumptions of the definition of a k-solvable group. Let $K = K_0 \subset \cdots \subset K_m = P$ denote the chain of invariant subfields corresponding to the actions of the groups $G_0, \ldots, G_m$.

Suppose that the group G_i is a normal subgroup in the group G_{i-1} and that the quotient G_{i-1}/G_i is abelian. The abelian quotient group G_{i-1}/G_i acts on the invariant subfield K_i, leaving the invariant subfield K_{i-1} pointwise fixed. Therefore, every element of the field K_i is expressible through the elements of the subfield K_{i-1} by means of summation and root extraction (see Theorem 2.3).

Suppose that the group G_i is a subgroup of index $m \leq k$ in the group G_{i-1}. There exists an element $a \in P$ whose stabilizer is equal to G_i (Theorem 2.13). The field K_i carries an action of the group G_{i-1} of automorphisms with invariant subfield K_{i-1}. Since the index of the stabilizer G_i of the element a in the group G is equal to m, the element a satisfies an algebraic equation of degree $m \leq k$ over the field K_{i-1}. By Theorem 2.21, every element of the field K_i is a polynomial in a with coefficients from the field K_{i-1}.

Repeating the same argument, we will be able to express every element of the field P in terms of the elements of the field K with the help of arithmetic operations, root extractions, and solving auxiliary algebraic equations of degree k or less. □

2.5 Galois Equations

An algebraic equation over a field K is called a *Galois equation* if the extension of the field K obtained by adjoining any single root of that equation to K contains all other roots. In this section, we prove that for every algebraic equation over a field K, there exists a Galois equation for which the extension of the field K obtained by adjoining all roots of the initial equation coincides with the extension obtained by adjoining a single root of the Galois equation. The proof is based on Theorem 2.21. Galois equations provide convenient tools for constructing Galois groups (see Sects. 2.6 and 2.7).

Let K be any field. Let $\tilde{P}$ denote the algebra $K[x_1, \ldots, x_m]$ of polynomials over the field K in the variables $x_1, \ldots, x_m$. The algebra $\tilde{P}$ carries an action of a group π of automorphisms isomorphic to the permutation group $S(m)$ of m elements: the action of the group consists of simultaneous permutations of the variables in all polynomials from the ring $K[x_1, \ldots, x_m]$. The invariant subalgebra $\tilde{K}$ with respect to this action consists of all symmetric polynomials in the variables $x_1, \ldots, x_m$.

Let $y \in \tilde{P}$ be some polynomial in m variables whose orbit under the action of the group $S(m)$ contains exactly $n = m!$ distinct elements $y = y_1, \ldots, y_n$.[3] Let Q denote a polynomial over the algebra $\tilde{K}$ whose roots are the elements $y_1, \ldots, y_n \in \tilde{P}$ (see Theorem 2.16). We may assume that the derivative of the polynomial Q does

[3] The existence of such a polynomial will be established in Lemma 2.27.

not vanish at its roots $y_1, \dots, y_n$. Applying Proposition 2.22 to the action of the group $S(m)$ on the algebra $\tilde{P}$ with invariant subalgebra $\tilde{K}$, we obtain the following corollary.

Corollary 2.25 *For every element $F \in \tilde{P} = K[x_1, \dots, x_m]$, there exists a polynomial T whose coefficients are symmetric polynomials in the variables $x_1, \dots, x_m$ such that the following identity holds:*

$$FQ'(y) = T(y).$$

Let $b_0 + b_1x + \cdots + b_mx^m = 0$ be an algebraic equation over the field K, $b_i \in K$, whose roots $x_1^0, \dots, x_m^0$ are distinct. Let P be the field obtained from K by adjoining all these roots. Consider the map $K[x_1, \dots, x_m] \to P$ assigning to each polynomial its value at the point $(x_1^0, \dots, x_m^0) \in P^m$.

Corollary 2.26 *Let $y \in K[x_1, \dots, x_m]$ be a polynomial such that all $n = m!$ polynomials obtained from y by all possible permutations of the variables assume different values at the point $(x_1^0, \dots, x_m^0) \in P^m$. Then the value of the polynomial y at this point generates the field P over the field K.*

Proof Indeed, the algebraic elements $x_1^0, \dots, x_m^0$ generate the field P over the field K. Therefore, every element of the field P is the value of some polynomial from the ring $K[x_1, \dots, x_m]$ at the point $(x_1^0, \dots, x_m^0)$. However, by Corollary 2.25, every polynomial F of $x_1, \dots, x_m$ multiplied by $Q'(y)$ is representable as a polynomial T of y with coefficients from the algebra $\tilde{K}$. We substitute into the corresponding identity $F(x_1, \dots, x_m) = Q'(y)T(y)$ the point $(x_1^0, \dots, x_m^0)$. By our assumption, all $n = m!$ roots of the polynomial Q assume different values at the point $(x_1^0, \dots, x_m^0)$. Therefore, the function $Q'(y)$ is different from 0 at this point, and the values of all symmetric polynomials at the point $(x_1^0, \dots, x_m^0)$ belong to the field K (since symmetric polynomials of the roots of an equation can be expressed through the coefficients of that equation). □

Lemma 2.27 *For every set of m distinct elements $x_1^0, \dots, x_m^0$ of the field $P \supseteq K$, there exists a linear polynomial $y = \lambda_1x_1 + \cdots + \lambda_mx_m$ with coefficients $\lambda_1, \dots, \lambda_m$ from the field K such that all $n = m!$ polynomials obtained from y by permutations of the variables assume different values at the point $(x_1^0, \dots, x_m^0) \in P^m$.*

Proof Consider the $n = m!$ points obtained from the point $(x_1^0, \dots, x_m^0)$ by all possible permutations of the coordinates. For every pair of points, the linear polynomials assuming the same values at those points form a proper vector subspace in the vector space of all linear polynomials with coefficients in the field K. The proper subspaces corresponding to pairs of points cannot cover the entire space (Proposition 2.10). Every linear polynomial y not lying in the union of the subspaces described above has the desired property. □

Definition 2.28 An equation $a_0 + a_1x + \cdots + a_mx^m = 0$ over a field K is called a *Galois equation* if its roots $x_1^0, \dots, x_m^0$ have the following property: for every pair of roots x_i^0, x_j^0, there exists a polynomial $P_{i,j}(t)$ over the field K such that $P_{i,j}(x_i^0) = x_j^0$.

Theorem 2.29 *Suppose that a field P is obtained from the field K by adjoining all roots of an algebraic equation over the field K with no multiple roots. Then the same field P can be obtained from the field K by adjoining a single root of some (in general, different) irreducible Galois equation over the field K.*

Proof By the assumption of the theorem, all roots $x_1^0, \dots, x_m^0$ of the equation are distinct. Consider a linear homogeneous polynomial y with coefficients in the field K such that all $n = m!$ linear polynomials obtained from y by permutations of the variables assume different values at the point $(x_1^0, \dots, x_m^0)$. Consider an equation of degree n over the field K whose roots are these values. By the corollary proved above, the equation thus obtained is a Galois equation, and its roots generate the field P. The Galois equation we have obtained may turn out to be reducible. Equating any irreducible component of it to zero, we obtain the required irreducible Galois equation. □

2.6 Automorphisms Related to a Galois Equation

In this section, we construct a group of automorphisms of an extension obtained from a base field (field of coefficients) by adjoining all roots of some Galois equation. We will show (Theorem 2.31) that the invariant subfield of this group coincides with the field of coefficients.

Let $Q = b_0 + b_1 x + \dots + b_n x^n$ be an irreducible polynomial over the field K. Then all fields generated over the field K by a root of the polynomial Q are isomorphic to each other and admit the following abstract description: every such field is isomorphic to the quotient of the ring $K[x]$ by the ideal I_Q generated by the irreducible polynomial Q. We let $K[x]/I_Q$ denote this field.

Let M be an extension of the field K containing all n roots $x_1^0, \dots, x_m^0$ of the equation $Q(x) = 0$. To every root x_i^0, we associate the field K_i obtained by adjoining the root x_i^0 to the field K. All the fields $K_i, i = 1, \dots, n$, are isomorphic to each other and are isomorphic to the field $K[x]/I_Q$. Let σ_i denote the isomorphism of the field $K[x]/I_Q$ to the field K_i that fixes all elements of the coefficient field K and takes the polynomial x to the element x_i^0.

Lemma 2.30 *Suppose that the equation $Q = b_0 + b_1 x + \dots + b_n x^n = 0$ is irreducible over the field K. Then the images $\sigma_i(a)$ of an element a of the field $K[x]/I_Q$ in the field M under all isomorphisms σ_i, $i = 1, \dots, n$, coincide if and only if the element a lies in the field of coefficients K.*

Proof If $b = \sigma_1(a) = \dots = \sigma_n(a)$, then $b = (\sigma_1(a) + \dots + \sigma_n(a))n^{-1}$. Therefore, the element b is the value of a symmetric polynomial of the roots $x_1^0, \dots, x_m^0$ of the equation $Q(x) = 0$; hence it belongs to the field K. □

We are now ready for the main theorem of this section.

Theorem 2.31 *Suppose that the field P is obtained from the field K by adjoining all roots of an irreducible algebraic equation over the field K. Then an element $b \in P$ is fixed by all automorphisms of P fixing all elements of K if and only if $b \in K$.*

Proof By Theorem 2.29, we can assume that the field P is obtained from the field K by adjoining all roots (or equivalently, a single root) of some irreducible Galois equation. By the definition of a Galois equation, all the fields K_i mentioned in Lemma 2.30 coincide with the field P. The isomorphism $\sigma_j\sigma_i^{-1}$ between the field K_i and the field K_j is an automorphism of the field P fixing all elements of the field K. By the lemma, an element b is fixed under all such automorphisms if and only if $b \in K$. □

2.7 The Fundamental Theorem of Galois Theory

In Sects. 2.2, 2.3, 2.5, and 2.6, we have, in fact, proved the central theorems of Galois theory. In this section, we give a summary. We define Galois extensions in Sect. 2.7.1 and Galois groups in Sect. 2.7.2. We prove the fundamental theorem of Galois theory in Sect. 2.7.3 and discuss in Sect. 2.7.4 the properties of the Galois correspondence and the behavior of the Galois group under extensions of the field of coefficients.

2.7.1 *Galois Extensions*

We give two equivalent definitions:

Definition 2.32 A field P obtained from a field K by adjoining all roots of an algebraic equation over the field K is called a *Galois extension* of the field K.

Definition 2.33 A field P is a *Galois extension* of its subfield K if there exists a finite group G of automorphisms of the field P whose invariant subfield is the field K.

Proposition 2.34 *Definitions 2.32 and 2.33 are equivalent. The group G from Definition 2.33 coincides with the group of all automorphisms of the field P over the field K. It follows that the group G is uniquely defined.*

Proof If the field P is a Galois extension of the field K in the sense of Definition 2.32, then by Theorem 2.31, the field P is also a Galois extension of the field K in the sense of Definition 2.33. Suppose now that the field P is a Galois extension of the field K in the sense of Definition 2.33. By Corollary 2.11, there exists an element $a \in P$ that is displaced (is not fixed) under the action of every nonidentity element of the group G. Consider the orbit O of the element a under the action of G. By Theorem 2.16, there exists an algebraic equation over the field K whose set of roots coincides with O. By Theorem 2.21, every element of the orbit, i.e., every root of this algebraic equation, generates the field P over the field K. Therefore, the field P is a Galois extension of the field K in the sense of Definition 2.32.

Every automorphism σ of the field P over the field K takes the element a to some element of the set O, since the set O is the set of all solutions of an algebraic equation with coefficients in the field K. Hence σ defines an element g of the group G such that $\sigma(a) = g(a)$. The automorphism σ must coincide with g, since a generates the field P over the field K. Therefore, the group G coincides with the group of all automorphisms of the field P over the field K. □

2.7.2 Galois Groups

We now proceed to a discussion of Galois groups, which are central objects in Galois theory.

Definition 2.35 The *Galois group of a Galois extension P of the field K* (or simply the Galois group of P over K) is defined as the group of all automorphisms of the field P over the field K. The *Galois group of an algebraic equation over the field K* is defined as the Galois group of the Galois extension P of K obtained by adjoining all roots of this algebraic equation to the field K.

Suppose that the field P is obtained by adjoining to K all roots of the equation

$$a_0 + a_1 x + \cdots + a_n x^n = 0 \tag{2.4}$$

over the field K. Every element σ of the Galois group of P over K permutes the roots of (2.4). Indeed, acting by σ on both parts of (2.4) yields

$$\sigma(a_0 + a_1 x + \cdots + a_n x^n) = a_0 + a_1 \sigma(x) + \cdots + a_n (\sigma(x))^n = 0.$$

Thus, the Galois group of the field P over the field K admits a representation in the permutation group of the roots of (2.4). This representation is faithful: if an automorphism fixes all the roots of (2.4), then it fixes all elements of the field P and hence is trivial.

Definition 2.36 A *relation* between the roots of (2.4) over the field K is defined as any polynomial Q belonging to the ring $K[x_1, \ldots, x_n]$ that vanishes at the point $(x_1^0, \ldots, x_n^0)$, where $x_1^0, \ldots, x_n^0$ is the collection of all roots of (2.4).

Proposition 2.37 *Every automorphism in the Galois group preserves all relations over the field K between the roots of* (2.4). *Conversely, every permutation of the roots preserving all relations between the roots over the field K extends to an automorphism from the Galois group.*

Thus the Galois group of the field P over the field K can be identified with the group of all permutations of the roots of (2.4) that preserve all relations between the roots defined over the field K.

Proof If a permutation $\sigma \in S(n)$ corresponds to an element of the Galois group, then the polynomial σQ obtained from a relation Q by permuting the variables $x_1, \dots, x_n$ according to σ also vanishes at the point $(x_1^0, \dots, x_n^0)$. Conversely, suppose that a permutation σ preserves all relations between the roots over the field K. Extend the permutation σ to an automorphism of the field P over the field K. Every element of the field P is the value of some polynomial Q_1 belonging to the ring $K[x_1, \dots, x_n]$ at the point $(x_1^0, \dots, x_n^0)$. It is natural to define the value of the automorphism σ at this element as the value at the point $(x_1^0, \dots, x_n^0)$ of the polynomial σQ_1 obtained from Q_1 by permuting the variables according to σ. We need to verify that the automorphism σ is well defined. Let Q_2 be a different polynomial in the ring $K[x_1, \dots, x_n]$ whose value at the point $(x_1^0, \dots, x_n^0)$ coincides with the value of Q_1 at this point. But then the polynomial $Q_1 - Q_2$ is a relation between the roots over the field K. Therefore, the polynomial $\sigma Q_1 - \sigma Q_2$ must also vanish at the point $(x_1^0, \dots, x_n^0)$, but this means exactly that the automorphism σ is well defined. □

2.7.3 The Fundamental Theorem

Suppose that the field P is a Galois extension of a field K. Galois theory describes all intermediate fields, i.e., all fields lying in the field P and containing the field K. To every subgroup H of the Galois group of the field P over the field K, we assign the subfield P_H consisting of all elements of P that are fixed under the action of H. This correspondence is called the *Galois correspondence*.

Theorem 2.38 (The fundamental theorem of Galois theory) *The Galois correspondence of a Galois extension is a one-to-one correspondence between all subgroups of the Galois group and all intermediate fields.*

Proof First, by Theorem 2.13, different subgroups of the Galois group have different invariant subfields. Second, if a field P is a Galois extension of the field K, then it is also a Galois extension of every intermediate field. This is obvious if we use Definition 2.32 of a Galois extension. From Definition 2.33 of a Galois extension, it can be seen that every intermediate field is the invariant subfield for some group of automorphisms of the field P over the field K. The theorem is proved. □

2.7.4 Properties of the Galois Correspondence

We now discuss the simplest properties of the Galois correspondence.

Proposition 2.39 *An intermediate field is a Galois extension of the field of coefficients if and only if under the Galois correspondence, this field is assigned to a normal subgroup of the Galois group. The Galois group of an intermediate Galois extension over the field of coefficients is isomorphic to the quotient of the Galois group of the initial extension by the normal subgroup corresponding to the intermediate Galois extension.*

Proof Let H be a normal subgroup of the Galois group G, and L_H the intermediate field corresponding to the subgroup H. The field L_H is mapped to itself under the automorphisms in the group G, since the fixed-point set of a normal subgroup is invariant under the action of the group (Proposition 2.4). The group of automorphisms of the field L_H induced by the action of the group G is isomorphic to the quotient group G/H. The invariant subfield of this induced group of automorphisms of L_H coincides with the field K. Thus if H is a normal subgroup of the group G, then L_H is a Galois extension of the field K with Galois group G/H.

Let K_1 be an intermediate Galois extension of the field K. The field K_1 is obtained from the field K by adjoining all roots of some algebraic equation over K. Every automorphism in the Galois group G can only permute the roots of this equation, and hence it maps the field K_1 to itself. Suppose that the field K_1 corresponds to a subgroup H, i.e., $K_1 = L_H$. An element g of the group G takes the field L_H to the field $L_{gHg^{-1}}$. Thus, if an intermediate Galois extension K_1 corresponds to a subgroup H, then for every element $g \in G$, we have $H = gHg^{-1}$. In other words, the subgroup H is a normal subgroup of the Galois group G. □

Proposition 2.40 *The smallest algebraic extension of a field K containing two given Galois extensions of K is a Galois extension of the field K.*

Proof The smallest field P containing both Galois extensions can be constructed in the following way. Suppose that the first field is obtained from the field K by adjoining all roots of a polynomial Q_1, and the second field by adjoining all roots of a polynomial Q_2. The field P can be obtained by adjoining all roots of the polynomial $Q = Q_1Q_2$ to the field K and is therefore a Galois extension of the field K. □

Proposition 2.41 *The intersection of two Galois extensions is a Galois extension. The Galois group of the intersection is a quotient group of the Galois group of each initial Galois extension.*

Proof Let P be the smallest field containing both Galois extensions. As we have proved, P is a Galois extension of the field K. The Galois group G of the field P over the field K preserves the first as well as the second extension of K. We conclude that the intersection of the two Galois extensions is also mapped to itself under the action of the group G. Therefore, by Propositions 2.4 and 2.39, the intersection of two Galois extensions is also a Galois extension. From the same proposition, it follows that the Galois group of the intersection is a quotient group of the Galois group of each initial Galois extension. □

2.7.5 Changing the Field of Coefficients

Let

$$a_0 + a_1x + \cdots + a_nx^n = 0 \tag{2.5}$$

be an algebraic equation over the field K, and P a Galois extension of the field K obtained from K by adjoining all roots of (2.5). Consider a larger field $\tilde{K} \supset K$ and its Galois extension $\tilde{P}$ obtained from the field $\tilde{K}$ by adjoining all roots of (2.5). What is the relation between the Galois group of $\tilde{P}$ over $\tilde{K}$ and the Galois group G of P over K? In other words, what happens with the Galois group of (2.5) if we change the base field (i.e., pass from the field K to the field $\tilde{K}$)?

Generally speaking, as the field of coefficients becomes bigger, the Galois group of the same equation becomes smaller, i.e., it is replaced with some subgroup. Indeed, there may be more relations between the roots of (2.5) over the bigger field. We now give a more precise statement.

Let K_1 denote the intersection of the fields P and $\tilde{K}$. The field K_1 includes the field K and lies in the field P, i.e., we have $K \subset K_1 \subset P$. By the fundamental theorem of Galois theory, the field K_1 corresponds to a subgroup G_1 of the Galois group G.

Theorem 2.42 *The Galois group $\tilde{G}$ of the field $\tilde{P}$ over the field $\tilde{K}$ is isomorphic to the subgroup G_1 of the Galois group G of the field P over the field K.*

Proof The Galois group $\tilde{G}$ fixes all elements of the field K (since $K \subset \tilde{K}$) and permutes the roots of (2.5). Hence the field P is mapped to itself under all automorphisms from the group $\tilde{G}$. The fixed-point set of the induced group of automorphisms of the field P consists precisely of all elements in the field P that lie in the field $\tilde{K}$, i.e., of all elements of the field $K_1 = P \cap \tilde{K}$. Therefore, the induced group of automorphisms of the field P coincides with the subgroup G_1 of the Galois group G. It remains to show that the homomorphism of the group $\tilde{G}$ into the group G_1 described above has trivial kernel. Indeed, the kernel of this homomorphism fixes all roots of (2.5), i.e., contains only the identity element of the group $\tilde{G}$. The theorem is proved. □

Suppose now that under the assumptions of the preceding theorem, the field $\tilde{K}$ is itself a Galois extension of the field K with Galois group Γ. By Proposition 2.40, the field K_1 is also a Galois extension of the field K in this case. Let Γ_1 denote the Galois group of the extension K_1 of the field K.

Theorem 2.43 (How the Galois group changes as the field of coefficients changes) *When the field of coefficients is replaced with a Galois extension, the Galois group G of the initial equation is replaced with a normal subgroup G_1. The quotient group G/G_1 of the group G by this normal subgroup is isomorphic to a quotient group of the Galois group of the new field of coefficients $\tilde{K}$ over the old field of coefficients K.*

Proof Indeed, the group G_1 corresponds to the field $P \cap \tilde{K}$, which is a Galois extension of the field K. Hence the group G_1 is a normal subgroup of the group G, and its quotient group G/G_1 is isomorphic to the Galois group of the field K_1 over the field K. But the Galois group of the field K_1 over the field K is isomorphic to the quotient group Γ/Γ_1. The theorem is proved. □

2.8 A Criterion for Solvability of Equations by Radicals

An algebraic equation over a field K is said to be *solvable by radicals* if there exists a chain of extensions $K = K_0 \subset K_1 \cdots \subset K_n$ in which every field K_{j+1} is obtained from the field K_j, $j = 0, \dots, n-1$, by adjoining some radical (i.e., some element y such that $y^m \in K_j$ for some $m > 0$), and the field K_n contains all roots of this algebraic equation. Galois theory was created to tell when a given equation is solvable by radicals.

In Sect. 2.8.1, we consider the group of all nth roots of unity that lie in a given field K. In Sect. 2.8.2, we consider the Galois group of the equation $x^n = a$. In Sect. 2.8.3, we give a criterion of solvability of an algebraic equation by radicals (in terms of the Galois group of the given equation).

2.8.1 *Roots of Unity*

Let K be a field. Let K_E^* denote the multiplicative group of all roots of unity lying in the field K (i.e., $a \in K_E^*$ if and only if $a \in K$ and for some positive integer n, we have $a^n = 1$).

Proposition 2.44 *If there is a subgroup of the group K_E^* consisting of ℓ elements, then the equation $x^\ell = 1$ has exactly ℓ solutions in the field K, and the subgroup under consideration is formed by all these solutions.*

Proof Every element in a group of order ℓ satisfies the equation $x^\ell = 1$. The field contains no more than ℓ roots of this equation, and the subgroup has exactly ℓ elements by assumption. □

From Proposition 2.44, it follows in particular that the group K_E^* has at most one subgroup of any given finite order.

Proposition 2.45 *A finite abelian group that has at most one cyclic subgroup of any given finite order is cyclic. In particular, every finite subgroup of the group K_E^* is cyclic.*

Proof From the classification theorem for finite abelian groups, it follows that an abelian group satisfying the assumptions of the proposition is determined by the number m of its elements up to isomorphism: if $m = p_1^{k_1} \cdots p_n^{k_n}$ is a factorization of m, then $G = (\mathbb{Z}/p_1^{k_1}\mathbb{Z}) \times \cdots \times (\mathbb{Z}/p_n^{k_n}\mathbb{Z})$. But this group is cyclic by the Chinese remainder theorem. □

A cyclic group with m elements can be identified with the group of residues modulo m.

Proposition 2.46 *The full automorphism group of the group $\mathbb{Z}/m\mathbb{Z}$ is isomorphic to the multiplicative group of all invertible elements in the ring of residues modulo m. In particular, this automorphism group is abelian.*

Proof An automorphism F of the group $\mathbb{Z}/m\mathbb{Z}$ is uniquely determined by the element $F(1)$, which must obviously be invertible in the multiplicative group of the ring of residues. This automorphism coincides with multiplication by $F(1)$. □

Proposition 2.47 *Suppose that a Galois extension P of a field K is obtained from K by adjoining some roots of unity. Then the Galois group of the field P over the field K is abelian.*

Proof All roots of unity that lie in the field P form a cyclic group with respect to multiplication. A transformation from the Galois group defines an automorphism of this group and is uniquely determined by this automorphism, i.e., the Galois group embeds into the full automorphism group of a cyclic group. The result now follows from Proposition 2.46. □

2.8.2 The Equation $x^n = a$

Proposition 2.48 *Suppose that a field K contains all roots of unity of degree n. Then the Galois group of the equation $x^n - a = 0$ over the field K is a subgroup of the cyclic group with n elements, provided that $a \in K \setminus \{0\}$.*

Proof The group of all roots of unity of degree n is cyclic (see Proposition 2.45). Let ξ be any generator of this group. Fix any root x_0 of the equation $x^n - a = 0$. Then we can label all roots of the equation $x^n - a = 0$ with residues i modulo n by setting x_i to be $\xi^i x_0$.

Suppose that a transformation g in the Galois group takes the root x_0 to the root x_i. Then $g(x_k) = g(\xi^k x_0) = \xi^{k+i} x_0 = x_{k+i}$ (recall that by our assumption, $\xi \in K$, whence $g(\xi) = \xi$), i.e., every transformation in the Galois group defines a cyclic permutation of the roots. Therefore, the Galois group embeds into the cyclic group with n elements. □

Lemma 2.49 *The Galois group G of the equation $x^n - a = 0$ over the field K, where $a \in K \setminus \{0\}$, has an abelian normal subgroup G_1 such that the corresponding quotient G/G_1 is abelian. In particular, the group G is solvable.*

Proof Let P be the extension of the field K obtained by adjoining all roots of the equation $x^n = a$ to this field. The ratio of any two roots of the equation $x^n = a$

is a root of unity of degree n. This implies that the field P contains all nth roots of unity. Let K_1 denote the extension of the field K obtained by adjoining all roots of unity of degree n. We have the inclusions $K \subset K_1 \subset P$. Let G_1 denote the Galois group of the equation $x^n = a$ over the field K_1. By Proposition 2.48, the group G_1 is abelian. The group G_1 is a normal subgroup of the group G, since the field K_1 is a Galois extension of the field K. The quotient group G/G_1 is abelian, since by Proposition 2.47, the Galois group of the field K_1 over the field K is abelian. □

2.8.3 *Solvability by Radicals*

We shall prove the following criterion for the solvability of an algebraic equation by radicals.

Theorem 2.50 (A criterion for solvability of equations by radicals) *An algebraic equation over some field is solvable by radicals if and only if its Galois group is solvable.*

Proof Solvability of an equation by radicals over a field K implies the existence of a chain of extensions $K = K_0 \subset K_1 \cdots \subset K_n$ in which every field K_{j+1} is obtained from the field K_j, $j = 0, 1, \ldots, n-1$, by adjoining all roots of the polynomial $x^n - a$ for $a \in K_j$ and the field K_n contains all roots of the initial equation. Let G_j denote the Galois group of our equation over the field K_j. Let us see what happens to the Galois group when we pass from the field K_j to the field K_{j+1}. According to Theorem 2.42, the group G_{j+1} is a normal subgroup of the group G_j, and moreover, the quotient G_j/G_{j+1} is simultaneously a quotient of the Galois group of the field K_{j+1} over the field K_j. Since the field K_{j+1} is obtained from the field K_j by adjoining all roots of the polynomial $x^n - a$, we conclude by Lemma 2.49 that the Galois group of the field K_{j+1} over the field K_j is solvable. (When the field K contains all roots of unity, the Galois group of the field K_{j+1} over the field K_j is abelian.) Since all roots of the algebraic equation lie in the field K_n by assumption, the Galois group G_n of the algebraic equation over the field K_n is trivial.

Thus, if the equation can be solved by radicals, then its Galois group admits a chain of subgroups $G = G_0 \supset G_1 \supset \cdots \supset G_n$ in which every group G_{j+1} is a normal subgroup of the group G_j with a solvable quotient G_j/G_{j+1}, and the group G_n is trivial. (If the field K contains all roots of unity, then the quotients G_j/G_{j+1} are abelian.) Thus, if the equation is solvable by radicals, then its Galois group is solvable.

Suppose now that the Galois group G of an algebraic equation over the field K is solvable. Let $\tilde{K}$ denote the field obtained from the field K by adjoining all roots of unity. The Galois group $\tilde{G}$ of the algebraic equation over the larger field $\tilde{K}$ is a subgroup of the Galois group G. Hence the Galois group $\tilde{G}$ is solvable. Let $\tilde{P}$ denote the field obtained from the field $\tilde{K}$ by adjoining all roots of the algebraic equation. The solvable group $\tilde{G}$ acts by automorphisms of the field $\tilde{P}$ with invariant subfield $\tilde{K}$. By Theorem 2.3, every element of the field $\tilde{P}$ is expressible by radicals through the elements of the field $\tilde{K}$. By definition of the field $\tilde{K}$, every element of that field is expressible through the roots of unity and the elements of the field K. The theorem is proved. □

2.9 A Criterion for Solvability by k-Radicals

We say that an algebraic equation is *solvable by k-radicals* if there exists a chain of extensions $K = K_0 \subset K_1 \cdots \subset K_n$ in which for every j, $0 < j \le n$, either the field K_{j+1} is obtained from the field K_j by adjoining a radical, or the field K_{j+1} is obtained from the field K_j by adjoining a root of some equation of degree at most k, and the field K_n contains all roots of the initial equation. In this section, we explain how to determine when a given algebraic equation is solvable by k-radicals. In Sect. 2.9.1, we discuss the properties of k-solvable groups. In Sect. 2.9.2, we prove a criterion for solvability by k-radicals.

Let us begin with the following simple statement.

Proposition 2.51 *The Galois group of an equation of degree $m \le k$ is isomorphic to a subgroup of the symmetric group $S(k)$.*

Proof Every element of the Galois group permutes the roots of the equation and is uniquely determined by the permutation of roots thus obtained. Hence the Galois group of an equation of degree m is isomorphic to a subgroup of the group $S(m)$. For $m \le k$, the group $S(m)$ is a subgroup of the group $S(k)$. □

2.9.1 Properties of k-Solvable Groups

In this subsection, we show that k-solvable groups (see Sect. 2.4) have properties similar to those of solvable groups. We begin with a lemma that characterizes subgroups of the group $S(k)$.

Lemma 2.52 *A group is isomorphic to a subgroup of the group $S(k)$ if and only if it has a collection of m subgroups, $m \le k$, such that:*

1. *The intersection of these subgroups contains no nontrivial normal subgroups of the entire group.*
2. *The sum of the indexes of these subgroups in the group does not exceed k.*

Proof Suppose that G is a subgroup of the group $S(k)$. Consider a representation of the group G as a subgroup of permutations of a set M with k elements. Suppose that under the action of the group G, the set M splits into m orbits. Choose a single point x_i in every orbit. The collection of stabilizers of points x_i satisfies the conditions of the lemma. Indeed, the index of the stabilizer H_i of x_i equals the cardinality of the orbit of x_i; hence the sum of these indices is k. Let H be the intersection of all stabilizers H_i. Suppose that H contains a nontrivial normal subgroup F. Every element x of M has the form $x = gx_i$ for some $g \in G$ and i. It follows that x is a fixed point for all elements of gFg^{-1}, since x_i is a fixed point for all elements of F. We conclude that F acts trivially on M, a contradiction.

Conversely, let a group G have a collection of subgroups $G_1, \dots, G_n$ satisfying the conditions of the lemma. Let P denote the union of the sets P_i, where $P_i = G/G_i$ consists of all right cosets with respect to the subgroup G_i, $1 \le i \le n$. The group G acts naturally on the set P. The representation of the group G in the group $S(P)$ of all permutations of P is faithful, since the kernel of this representation lies in the intersection of the groups G_i. The group $S(P)$ embeds into the group $S(k)$, since the number of elements in the set P is the sum of the indices of the subgroups G_i. □

Corollary 2.53 *Every quotient group*[4] *of the symmetric group $S(k)$ is isomorphic to a subgroup of $S(k)$.*

Proof Suppose that a group G is isomorphic to a subgroup of the group $S(k)$ and that G_i are subgroups in G satisfying the conditions of Lemma 2.52. Let π be an arbitrary homomorphism of the group G (onto some other group). Then the collection of the subgroups $\pi(G_i)$ in the group $\pi(G)$ also satisfies the conditions of the lemma. □

We say that a normal subgroup H of a group G is of *depth at most k* if the group G has a subgroup G_0 of index at most k such that H is the intersection of all subgroups conjugate to G_0. We say that a group is of depth at most k if its identity subgroup is of depth at most k.

A *normal tower* of a group G is a nested chain of subgroups $G = G_0 \supset \cdots \supset G_n = \{e\}$ in which every succeeding group is a normal subgroup of the preceding group.

Corollary 2.54 *If a group G is a subgroup of the group $S(k)$, then the group G has a nested chain of subgroups $G = \Gamma_0 \supset \cdots \supset \Gamma_m = \{e\}$ in which the group Γ_m is trivial, and for every $i = 0, 1, \dots, m-1$, the group Γ_{i+1} is a normal subgroup of the group Γ_i of depth at most k.*

Proof Let G_i be a collection of subgroups of the group G satisfying the conditions of Lemma 2.52. Let F_i denote the normal subgroup of the group G obtained as the intersection of all subgroups conjugate to the subgroup G_i. The chain of subgroups $\Gamma_1 = F_1$, $\Gamma_2 = F_1 \cap F_2, \dots, \Gamma_m = F_1 \cap F_2 \cap \cdots \cap F_m$ satisfies the conditions of the corollary. □

Lemma 2.55 *A group G is k-solvable if and only if it admits a normal tower of subgroups $G = G_0 \supset \cdots \supset G_n = \{e\}$ in which for every i, $0 < i \le n$, either the normal subgroup G_i has depth at most k in the group G_{i-1}, or the quotient G_{i-1}/G_i is abelian.*

---

[4] And indeed every subquotient, i.e., a quotient group of a subgroup.

Proof 1. Suppose that the group G admits a normal tower $G = G_0 \supset \cdots \supset G_n = \{e\}$ satisfying the conditions of the lemma. If for some i, the normal subgroup G_i has depth at most k in the group G_{i-1}, then the group G_{i-1}/G_i has a chain of subgroups $G_{i-1}/G_i = \Gamma_0 \supset \cdots \supset \Gamma_m = \{e\}$ in which the index of every succeeding group in its preceding group does not exceed k. For every such number i, we can insert the chain of subgroups $G_{i-1} = \Gamma_{0,i} \supset \cdots \supset \Gamma_{m_i,i}$ between G_{i-1} and G_i, where $\Gamma_{j,i} = \pi^{-1}(\Gamma_j)$, and $\pi : G_{i-1} \to G_{i-1}/G_i$ is the canonical projection to the quotient group. We thus obtain a chain of subgroups satisfying the definition of a k-solvable group.

2. Suppose that a group G is k-solvable, and $G = G_0 \supset G_1 \supset \cdots \supset G_n = \{e\}$ is a chain of subgroups satisfying the assumptions listed in the definition of a k-solvable group. We will successively replace subgroups in the chain with smaller subgroups. Let i be the first number for which the group G_i is not a normal subgroup in the group G_{i-1} but rather a subgroup of index $\leq k$. In this case, the group G_{i-1} has a normal subgroup H lying in the group G_i such that the group G_{i-1}/H is isomorphic to a subgroup of $S(k)$. Indeed, we can take H to be the intersection of all subgroups in G_{i-1} conjugate to the group G_i. We can now modify the chain $G = G_0 \supset G_1 \supset \cdots \supset G_n = \{e\}$ in the following way: Every subgroup labeled by numbers less than i remains the same. Every group G_j with $i \leq j$ is replaced with the group $G_j \cap H$. Repeat the same procedure for the chain of subgroups thus obtained, and so on. Finally, we obtain a normal tower of subgroups satisfying the conditions of the lemma.

□

Theorem 2.56 *The following statements hold:*

1. *All subgroups and quotient groups of a k-solvable group are k-solvable.*
2. *If a group has a k-solvable normal subgroup such that the corresponding quotient group is k-solvable, then the group is also k-solvable.*

Proof The only nonobvious statement of this theorem is that about a quotient group. It follows easily from Lemma 2.55. □

2.9.2 Solvability by k-Radicals

We shall prove the following criterion for solvability by k-radicals.

Theorem 2.57 (A criterion for solvability of equations by k-radicals) *An algebraic equation over a field is solvable by k-radicals if and only if its Galois group is k-solvable.*

Proof 1. Suppose that the equation can be solved by k-radicals. We need to prove that the Galois group of the equation is k-solvable. This is proved in exactly the same way as the statement that the Galois group of an equation solvable by radicals is solvable.

Let $K = K_0 \subset K_1 \subset \cdots \subset K_n$ be a chain of fields that arises in the solution of the equation by k-radicals, and $G_0 \supset \cdots \supset G_n$ the chain of Galois groups of the equation over these fields. By the assumption, the field K_n contains all roots of the equation, and therefore, the group G_n is trivial and in particular, is k-solvable. Suppose that the group G_{i+1} is k-solvable. We need to prove that the group G_i is also k-solvable.

If the field K_{i+1} is obtained from the field K_i by adjoining all roots of the equation $x^n - a = 0$, where $a \in K_i$, then the Galois group of the field K_{i+1} over the field K_i is solvable, hence k-solvable. If the field K_{i+1} is obtained from the field K_i by adjoining all roots of an algebraic equation of degree at most k, then the Galois group of the field K_{i+1} over the field K_i is a subgroup of the group $S(k)$ (see Proposition 2.51), hence is k-solvable.

By Theorem 2.3, the group G_{i+1} is a normal subgroup of the group G_i; moreover, the quotient group G_i/G_{i+1} is simultaneously a quotient group of the Galois group of the field K_{i+1} over the field K_i. The group G_{i+1} is solvable by the induction hypothesis. The Galois group of the field K_{i+1} over the field K_i is k-solvable, as we have just proved. Using Theorem 2.56, we conclude that the group G_i is k-solvable.

2. Suppose that the Galois group G of an algebraic equation over the field K is k-solvable. Let $\tilde{K}$ denote the field obtained from the field K by adjoining all roots of unity. The Galois group $\tilde{G}$ of the same equation over the larger field $\tilde{K}$ is a subgroup of the group G. Therefore, the Galois group $\tilde{G}$ is k-solvable. Let $\tilde{P}$ denote the field obtained from the field $\tilde{K}$ by adjoining all roots of the given algebraic equation. The group $\tilde{G}$ acts by automorphisms on $\tilde{P}$ with invariant subfield $\tilde{K}$. By Theorem 2.24, every element of the field $\tilde{P}$ can be expressed through the elements of the field $\tilde{K}$ by taking radicals, performing arithmetic operations, and solving algebraic equations of degree at most k. By definition of the field $\tilde{K}$, every element of this field is expressible through elements of the field K and roots of unity. The theorem is proved. □

2.9.3 *Unsolvability of the General Equation of Degree $k + 1 > 4$ by k-Radicals*

Let K be a field. A *generic algebraic equation* of degree k with coefficients in the field K is an equation

$$x^k + a_1 x^{k-1} + \cdots + a_0 = 0 \tag{2.6}$$

whose coefficients are generic elements of the field K. Closely related to generic equations is the *general* equation (2.6), in which the coefficients a_i are formal variables. Do there exist formulas containing radicals (k-radicals) and the variables

$a_1, \dots, a_k$ that give solutions of the equation $x^k + a_1^0 x^{k-1} + \dots + a_0^0 = 0$ as one substitutes the particular elements $a_1^0, \dots, a_k^0$ of the field K for the variables?

This question can be formalized in the following way. The general algebraic equation can be viewed as an equation over the field $K\{a_1, \dots, a_k\}$ of rational functions in k independent variables $a_1^0, \dots, a_k^0$ with coefficients in the field K (in this interpretation, the coefficients of (2.6) are the elements $a_1^0, \dots, a_k^0$ of the field $K\{a_1, \dots, a_k\}$). We can now ask whether (2.6) is solvable over the field $K\{a_1, \dots, a_k\}$ by radicals (or by k-radicals).

Let us compute the Galois group of (2.6) over the field $K\{a_1, \dots, a_k\}$. Consider yet another copy $K\{x_1, \dots, x_k\}$ of the field of rational functions in k variables equipped with the group $S(k)$ of automorphisms acting by permutations of the variables $x_1, \dots, x_k$. The invariant subfield $K_S\{x_1, \dots, x_k\}$ consists of symmetric rational functions. By the fundamental theorem of symmetric functions, this field is isomorphic to the field of rational functions of $\sigma_1 = x_1 + \dots + x_k, \dots, \sigma_n = x_1 \cdots x_k$. Therefore, the map $F(a_1) = -\sigma_1, \dots, F(a_n) = (-1)^n \sigma_n$ extends to an isomorphism $F : K\{a_1, \dots, a_k\} \to K_S\{x_1, \dots, x_k\}$. Let us identify the fields $K\{a_1, \dots, a_k\}$ and $K_S\{x_1, \dots, x_k\}$ by the isomorphism F. From a comparison of Viète's formulas with the formulas defining the map F, it becomes clear that under this identification, the variables become the roots of (2.6), the field $K\{x_1, \dots, x_k\}$ becomes the extension of the field $K\{a_1, \dots, a_k\}$ by adjoining all roots of (2.6), and the automorphism group $S(k)$ becomes the Galois group of (2.6). Thus we have proved the following proposition.

Proposition 2.58 *The Galois group of* (2.6) *over the field* $K\{a_1, \dots, a_k\}$ *is isomorphic to the permutation group* $S(k)$.

Theorem 2.59 *The general equation of degree* $k + 1 > 4$ *is not solvable by taking radicals and by solving auxiliary algebraic equations of degree* k *or less.*

Proof The group $S(k + 1)$ has the following normal tower of subgroups: $\{e\} \subset A(k + 1) \subset S(k + 1)$, where $A(k + 1)$ is the alternating group. For $k + 1 > 4$, the group $A(k + 1)$ is simple. The group $A(k + 1)$ is not a subgroup of the group $S(k)$, since the group $A(k + 1)$ has more elements than the group $S(k)$. Thus, for $k + 1 > 4$, the group $S(k + 1)$ is not k-solvable. To conclude the proof, it remains to use Theorem 2.57. □

As a corollary, we obtain the following theorem.

Theorem 2.60 (Abel) *The general algebraic equation of degree* 5 *and higher is not solvable by radicals.*

Remark 2.61 Abel proved this theorem by a different method before Galois theory appeared. His approach was later developed by Liouville. Liouville's method makes it possible, for example, to prove that many elementary integrals cannot be computed by elementary functions (see Chap. 1).

2.10 Unsolvability of Complicated Equations by Solving Simpler Equations

Is it possible to solve a given complicated algebraic equation using the solutions of other, simpler, algebraic equations as admissible operations? We have considered two well-posed questions of this kind: the question of solvability of equations by radicals (in which the simpler equations are those of the form $x^n - a = 0$) and the question of solvability of equations by k-radicals (in which the simpler equations are those of the form $x^n - a = 0$ and all algebraic equations of degree k or less). In this section, we address the general question of solvability of complicated equation by solving simpler equations. In Sect. 2.10.1, we set up the problem of B-solvability of equations and discuss a necessary condition for their solvability. In Sect. 2.10.2, we discuss classes of groups related to the problem of B-solvability of equations.

2.10.1 A Necessary Condition for Solvability

Let B be a collection of algebraic equations. An algebraic equation defined over a field K is automatically defined over every larger field K_1, $K \subset K_1$. We will assume that the collection B of algebraic equations contains, together with any equation defined over a field K, that same equation considered as an equation over any larger field $K_1 \supset K$.

Definition 2.62 An algebraic equation over a field K is said to be *solvable by solving equations from the collection* B, or B*-solvable* for short, if there exists a chain of fields $K = K_0 \subset K_1 \subset \cdots \subset K_n$ such that all roots of the equation belong to the field K_n, and for every $i = 0, \ldots, n-1$, the field K_{i+1} is obtained from the field K_i by adjoining all roots of some algebraic equation from the collection B defined over the field K_i.

Is a given algebraic equation B-solvable? Galois theory provides a necessary condition for B-solvability of equations. In this subsection, we discuss this condition. We assign to the collection B of equations the set $G(B)$ of Galois groups of these equations.

Proposition 2.63 *The set $G(B)$ contains, together with any finite group, all of its subgroups.*

Proof Suppose that some equation defined over the field K belongs to the collection B. Let P be the field obtained from K by adjoining all roots of this equation, G the Galois group of the field P over the field K, and $G_1 \subset G$ any subgroup. Let K_1 denote the intermediate field corresponding to the subgroup G_1. The Galois group of our equation over the field K_1 coincides with G_1. By our assumption, the collection B contains, together with any equation defined over the field K, the same equation defined over the bigger field K_1. □

Theorem 2.64 (A necessary condition for B-solvability) *If an algebraic equation over a field K is B-solvable, then its Galois group G admits a normal tower $G = G_0 \supset G_1 \supset \cdots \supset G_1 = \{e\}$ of subgroups in which every quotient G_i/G_{i+1} is a quotient of some group from $G(B)$.*

Proof Indeed, the B-solvability of an equation over the field K means the existence of a chain of extensions $K = K_0 \subset K_1 \subset \cdots \subset K_n$ in which the field K_{i+1} is obtained from the field K_i by adjoining all roots of some equation from B, and the last field K_n contains all roots of the initial algebraic equation. Let $G = G_0 \supset \cdots \supset G_n = \{e\}$ be the chain of Galois groups of this equation over this chain of subfields. We will show that the chain of subgroups thus obtained satisfies the property stated in the theorem. Indeed, by Theorem 2.42, the group G_{i+1} is a normal subgroup of the group G_i; moreover, the quotient group G_i/G_{i+1} is simultaneously a quotient of the Galois group of the field K_{i+1} over the field K_i. Since the field K_{i+1} is obtained from the field K_i by adjoining all roots of some equation from B, the Galois group of the field K_{i+1} over the field K_i belongs to the set $G(B)$. □

2.10.2 Classes of Finite Groups

Let M be a set of finite groups.

Definition 2.65 Define the *completion* $\mathscr{K}(M)$ of the set M as the minimal class of finite groups containing all groups from M and satisfying the following properties:

1. Together with any group, the class $\mathscr{K}(M)$ contains all of its subgroups.
2. Together with any group, the class $\mathscr{K}(M)$ contains all of its quotients.
3. If a group G has a normal subgroup H such that the groups H and G/H are in the class $\mathscr{K}(M)$, then the group G is in the class $\mathscr{K}(M)$.

The theorem proved above suggests the following problem: for a given set M of finite groups, describe its completion $\mathscr{K}(M)$. Recall the Jordan–Hölder theorem. A normal tower $G = G_0 \supset \cdots \supset G_n = \{e\}$ of a group G is said to be *unrefinable* (or *maximal*) if all quotient groups G_i/G_{i+1} of this tower are simple groups. The Jordan–Hölder theorem asserts that for every finite group G, the set of quotient groups associated to any unrefinable normal tower of the group G does not depend on the choice of unrefinable tower (and hence is an invariant of the group).

Proposition 2.66 *A group G belongs to the class $\mathscr{K}(M)$ if and only if every quotient group G_i/G_{i+1} with respect to an unrefinable normal tower of the group G is a subquotient of a group from M.*

A subquotient is a quotient of a subgroup.

Proof First, by definition of the class $\mathscr{K}(M)$, every group G satisfying the assumptions of the proposition belongs to the class $\mathscr{K}(M)$. Second, it is not hard to verify that groups G satisfying the assumptions of the proposition have properties 1–3 listed in the definition of the completion of M. □

Corollary 2.67 *The following statements hold.*

1. *The completion of the class of all finite abelian groups is the class of all finite solvable groups.*
2. *The completion of the set consisting of all abelian groups and the group $S(k)$ is the class of all finite k-solvable groups.*

Remark 2.68 Necessary conditions for solvability of algebraic equations by radicals and by k-radicals are particular cases of Theorem 2.64.

Chapter 3
Solvability and Picard–Vessiot Theory

Picard discovered a similarity between linear differential equations and algebraic equations, and he initiated the development of a differential analogue of Galois theory. The culmination of this theory is the Picard–Vessiot theorem, in which the question of solvability of a linear differential equation relates to the question of solvability of a certain algebraic Lie group.

In this chapter, we discuss the simplest Picard–Vessiot extensions in detail, and describe a similarity between linear differential equations and algebraic equations. We state the main results of the Picard–Vessiot theory and also discuss the linear-algebraic part of the theory that is used for constructing explicit solutions of Fuchsian differential equations (see Sect. 6.1). A criterion of Kolchin will be stated that will allow us to establish explicit criteria for different kinds of solvability in the context of Fuchsian systems of differential equations with sufficiently small coefficients (see Sect. 6.2).

3.1 Similarity Between Linear Differential Equations and Algebraic Equations

We now recall the simplest properties of linear differential equations and their analogues for algebraic equations.

3.1.1 Division with Remainder and the Greatest Common Divisor of Differential Operators

A *linear differential operator of order n* over a differential field K is an operator $L = a_n D^n + \cdots + a_0$, where $a_i \in K$ and $a_n \neq 0$, acting on an element y of K by the formula

A. Khovanskii, *Topological Galois Theory*, Springer Monographs in Mathematics,
DOI 10.1007/978-3-642-38871-2_3

$$L(y) = a_n y^{(n)} + \cdots + a_0 y.$$

For operators L_1 and L_2 over K, their composition $L = L_1 \circ L_2 = L_1(L_2)$ is also an operator over K. The composition of operators is, in general, noncommutative, but this noncommutativity does not affect the leading terms. Namely, the leading term of the operator $L = L_1 \circ L_2$ equals the product of the leading terms of the operators L_1 and L_2. For operators L and L_2 of orders n and k over K, there exist unique operators L_1 and R over K such that $L = L_1 \circ L_2 + R$, and the order of R is strictly less than k. The operator R is called *the remainder on right division* of the operator L by the operator L_2. The operators L_1 and R can be constructed explicitly: the algorithm for the division of operators with remainder is based on the remark about the leading term of the composition that we made above and is very similar to the algorithm for division with remainder for polynomials in one variable.

For any two operators L_1 and L_2 over K, one can find their right greatest common divisor N explicitly, i.e., an operator N over K of maximal order that divides the operators L_1 and L_2 from the right, i.e., $L_1 = M_1 \circ N$ and $L_2 = M_2 \circ N$, where M_1 and M_2 are certain operators over K. Finding operators M_1, M_2, and N in terms of the operators L_1 and L_2 is very similar to the standard Euclidean algorithm for finding the greatest common divisor of two polynomials in one variable and is based on the algorithm for division of operators with remainder. As in the commutative case, the greatest common divisor N is representable in the form $N = A \circ L_1 + B \circ L_2$, where A and B are some operators over K.

It is clear that y is a solution of the equation $N(y) = 0$ if and only if $L_1(y) = 0$ and $L_2(y) = 0$.

3.1.2 *Reduction of Order for a Linear Differential Equation as an Analogue of Bézout's Theorem*

Let L be a linear differential operator over K, y_1 a nonzero element of the field K, $p = y_1'/y_1$ its logarithmic derivative, and $L_2 = D - p$ the first-order operator annihilating y_1. The remainder R on right division of L by L_2 is the operator of multiplication by c_0, where $c_0 = L(y_1)/y_1$. Indeed, the desired equality can be obtained by plugging $y = y_1$ into the identity $L(y) \equiv L_1 \circ L_2(y) + c_0 y$. The operator L is divisible by L_2 from the right if and only if the element y_1 satisfies the identity $L(y_1) \equiv 0$.

Using a nonzero solution y_1 of the equation $L(y) = 0$ of order n, one can reduce the order of this equation. To this end, one needs to represent the operator L in the form $L = L_1 \circ L_2$, where L_1 is an operator of order $(n - 1)$. The coefficients of the operator L_1 lie in an extension of the differential field K obtained by adjoining the logarithmic derivative p of the element y_1. If one knows some solution u of the equation $L_1(u) = 0$, then using it, one can construct a certain solution y of the initial equation $L(y) = 0$. To do so, it suffices to solve the equation $L_2(y) =$

$y' - py = u$. The procedure just described is called the *reduction of order* of a differential equation.

Remark 3.1 An operator annihilating y_1 is defined up to a factor, which can be an arbitrary function, and the procedure of order reduction depends on the choice of this function. It is easier to divide by the operator $\tilde{L}_2 = D \circ y_1^{-1}$, which is the composition of multiplication by the element y_1^{-1} and differentiation. To do so, it suffices to compute the operator $L_3 = L \circ y_1$ that is the composition of the multiplication by y_1 and the operator L. The operator L_3 is divisible by D from the right, i.e., $L_3 = \tilde{L}_1 \circ D$, since $L_3(1) = L \circ y_1(1) = 0$. It is readily seen that $L = \tilde{L}_1 \circ \tilde{L}_2$. The initial equation $L(y) = 0$ reduces to the equation $\tilde{L}_1(u) = 0$ of lower order. This very procedure of order reduction is usually given in textbooks on differential equations. Note that the coefficients of the operator $\tilde{L}_1$ lie in the extension of the differential field K obtained by adjoining the element y_1 itself, rather than its logarithmic derivative p, which sometimes makes the operator $\tilde{L}_1$ less convenient than the operator L_1.

In algebra, one has the following analogues of the facts just mentioned:

1. The remainder on division of a polynomial P in the variable x by $(x - a)$ equals the value of P at the point a (Bézout's theorem).
2. If one solution x_1 of the equation $P(x) = 0$ is known, then the degree of that equation can be reduced: the remaining roots of the polynomial P satisfy the equation $Q(x) = 0$ of lower degree, where $Q = P/(x - x_1)$.

Despite the analogy, there is a difference: solutions of a differential equation obtained by order reduction are not, in general, solutions of the initial equation.

Remark 3.2 The exponentials are eigenfunctions for differential operators $P(D)$ with constant coefficients. This fact is equivalent to Bézout's theorem. Indeed, if $P = Q(x - a) + P(a)$, then $P(D) = Q(D) \circ (D - a) + P(a)$. Thus a solution y_1 of the differential equation $(D - a)y = 0$ is an eigenvector of the operator $P(D)$ with the eigenvalue $P(a)$.

3.1.3 A Generic Linear Differential Equation with Constant Coefficients and Lagrange Resolvents

We will now show that solving a linear differential equation with constant coefficients is similar to solving an algebraic equation by radicals (see Sect. 2.1).

Consider a linear differential operator $L = T(D) = D^n - a_{n-1}D^{n-1} - \cdots - a_0E$ with constant complex coefficients $a_{n-1}, \ldots, a_0$. Suppose that the characteristic polynomial $T(t) = t^n - a_{n-1}t^{n-1} - \cdots - a_0$ of this operator has exactly n distinct complex roots $\lambda_1, \ldots, \lambda_n$. Then the linear differential equation $L(y) = 0$ can be solved explicitly with the help of generalized Lagrange resolvents. Lagrange resolvents are used here in exactly the same way as in solving an algebraic equation by radicals. We now proceed with a detailed discussion.

Consider the vector space V of all solutions of the equation

$$L(y) = y^{(n)} - a_{n-1}y^{(n-1)} - \cdots - a_0 y = 0. \tag{3.1}$$

It is clear that the differentiation D takes the space V to itself. The operator $D : V \to V$ satisfies the equation $T(D) = 0$. Let $L_i = T_i(D)$ denote the generalized Lagrange resolvent of the operator $D : V \to V$ corresponding to the root λ_i (see Sect. 2.1.3).

Theorem 3.3 *Every solution y of* (3.1) *is the sum of its generalized Lagrange resolvents: $y = y_1 + \cdots + y_n$. The generalized Lagrange resolvent $y_i = L_i y$ satisfies the differential equation $y_i' = \lambda_i y_i$.*

Proof The result follows immediately from Proposition 2.5. □

We see that generalized Lagrange resolvents help to reduce the general equation (3.1), for which the characteristic equation has only simple roots, to equations of the form $y_i' = \lambda_i y_i$.

We now introduce some convenient notation. Let $Q(x) = b_0 + b_1 x + \cdots + b_k x^k$ be a polynomial over the field $\mathbb{C}$, and $u_0, \ldots, u_k$ a sequence of complex numbers. It will sometimes be convenient to write $Q\langle u_0, \ldots, u_k\rangle$ for the complex number $(b_0 u_0 + b_1 u_1 + \cdots + b_k u_k) \in \mathbb{C}$.

Using this notation, we can give a formula for the solution y of Eq. (3.1) with the following initial values: $y(t_0) = y_0^{(0)}, \ldots, y^{n-1}(t_0) = y_0^{(n-1)}$.

Corollary 3.4 *The solution of the Cauchy problem stated above is given by the following formula:*

$$y(t) = \sum_{1 \le i \le n} T_i \left\langle y_0^{(0)}, \ldots, y_0^{(n-1)} \right\rangle \exp(\lambda_i (t - t_0)).$$

Using interpolating polynomials with multiple nodes, one can give explicit formulas for a solution of a linear differential equation with constant coefficients whose characteristic equation has multiple roots (see [60]).

3.1.4 Analogue of Viete's Formulas for Differential Operators

If one knows all roots $x_1, \ldots, x_n$ of a degree-n polynomial P with leading coefficient 1, then the polynomial P can be recovered: by Viete's formulas,

$$P(x) = x^n + p_1 x^{n-1} + \cdots + p_n,$$

where

$$p_1 = -\sigma_1, \ldots, p_n = (-1)^n \sigma_n \quad \text{and} \quad \sigma_1 = x_1 + \cdots + x_n, \quad \ldots, \quad \sigma_n = x_1 \cdots x_n.$$

The functions $\sigma_1, \ldots, \sigma_n$ are unchanged under permutation of the roots and are called the *elementary symmetric functions*.

Similarly, if one knows n linearly independent solutions $y_1, \ldots, y_n$ of a linear differential equation $Ly = 0$ of order n, where L is an operator whose coefficient of the highest derivative equals 1, then the operator L can be recovered. Indeed, first of all, there is at most one such operator: the difference $L_1 - L_2$ of two operators satisfying these properties is an operator of order less than n having n linearly independent solutions, which is possible only if L_1 coincides with L_2.

The Wronskian (determinant) W of n independent solutions $y_1, \ldots, y_n$ of a linear differential equation is nonzero. Consider the equation $W(y, y_1, \ldots, y_n) = 0$, where $W(y, y_1, \ldots, y_n)$ is the Wronskian of an unknown function y and the functions $y_1, \ldots, y_n$. Expanding the Wronskian

$$W(y, y_1, \ldots, y_n) = \begin{vmatrix} y & y_1 & \ldots & y_n \\ \vdots & \vdots & & \vdots \\ y^{(n)} & y_1^{(n)} & \ldots & y_n^{(n)} \end{vmatrix}$$

with respect to the first column and dividing it by W, we obtain the equation

$$y^{(n)} + p_1 y^{(n-1)} + \cdots + p_n y = 0, \tag{3.2}$$

where $p_1 = -\varphi_1, \ldots, p_n = (-1)^n \varphi_n$ and

$$\varphi_1 = \frac{1}{W} \begin{vmatrix} y_1 & \ldots & y_n \\ \vdots & & \vdots \\ y_1^{(n-2)} & \ldots & y_n^{(n-2)} \\ y_1^{(n)} & \ldots & y_n^{(n)} \end{vmatrix}, \quad \ldots, \quad \varphi_n = \frac{1}{W} \begin{vmatrix} y_1' & \ldots & y_n' \\ \vdots & & \vdots \\ y_1^{(n-1)} & \ldots & y_n^{(n-1)} \\ y_1^{(n)} & \ldots & y_n^{(n)} \end{vmatrix}. \tag{3.3}$$

The functions $y_1, \ldots, y_n$ and their linear combinations are solutions of Eq. (3.2). Formulas (3.2) and (3.3) are similar to Viete's formulas. The functions $\varphi_1, \ldots, \varphi_n$ are rational functions of $y_1, \ldots, y_n$ and their derivatives up to order n. These functions depend only on the vector space V spanned by the functions $y_1, \ldots, y_n$, and they do not depend on the choice of a particular basis $y_1, \ldots, y_n$ in the space V. In other words, the functions $\varphi_1, \ldots, \varphi_n$ are $\mathrm{GL}(V)$-invariant functions of $y_1, \ldots, y_n$ and of their derivatives. We will call the functions $\varphi_1, \ldots, \varphi_n$ the *elementary differential invariants* of $y_1, \ldots, y_n$.

3.1.5 An Analogue of the Theorem on Symmetric Functions for Differential Operators

As is known from algebra, every rational function in the variables $x_1, \dots, x_n$ that does not change under permutations of the variables is, in fact, a rational function of the elementary symmetric polynomials $\sigma_1, \dots, \sigma_n$ in the variables $x_1, \dots, x_n$. In other words, every rational expression that depends symmetrically on the roots of a polynomial of degree n can be expressed rationally through the coefficients of that polynomial. A similar theorem for linear differential equations is due to Picard.

Theorem 3.5 (Picard) *Every rational function R of linearly independent functions $y_1, \dots, y_n$ and their derivatives that is* $\mathrm{GL}(V)$*-invariant (i.e., that does not change if the functions $y_1, \dots, y_n$ are replaced by their linear combinations $z_1 = a_{11}y_1 + \dots + a_{1n}y_n, \dots, z_n = a_{n1}y_1 + \dots + a_{nn}y_n$, where the matrix $A = \{a_{ij}\}$ is constant and nondegenerate) is in fact a rational function of the elementary differential invariants $\varphi_1, \dots, \varphi_n$ of the functions $y_1, \dots, y_n$ and the derivatives of those invariants.*

Proof Suppose that R depends on the vector function $\mathbf{y} = (y_1, \dots, y_n)$ and its derivatives of orders up to $n + k$. Differentiating the identity

$$\mathbf{y}^{(n)} - \varphi_1 \mathbf{y}^{(n-1)} + \dots + (-1)^n \varphi_n \mathbf{y} = 0,$$

we can express the derivative $\mathbf{y}^{(n+i)}$ through $\mathbf{y}$, its derivatives of orders less than n, elementary differential invariants, and their derivatives of orders at most i. Substituting the thus obtained expressions for the higher derivatives of the vector function $\mathbf{y}$ into the rational function R, we obtain a rational function $\tilde{R}$ of the vector function $\Phi = (\varphi_1, \dots, \varphi_n)$, its derivatives of orders at most k, and the entries of the fundamental matrix Y, where

$$Y = \begin{vmatrix} y_1 & \dots & y_n \\ \vdots & & \vdots \\ y_1^{(n-1)} & \dots & y_n^{(n-1)} \end{vmatrix}.$$

Fix any nondegenerate matrix Y_0. Then

$$\tilde{R}\left(\Phi(t), \dots, \Phi^{(k)}(t), Y(t)\right) \equiv \tilde{R}\left(\Phi(t), \dots, \Phi^{(k)}(t), Y_0\right).$$

Indeed, prior to computing the value of the function at the point $t = t_0$, we can apply a linear transformation mapping the fundamental matrix $Y(t_0)$ to the matrix Y_0. By our assumptions, such a linear transformation does not change the value of the function. □

Corollary 3.6 *Every rational function of independent solutions $y_1, \dots, y_n$ of a linear differential equation and their derivatives that does not change as we pass* *to any other basis $z_1, \ldots, z_n$ in the space of solutions is in fact a rational function of the coefficients of the differential equation and their derivatives.*

3.2 A Picard–Vessiot Extension and Its Galois Group

Consider a linear differential equation

$$y^{(n)} + p_1 y^{(n-1)} + \cdots + p_n y = 0 \tag{3.4}$$

with coefficients in some functional differential field K. (Recall that we always assume that the field K contains all complex constants.)

Definition 3.7 A functional differential field P is called a *Picard–Vessiot extension* of the field K if there is a linear differential equation (3.4) with coefficients in K such that P is obtained from K by adjoining all solutions of (3.4). The *Galois group of the Picard–Vessiot extension P over the field K* is defined as the group G_P of all automorphisms of P that fix all elements of K.

Let V denote the solution space of (3.4). Every element τ of the Galois group G_P maps a solution of (3.4) to a solution of (3.4) and preserves the linear relations among the solutions. Therefore, the element τ gives rise to a linear transformation $A(\tau) : V \to V$ of the solution space. The map $\tau \mapsto A(\tau)$ is a group homomorphism $\pi : G_P \to \mathrm{GL}(V)$. The group $G = \pi(G_P) \subset \mathrm{GL}(V)$ is called the *Galois group of* (3.4). The homomorphism τ has trivial kernel, since the field P is generated by the solutions of (3.4). Thus, the group G_P is isomorphic to the Galois group G of (3.4).

In general, the group G is different from $\mathrm{GL}(V)$, the reason being that elements $\tau \in G$ preserve not only algebraic linear relations but also differential relations over the field K (see below) among the solutions of (3.4).

In this section, we deal with abstract (not only functional) differential rings and fields. Given a differential field K, we define the ring $K\{x_1, \ldots, x_n\}$ of differential polynomials consisting of polynomials over K in the elements x_i and their derivatives. In other words, the elements of the ring $K\{x_1, \ldots, x_n\}$ are polynomials over K in an infinite set of variables $\{x_i, z_{i,j}\}$, where $1 \leq i \leq n$, $1 \leq j < \infty$, and the differentiation is defined as the extension of the differentiation in the field K given by the identifications $x_i^{(k)} = z_{i,k}$.

A differential relation over the field K between the solutions of (3.4) is by definition a differential polynomial $Q \in K\{x_1, \ldots, x_n\}$ that vanishes under the substitution $x_1 = y_1, \ldots, x_n = y_n$. The set J of all such relations is obviously a prime ideal of the ring $K\{x_1, \ldots, x_n\}$ that is stable under the differentiation (i.e., this ideal is a *differential ideal*). The initial Picard–Vessiot extension P is the field of fractions of the quotient ring of $K\{x_1, \ldots, x_n\}$ by the ideal J.

The action of a group $A \in \mathrm{GL}(\tilde{V})$ on the space $\tilde{V}$ spanned by $x_1, \dots, x_n$ extends to its representations in differential automorphism groups of the ring $K\{x_1, \dots, x_n\}$ and its field of fractions $\mathscr{F}$.

Remark 3.8 Elements of the space $\tilde{V}$ satisfy the linear differential equation

$$x^n - \varphi_1 x^{(n-1)} + \dots + (-1)^n \varphi_n x = 0,$$

in which $\varphi_1, \dots, \varphi_n$ are elementary differential invariants (the definition of these invariants from Sect. 3.1.4 carries over verbatim to abstract differential fields) of the elements $x_1, \dots, x_n$, which belong to the field $\mathscr{F}_0 \subset \mathscr{F}$, which is invariant under the action of the group $\mathrm{GL}(\tilde{V})$ on the field $\mathscr{F}$.

Part 1 of the following proposition is similar to Proposition 2.37 on the Galois group of an algebraic equation.

Proposition 3.9 1. *An element $A \in \mathrm{GL}(V)$ belongs to the Galois group G of the differential equation* (3.4) *over a differential field K if and only if A preserves all differential relations over K among the solutions of the equation.*
2. *The Galois group G of a linear differential equation* (3.4) *is an algebraic subgroup of* $\mathrm{GL}(V)$.

Proof 1. Automorphisms from the group G_P obviously preserve all differential relations over the field K among the solutions. Conversely, the substitution $x_1 = y_1, \dots, x_n = y_n$ defines an identification of the solution space V with the space $\tilde{V}$ spanned by $x_1, \dots, x_n$. This allows us to regard A as an element of $\mathrm{GL}(\tilde{V})$ that extends to an automorphism of the ring $K\{x_1, \dots, x_n\}$. By our assumption, this automorphism maps the relations ideal J to itself, i.e., defines an automorphism of the field of fractions of the quotient ring $K\{x_1, \dots, x_n\}/J$ isomorphic to the field P.
2. The equality $Q(Ay_1, \dots, Ay_n) = 0$ for $Q \in K\{x_1, \dots, x_n\}$ can be regarded as a polynomial relation $\tilde{Q}(\{a_{i,j}\}) = 0$ among the entries $\{a_{i,j}\}$ of the complex $n \times n$ matrix $A = \{a_{i,j}\}$, where $\tilde{Q}$ is a polynomial with coefficients from the field P that is uniquely determined by Q. The set of equalities $Q(Ay_1, \dots, Ay_n) = 0$, for $Q \in J$, can be regarded as a set of polynomial relations over the field P between the entries $\{a_{i,j}\}$. The field P is an (infinite-dimensional) vector space over $\mathbb{C}$. Therefore, polynomial relations over the field P can be regarded as polynomial relations over $\mathbb{C}$ with vector coefficients. Such relations are equivalent to the vanishing of all coordinates of all the vectors in some fixed basis, i.e., equivalent to a set of polynomial relations over the field $\mathbb{C}$. The intersection of any collection of algebraic sets is an algebraic set. □

Using the isomorphism between the Galois group G_P of the Picard–Vessiot extension P and the Galois group G of a linear differential equation, one can define a structure of a linear algebraic group on G_P. If two different linear differential equations over the field K define the same Picard–Vessiot extension, then the Galois groups of these equations are isomorphic not only as abstract groups but also as algebraic groups. Thus the algebraic group structure on the Galois group of a Picard–Vessiot extension is well defined.

3.3 The Fundamental Theorem of Picard–Vessiot Theory

Let P be a Picard–Vessiot extension of a differential field K, and G its Galois group. Picard–Vessiot theory describes all intermediate differential fields, i.e., all differential fields containing the field K that are contained in the field P. To each subgroup Γ of the Galois group G, we assign the differential field $\mathrm{Fix}(\Gamma)$ consisting of all elements of the field P that are fixed under the action of the group Γ (it is clear that $K \subseteq \mathrm{Fix}(\Gamma)$). To each intermediate differential field F, $K \subseteq F \subseteq P$, we assign a subgroup $G_{\mathrm{PV}}(F) \subseteq G$, which is the Galois group of the Picard–Vessiot extension P of the field F (P is a Picard–Vessiot extension of the field K, and hence it is automatically a Picard–Vessiot extension of the intermediate field F, $K \subset F \subset P$). The maps Fix and G_{PV} establish the *Galois correspondence* between the subgroups of the Galois group and the intermediate differential fields in the Picard–Vessiot extension. We state the following theorem without proof.

Theorem 3.10 (Fundamental theorem of Picard–Vessiot theory) *The Galois correspondence is a one-to-one correspondence between the set of algebraic subgroups of the Galois group and the set of intermediate differential fields in the Picard–Vessiot extension. More precisely, the following statements hold:*

1. *The composition* $\mathrm{Fix} \circ G_{\mathrm{PV}}$ *of the maps* Fix *and* G_{PV} *is the identity transformation on the set of intermediate fields: if F is a differential field and $K \subseteq F \subseteq P$, then* $\mathrm{Fix}(G_{\mathrm{PV}}(F)) = F$.
2. *The composition* $G_{\mathrm{PV}} \circ \mathrm{Fix}$ *of the maps* G_{PV} *and* Fix *takes every subgroup Γ of the Galois group G to its algebraic closure $\overline{\Gamma}$ in the group G: if Γ is a subgroup of the Galois group, $\Gamma \subset G$, then* $G_{\mathrm{PV}}(\mathrm{Fix}(\Gamma)) = \overline{\Gamma}$.
3. *An intermediate differential field F, $K \subseteq F \subseteq P$, is a Picard–Vessiot extension of the field K if and only if the group $G_{\mathrm{PV}}(F)$ is a normal subgroup of the group G. Furthermore, the Galois group of the Picard–Vessiot extension F of the field K is the quotient group of G by the normal subgroup $G_{\mathrm{PV}}(F)$.*

Let us prove a useful characteristic property of Picard–Vessiot extensions, which follows directly from the fundamental theorem.

Corollary 3.11 *A differential field P is a Picard–Vessiot extension of a differential field K, $K \subseteq P$, if and only if there exists a group Γ of automorphisms of P such that:*

1. *It fixes all elements of K and no other elements.*
2. *There exists a finite-dimensional vector subspace V of P over the field of constants such that V is invariant under the group Γ, and the field P is the minimal differential field containing both V and K.*

Proof A Picard–Vessiot extension satisfies these conditions. This follows from part 1 of the fundamental theorem applied to the field $F = K$. Conversely, let $y_1, \dots, y_n$ be a basis of the vector space V from part 2 of this corollary. The coefficients of the order-n linear differential equation in the functions $y_1, \dots, y_n$ are invariant under all linear transformations of the space V. Hence they are invariant under the action of the group Γ and as a consequence, lie in K. Therefore, P is obtained from K by adjoining all solutions of the above-mentioned differential equation, and P is a Picard–Vessiot extension of the field K. □

What happens to the Galois group of a linear differential equation if the differential coefficient field K is replaced with a bigger differential field K_1? This question is of particular interest when the field K_1 is a Picard–Vessiot extension of the field K. Let G_1 denote the Galois group of the extension K_1 of the differential field K. Results on unsolvability of linear differential equations are based on the following theorem of Picard–Vessiot theory, which we state without proof and whose formulation is very similar to that of Theorem 2.42.

Theorem 3.12 (How the Galois group of a linear differential equation changes when the coefficient field is replaced with a Picard–Vessiot extension) *If the coefficient field K is replaced with a Picard–Vessiot extension K_1, then the Galois group G of the equation is replaced with a certain algebraic normal subgroup H of G. The quotient group G/H of the group G by this normal subgroup is isomorphic to some algebraic quotient group of the Galois group G_1 of the new differential field K_1 over the old differential field K.*

3.4 The Simplest Picard–Vessiot Extensions

In this section, we consider the following simplest Picard–Vessiot extensions: an algebraic extension and extensions obtained by adjoining an integral and by adjoining an exponential of integral.

3.4.1 Algebraic Extensions

Consider an algebraic equation

$$Q(x) = x^n + a_{n-1}x^{n-1} + \cdots + a_0 = 0 \tag{3.5}$$

over a functional differential field K, and consider its Galois extension P obtained from the field K by adjoining all solutions of (3.5).

Lemma 3.13 *The field P is a differential field. If a transformation of P fixing K pointwise preserves the arithmetic operations in P, then it automatically preserves differentiation.*

Proof Changing the algebraic equation (3.5) if necessary, we may assume that it is irreducible over the field K and that each root x_i of (3.5) generates the field P over K. Differentiating the identity $Q(x_i) = 0$, we obtain that

$$\frac{\partial Q}{\partial x}(x_i)x_i' + \frac{\partial Q}{\partial t}(x_i) = 0,$$

where

$$\frac{\partial Q}{\partial t} = \sum_{i=1}^{n-1} a_i' x^i.$$

The polynomial $\partial Q/\partial t$ cannot vanish at the point x_i, since the equation $Q = 0$ is irreducible. Thus we obtain the algebraic expression

$$x_i' = -\frac{\partial Q}{\partial x}(x_i) \Big/ \frac{\partial Q}{\partial t}(x_i)$$

for the derivative of the root x_i, which is the same for all roots x_i of the polynomial Q. Both statements of the lemma now follow. □

The Galois group Γ of the Galois extension P over the field K fixes only the elements of K. The vector space V over the field of constants spanned by the roots $x_1, \dots, x_n$ of (3.5) is invariant under the action of the group Γ (although not fixed pointwise). By Corollary 3.11, the differential field P is a Picard–Vessiot extension. The Galois group of the Picard–Vessiot extension P of the field K coincides with the Galois group of the algebraic equation (3.5).

The fundamental theorem of Picard–Vessiot theory for the Picard–Vessiot extension P of the differential field K coincides with the fundamental theorem of Galois theory for the Galois extension P of the field K.

3.4.2 Adjoining an Integral

Let y_1 be an integral of a nonzero element from a functional differential field K, so that $y_1' = a$, $a \in K \setminus \{0\}$. The homogeneous differential equation $ay'' - a'y' = 0$ has independent solutions y_1 and 1. The extension of the differential field K obtained from K by adjoining the element y_1 is therefore a Picard–Vessiot extension (recall that we always assume that the field K contains all complex constants).

Lemma 3.14 *The integral y_1 either belongs to the field K or is transcendental over K.*

Proof Suppose that the integral y_1 is algebraic over the field K. Let $Q(y) = a_n y^n + \cdots + a_0 = 0$ be an equation irreducible over K satisfied by y_1. We can assume that $n > 1$ and that $a_n = 1$. Differentiating the identity $Q(y) = 0$, we obtain the equation $(na + a'_{n-1})y^{n-1} + \cdots + a'_0 = 0$ in y_1. If $(na_n + a'_{n-1}) \neq 0$, then we obtain an equation of smaller degree in y_1, which contradicts the irreducibility of the polynomial Q. If $na + a'_{n-1} = 0$, then $(-a_{n-1}/n)' = a$. In this case, $y_1 = -a_{n-1}/n + C$ for some $C \in \mathbb{C}$, i.e., we have $y_1 \in K$. This contradiction proves the lemma. □

Suppose that y_1 is transcendental over K. Let us show that the only independent differential relation on y_1 over K is $y'_1 = a$. Indeed, using this relation, we can reduce every differential polynomial in y_1 over K to a polynomial in y_1 with coefficients from K. But no such nontrivial polynomial can vanish at y_1, since the element y_1 is transcendental over K. Therefore, the Galois group of the equation $ay'' - a'y' = 0$ consists only of the linear transformations A such that $Ay_1 = y_1 + C$, $A(1) = 1$, where C is any complex number. Thus the Galois group of a nontrivial integral extension is isomorphic to the additive group of complex numbers.

An algebraic group is said to be *unipotent* if it does not have elements of finite order other than the identity (this notion was introduced by Kolchin [64], who initially called such groups *anticompact*). The Galois group of a nontrivial integral extension is obviously unipotent.

Proposition 3.15 *There are no differential fields between the field K and $K\langle y\rangle$, where y is an integral over K not belonging to K.*

Proof Indeed, let F be a differential field such that $K \subset F \subset K\langle y\rangle$. Suppose that $b \in F$ and $b \notin K$. Then the element b can be represented as a nontrivial rational function of y with coefficients from K. The existence of such a function means that the element y is algebraic over F. But the element y is an integral over F, since $y' = a \in K$. An integral is algebraic over a differential field if and only if it belongs to this field (see Lemma 3.14), i.e., $F = K\langle y\rangle$. □

This proposition proves the fundamental theorem of Picard–Vessiot theory for adjoining an integral. Indeed, the Galois group of the field $K\langle y\rangle$ over the field K has no nontrivial algebraic subgroups, and the pair of differential fields $K \subset K\langle y\rangle$ has no proper intermediate differential fields.

3.4.3 Adjoining an Exponential of Integral

Let y_1 be the exponential of integral of an element from a functional differential field K, i.e., $y'_1 = ay_1$, where $a \in K$. The extension of the field K by the element y_1 is by definition a Picard–Vessiot extension.

Lemma 3.16 *Let the exponential of integral y_1 be algebraic over the field K. Then y_1 is a radical over the field K.*

Proof Let $Q(y) = a_n y^n + \cdots + a_0 = 0$ be an equation irreducible over K such that y_1 satisfies this equation. We can assume that $a_n \neq 0$ for $n > 1$ and $a_0 = 1$. Differentiating the identity $Q(y_1) = 0$, we obtain the equation $\sum(a_k' + ka_k a)y^k = 0$ on y_1. This equation is of degree $\leq n$, but it has no constant term. All coefficients of this equation must vanish identically, since otherwise, we would have a contradiction to the irreducibility of the polynomial Q. The equality $a_n' + na_n a = 0$ means that the quotient $a_n/y_1^n = c$ is constant. Indeed, from the relation $y_1' = ay_1$, it follows that $\left(y_1^{-n}\right)' + na\left(y_1^{-n}\right) = 0$, i.e., that y_1^{-n} and a_n satisfy the same equation. Thus $y^n = a_n/c$. The lemma is proved. □

Suppose that the element y_1 is transcendental over K. Let us show that in this case, the unique independent differential relation on y_1 over K is $y_1' = ay_1$. Indeed, using this relation, we can reduce every differential polynomial of y_1 over K to a polynomial of y_1 with coefficients in K. But no such nontrivial polynomial can vanish on y_1, since y_1 is transcendental over K. Therefore, the Galois group of the equation $y' = ay$ consists of linear transformations of the form $Ay_1 = Cy_1$, where $C \neq 0$ is any nonzero complex number. Thus the Galois group of a nonalgebraic extension obtained by adjoining an exponential of integral coincides with the multiplicative group $\mathbb{C}^*$ of nonzero complex numbers.

An exponential of integral over K is an algebraic element y over K if and only if y is a radical over K. Hence, if adjoining an exponential of integral yields an algebraic extension, then the Galois group of this extension is a finite multiplicative subgroup in $\mathbb{C}^*$. An algebraic group is called *semisimple* if each of its nontrivial algebraic subgroups contains elements of finite order different from the identity. This notion was introduced by Kolchin [64] under the name quasicompact group. The Galois group of a nonalgebraic extension obtained by adjoining an exponential of integral is obviously quasicompact.

Proposition 3.17 *Let y be an exponential of integral over K, and suppose that the element y is transcendental over K. Then to each nonnegative integer n, one can associate a differential subfield lying between the fields K and $K\langle y\rangle$. Namely, this is the differential field K_n consisting of rational functions of y^n with coefficients in K. For different n, the fields K_n are different. Every intermediate differential field coincides with some field K_n.*

Proof Let F be a differential field properly containing the field K and properly contained in the field $K\langle y\rangle$. Repeating the arguments of Proposition 3.15, we obtain that the element y is algebraic over F. The element y is an exponential of integral over F. Thus the algebraic equation in y, irreducible over F, has the form $y^n - a = 0$, where $a \in F$ (see Lemma 3.16); hence $K_n \subseteq F$.

The field K_n must coincide with F, for otherwise, there would exist an element $b \in F$ such that $b \notin K_n$. The element b is some rational function R of y, and the relation $R(y) = b$ is not a consequence of the equation $y^n = a$. This contradicts the irreducibility of the equation $y^n = a$ over the field F, which proves that $K_n = F$. The fields K_n are different for different n, since y is transcendental over K. □

This proposition proves the fundamental theorem of the Picard–Vessiot theory for adjoining an exponential of integral. Indeed, every proper algebraic subgroup of the group $\mathbb{C}^*$ is the group of nth roots of unity for some n. An intermediate differential field between K and $K\langle y\rangle$ consists precisely of elements of the field $K\langle y\rangle$ that are fixed under the action of the group of nth roots of unity on $K\langle y\rangle$.[1]

3.5 Solvability of Differential Equations

We say that an algebraic group G is *solvable*, *k-solvable*, or *almost solvable in the category of algebraic groups* if it has a normal tower of algebraic subgroups $G = G_0 \supset \cdots \supset G_m = e$ with the following properties:

1. For solvable groups: for every $i = 1, \dots, m$, the quotient group G_{i-1}/G_i is abelian.
2. For k-solvable groups: for every $i = 1, \dots, m$, either the depth of the group G_i in the group G_{i-1} is at most k, or the group G_{i-1}/G_i is abelian.
3. For almost solvable groups: for every $i = 1, \dots, m$, either the index of the group G_i in the group G_{i-1} is finite, or the group G_{i-1}/G_i is abelian.

Theorem 3.18 (Picard–Vessiot) *A linear differential equation over a differential field K is solvable by quadratures, by k-quadratures, or by generalized quadratures if and only if the Galois group of the equation over K is a solvable, k-solvable, or almost solvable group, respectively, in the category of algebraic groups.*

Remark 3.19 In the classical Picard–Vessiot theorem, the question of solvability of equations by k-quadratures is not discussed. We have added this case to the theorem, firstly, because the result is similar, and secondly, because it carries over to the topological version of Galois theory.

In this section, we prove only the necessity of the conditions on the Galois group for the solvability of an equation. We postpone the proof of sufficiency to Sect. 3.7 and continue with the following theorem.

Theorem 3.20 *If a linear differential equation is solvable by quadratures, by k-quadratures, or by generalized quadratures, then the Galois group G of this equation is a solvable, k-solvable, or almost solvable group in the category of algebraic groups.*

[1]The group of nth roots of unity acts on $K\langle y\rangle$ as follows: the transformation of $K\langle y\rangle$ associated with a root of unity ζ sends a polynomial $P(y)$ to the polynomial $P(\zeta y)$.

Proof The solvability of equations by generalized quadratures over a field K means the existence of a chain of differential fields $K = K_0 \subset \cdots \subset K_N$ such that the first field of this chain coincides with the initial field K, the last field K_N contains all solutions of the differential equation, and for every $i = 1, \ldots, N$, the field K_i is obtained from the field K_{i-1} by adjoining an integral, or an exponential of integral, or all solutions of an algebraic equation. (For solvability by quadratures, the last of these types of extension is forbidden. For solvability by k-quadratures, one can adjoin the roots of algebraic equations only of degree at most k.)

Let $G = G_0 \supset \cdots \supset G_m = e$ be the nested chain of groups in which the group G_i is the Galois group of the initial equation over the field K_i. According to the fundamental theorem, Theorem 3.10, the quotient group G_{i-1}/G_i is a quotient of the Galois group of the Picard–Vessiot extension K_i of K_{i-1}. If this extension is obtained by adjoining an integral or an exponential of integral, then the group G_{i-1}/G_i is abelian as the quotient group of an abelian group (see Sects. 3.4.2 and 3.4.3). If the extension K_i of the field K_{i-1} is obtained by adjoining all roots of an algebraic equation, then the quotient group G_{i-1}/G_i is finite. If this algebraic equation has degree $\leq k$, then between the groups $G_i \supset G_{i-1}$, one can insert a chain of normal subgroups $G_i = G_{i1} \supset \cdots \supset G_{ip} = G_{i-1}$ such that the depth of the group G_{ij} in the group $G_{i,j-1}$ is at most k (see Sect. 2.9). This concludes the proof of the theorem. □

The theorem just proved can be reformulated in the following way: If a Picard–Vessiot extension is a Liouville extension, a k-Liouville extension, or a generalized Liouville extension, then its Galois group is respectively a solvable, k-solvable, or almost solvable group in the category of algebraic groups.

In this reformulation, the theorem becomes applicable to algebraic equations over differential fields as well. It gives stronger results on the unsolvability of algebraic equations.

Theorem 3.21 *If the Galois group of an algebraic equation over a differential field K is unsolvable, then this algebraic equation is unsolvable not only by radicals, but by quadratures as well. If the Galois group is not k-solvable, then the algebraic equation is unsolvable by k-quadratures over K.*

3.6 Linear Algebraic Groups and Necessary Conditions of Solvability

The Galois group of a linear differential equation is a linear algebraic group. Such groups have general properties that make it possible to reformulate the conditions of solvability, k-solvability, and almost solvability of the Galois group and to prove that these conditions are sufficient for the solvability of the equation (see Sect. 3.7).

First of all, let us point out that every linear algebraic group is a Lie group. Indeed, the set of singular points of every algebraic variety has codimension ≥ 1.

But the left (or right) multiplications in the group map every point of the group to every other point. Thus the group looks the same at each of its points, and therefore, the set of its singular points is empty. The connected component of the identity in an algebraic group is a normal subgroup of finite index. Indeed, the connected component of the identity is a normal subgroup in every Lie group, and every algebraic variety has only finitely many connected components. In the sequel, the following famous theorem of Lie, which will be stated without proof, plays a key role.

Theorem 3.22 (Lie) *A connected solvable linear algebraic Lie group reduces to triangular form in some basis.*

This theorem of Lie implies the following proposition.

Proposition 3.23 *A linear algebraic group is an almost solvable group in the category of algebraic groups if and only if all elements in its connected component of the identity reduce simultaneously to triangular form in some basis.*

Proof Every group of triangular matrices is solvable. This proves the claim in one direction. Let $G = G_0 \supset \cdots \supset G_n = e$ be a normal tower of algebraic subgroups of the group G such that each quotient group G_i/G_{i-1} is either abelian or finite. Consider the connected components of the identity in these groups. They form a normal tower $G^0 = G_0^0 \supset \cdots \supset G_n^0 = e$ of algebraic subgroups in the connected component G^0 of the identity of the group G. Furthermore, if the quotient G_{i-1}/G_i is abelian, then the quotient G_{i-1}^0/G_i^0 is also abelian. If the quotient group G_{i-1}/G_i is finite, then the groups G_{i-1}^0 and G_i^0 coincide. The claim is proved. □

Proposition 3.24 *A linear algebraic group G is a solvable or a k-solvable group in the category of algebraic groups if and only if all elements in its connected component G^0 of the identity reduce to triangular form in some basis, and the finite quotient group G/G_0 is, respectively, solvable or k-solvable.*

Proof It suffices to prove the "only if" part. Suppose that G is a solvable or a k-solvable group in the category of algebraic groups. By Proposition 3.23, the group G^0 is triangular in some basis. Furthermore, the group G^0 is a normal subgroup of finite index in the group G. The finite quotient group G/G_0 is respectively solvable or k-solvable. □

The group $\mathrm{GL}(n)$ has a remarkable Zariski topology. We now recall its definition. Fixing a basis in an n-dimensional vector space V establishes an identification between the linear maps $A : V \to A$ and the $n \times n$ matrices, which, in turn, can be identified with points of the n^2-dimensional space $\mathbb{C}^N$, where $N = n^2$. Under this identification, elements $A \in \mathrm{GL}(V)$ correspond to points of the open subset $U \subset \mathbb{C}^N$ given by the condition $\det A \neq 0$. A Zariski closed (i.e., closed in the Zariski topology) subset of U is the solution set in U of any system of polynomial equations. The Zariski closure of a set $X \subset U$ coincides with the set of all common zeros in U of all polynomials vanishing on X. The Zariski topology on $\mathrm{GL}(V)$ is induced by the topology in U just described (this topology does not depend on the choice of an iden210tification between $\mathrm{GL}(V)$ and U).

The Zariski topology on $\mathrm{GL}(V)$ makes it possible to associate with every subgroup $\Gamma \subset \mathrm{GL}(V)$ the algebraic group $\bar{\Gamma} \subset \mathrm{GL}(V)$ that is the Zariski closure of Γ. This operation allows us to generalize Propositions 3.23 and 3.24 to arbitrary linear algebraic groups.

Proposition 3.25 *The following statements hold:*

1. *A linear algebraic group is almost solvable if and only if it has a triangular normal subgroup H of finite index. A linear algebraic group is k-solvable or solvable if and only if a finite quotient group G/H of the group G by some triangular normal subgroup H of finite index is respectively k-solvable or solvable.*
2. *A linear algebraic group G is an almost solvable, k-solvable, or solvable group in the category of algebraic groups if and only if it is respectively an almost solvable, k-solvable, or solvable group.*

Proof Let $G = G_0 \supset \cdots \supset G_n = e$ be a normal tower of subgroups in the group G. Then the closures in the Zariski topology of the groups in this tower form a normal tower of algebraic groups $\overline{G} = \overline{G}_0 \supset \cdots \supset \overline{G}_n = e$. Furthermore, if the group G_{i-1}/G_i is abelian or finite or if the group G_i has depth $\leq k$ in G_{i-1}, then the group $\overline{G}_{i-1}/\overline{G}_i$ is respectively abelian or finite, or the group $\overline{G}_i$ has depth $\leq k$ in the group $\overline{G}_{i-1}$. This proves all parts of the proposition in one direction. In the other direction, all the assertions are obvious. □

3.7 A Sufficient Condition for the Solvability of Differential Equations

A group Γ of automorphisms of a differential field F with differential invariant subfield K is called an *admissible group of automorphisms* if there exists a finite-dimensional vector space V over the field of constants such that V is invariant under the action of Γ, and $K\langle V\rangle = F$. According to Picard–Vessiot theory (see Corollary 3.11), a differential field F is a Picard–Vessiot extension of the differential field K if and only if there exists an admissible group of automorphisms of the differential field F with differential invariant subfield K. In general, the existence of an admissible group of transformations of a Picard–Vessiot extension is not at all obvious, and it constitutes a significant part of the fundamental theorem of this theory. However, in many cases, the existence of an admissible group of automorphisms is known a priori. An example is provided by extensions of the field of rational functions by all solutions of any Fuchsian linear differential equation (see Sect. 6.1). In these cases, the monodromy group of the equation plays the role of the group Γ.

If the group Γ is solvable, then the elements of the field F are representable by quadratures through the elements of the field K. A construction of such a representation belongs essentially to linear algebra and does not use the main theorems of Picard–Vessiot theory. The admissible group Γ of automorphisms is isomorphic to the induced group of linear transformations of the space V, and it can be viewed as a linear algebraic group.

Theorem 3.26 (Liouville) *If all transformations in the admissible group Γ can be reduced to triangular form in the same basis, then the differential field F is a Liouville extension of the differential field K.*

Proof Let $e_1, \dots, e_n$ be a basis of the space V such that every transformation $\mu \in \Gamma$ has the form $\mu(e_i) = \sum_{j \le i} a_{i,j} e_j$. Recall that V is a vector space over the field of constants, and in particular, $a_{i,j}$ are constants. Consider the vector space $\tilde{V}$ spanned by the vectors $\tilde{e}_i = (e_i/e_1)'$, where $i = 2, \dots, n$. The space $\tilde{V}$ is invariant under the action of the group, and every transformation μ from Γ has triangular form in the basis $\tilde{e}_i$. Indeed,

$$\mu(\tilde{e}_i) = \mu\left(\frac{e_i}{e_1}\right)' = \left(\frac{a_{i,1}}{a_{1,1}} + \sum_{2 \le j \le i} \frac{a_{i,j}}{a_{1,1}} \frac{e_j}{e_1}\right)' = \sum_{2 \le j \le i} \frac{a_{i,j}}{a_{1,1}} \tilde{e}_j.$$

(Recall that by the definition of differential field automorphism, μ commutes with differentiation.) The space $\tilde{V}$ has a smaller dimension than the space V. We can assume, therefore, that the differential field $K\langle \tilde{V} \rangle$ is a Liouville extension of the differential field K. For $\mu \in \Gamma$, we have

$$\mu\left(\frac{e_1'}{e_1}\right) = \frac{a_{11} e_1'}{a_{11} e_1} = \frac{e_1'}{e_1},$$

and therefore, the element $e_1'/e_1 = a$ lies in the differential invariant subfield K. The differential field F is obtained from the differential field K by adjoining the exponential of integral e_1 of a and by adjoining the integrals e_i/e_1 of $\tilde{e}_i$ for $i = 2, \dots, n$. □

Proposition 3.27 *If a group Γ of admissible automorphisms of a field F with the invariant subfield K is almost solvable, then there exists a field K_0 invariant under the group Γ such that:*

1. *The field F is a Liouville extension of the field K_0.*
2. *The induced group of automorphisms of the field K_0 is finite, and each element of the field K_0 is algebraic over the field K.*
3. *If the group Γ is solvable, then each element of the field K_0 is representable by radicals over the field K.*

Proof Let V be a finite-dimensional vector subspace in F over the field of constants such that V is stable under the action of Γ, and $K\langle V \rangle = F$.

From Proposition 3.23, it follows that the group Γ has a normal subgroup Γ_0 of finite index that can be reduced to triangular form in some basis of the space V. Let K_0 be the differential invariant subfield of the group Γ_0. According to Theorem 3.36, the differential field F is a Liouville extension of the differential field K_0.

Obviously (see Proposition 2.4), the field K_0 is invariant under the action of the group Γ, and the induced group of automorphisms $\tilde{\Gamma}_0$ of this field is a finite quotient group of the group Γ. Thus every element of the field K_0 is algebraic over K (see Theorem 2.16). If the initial group Γ is solvable, then its finite quotient group $\tilde{\Gamma}_0$ is also solvable. In this case, every element of the field K_0 is expressible by radicals through the elements of the field K (see Theorem 2.3). □

The proof of the following statement is based on Galois theory.

Proposition 3.28 *If under the assumptions of Proposition 3.27, the group Γ is k-solvable, then every element of the field K_0 is expressible through the elements of the field K by radicals and solutions of algebraic equations of degree at most k.*

Proof Since the group $\tilde{\Gamma}_0$ is finite, the extension K_0 of the field K is a Galois extension of K. If the group Γ_0 is k-solvable, then its finite quotient group is also k-solvable. Proposition 3.28 now follows from Theorem 2.24. □

Let us now finish the proof of the Picard–Vessiot theorem (see Sect. 3.5).

Proof (Completion of the proof of Theorem 3.18) According to the fundamental theorem, Theorem 3.10, for every linear differential equation over a differential field K, its Galois group fixes the elements of the field K only. Thus Propositions 3.27 and 3.28 just proved are applicable, which establishes the sufficiency of the conditions on the Galois group in the Picard–Vessiot theorem.

The Picard–Vessiot theorem not only proves the Liouville–Mordukhai-Boltovskii criterion (see Sect. 1.8), but also helps to generalize it to the case of solvability by quadratures and k-quadratures. Namely, a linear differential equation of order n is solvable by generalized quadratures over a differential field K if and only if firstly, it has a solution y_1 satisfying the equation $y_1' = ay_1$, where a is an element of some algebraic extension K_1, and secondly, if the differential equation of order $(n-1)$ on the function $z = y' - ay$ with coefficients from the field K_1 obtained from the original equation by reduction of order (see Sect. 3.1.2) is solvable by generalized quadratures. There are similar statements for the solvability of a linear differential equation by quadratures and by k-quadratures. For solvability by quadratures, one also needs to require that the algebraic extension K_1 be obtained from K by adjoining radicals, and for the solvability by k-quadratures, that the extension K_1 be obtained from K by adjoining radicals and roots of algebraic equations of degree $\leq k$. For the proof of these statements, it suffices to look at the construction of solutions of differential equations.

Differential algebra helps to make this criterion significantly more precise. For linear differential equations whose coefficients are rational functions, there is a finite algorithm that determines whether an equation is solvable by generalized quadratures, and that makes it possible, if the equation is solvable, to find its solution [92]. The algorithm uses the following:

1. An estimate of the degree of the extension K_1 of the field K that depends only on the order of the equation and that follows from general group-theoretic arguments (see Sect. 6.2.2).
2. The theory of normal forms of linear differential equations in neighborhoods of their singular points.
3. Elimination theory for differential equations and inequalities depending on several functions (which is due to Seidenberg and generalizes the Seidenberg–Tarski theorem to the case of differential fields).

3.8 Other Kinds of Solvability

Kolchin extended the Picard–Vessiot theorem [64]. He considered the problems of solvability of linear differential equations by integrals and by exponentials of integrals separately, together with versions of these problems in which algebraic extensions are allowed.

The definition of Liouville extensions used three kinds of extensions: algebraic extensions, adjoining integrals, and adjoining exponentials of integrals. One can define more specific kinds of solvability using only some of these extensions as the building blocks (and using only special algebraic extensions). Let us list the main versions:

1. Solvability by integrals
2. Solvability by integrals and by radicals
3. Solvability by integrals and by algebraic functions
4. Solvability by exponentials of integrals
5. Solvability by exponentials of integrals and by algebraic functions

Let us decipher the third of these definitions. Consider an arbitrary chain of differential fields $K = K_0 \subseteq \cdots \subseteq K_n$ such that each subsequent field K_i, $i = 1, \dots, n$, is obtained from the preceding field K_{i-1} either by adjoining an integral over K_{i-1} or as an algebraic extension of the field K_{i-1}. Each element of the differential field K_n is, by definition, representable by integrals and algebraic functions over the field K. An equation is solvable over the field K by integrals and algebraic functions if each of its solutions is represented by integrals and algebraic functions. The other kinds of solvability in items 1–5 above are deciphered similarly.

Remark 3.29 There is no need to consider solvability by radicals and solvability by exponentials of integrals separately, since each radical is an exponential of integral.

Remark 3.30 Up to now, we have been dealing with special algebraic extensions obtained by adjoining all roots of algebraic equations of degrees at most k. We could possibly define, say, k-solvability by integrals combining such algebraic extensions with adjoining integrals. But we do not implement this to avoid overloading the text and because of the lack of interesting examples.

Definition 3.31 We say that a linear algebraic group G is a *special triangular group* if there exists a basis such that all elements in G are simultaneously reduced to triangular form in this basis, and all eigenvalues of all elements in G are equal to 1.

Definition 3.32 We say that a linear algebraic group is *diagonal* if there exists a basis such that all elements of the group are diagonal in that basis.

Theorem 3.33 (Kolchin's theorem on solvability by integrals) *A linear differential equation over a differential field K is solvable by integrals, by integrals and radicals, or by integrals and algebraic functions if and only if the Galois group of the equation over K is respectively a special triangular group, solvable and contains a special triangular normal subgroup of finite index, or contains a special triangular normal subgroup of finite index.*

Theorem 3.34 (Kolchin's theorem on solvability of integrals by exponentials) *A linear differential equation over a differential field K is solvable by exponentials of integrals or by the exponentials of integrals and algebraic functions if and only if its Galois group over K is respectively solvable and contains a diagonal normal subgroup of finite index or contains a diagonal normal subgroup of finite index.*

A couple of words about the proof of these theorems. The Galois group of an extension by an integral is unipotent (see Sect. 3.4.2). The Galois group of an extension by an exponential of integral is semisimple (see Sect. 3.4.3). Kolchin developed the theory of unipotent and semisimple linear algebraic groups. Here is one simple statement from this theory.

Proposition 3.35 ([64]) *The following statements hold:*

1. *A linear algebraic group is semisimple if and only if each element of the group can be reduced to diagonal form.*
2. *A linear algebraic group is unipotent if and only if all eigenvalues of all elements of the group are equal to* 1.

The theory of semisimple and uniipotent groups together with the fundamental theorem of Picard–Vessiot theory allowed Kolchin to prove his theorems on solvability by integrals and solvability by exponentials of integrals. Of course, the theorems of Kolchin, like the Picard–Vessiot theorem, are true not only for linear differential equations but also for Picard–Vessiot extensions (every such extension is generated by solutions of linear differential equations). Let us formulate criteria for different kinds of representability of all elements of a Picard–Vessiot extension with a triangular Galois group. This criterion follows easily from the theorems of Kolchin and Picard–Vessiot. We will apply this criterion in Sect. 6.2.3 when we discuss different kinds of solvability of Fuchsian systems with small coefficients.

Theorem 3.36 (Criteria for extensions with triangular Galois groups (cf. [64])) *Suppose that a Picard–Vessiot extension F of a differential field K has a triangular Galois group. Then every element of the field F:*

1. *Is representable by quadratures over the field K.*
2. *Is representable by integrals and by algebraic functions or by integrals and by radicals over the field K if and only if all eigenvalues of all operators from the Galois group are roots of unity. (These kinds of solvability are different in general if the Galois group is not triangular.)*
3. *Is representable by integrals over the field K if and only if all eigenvalues of all operators from the Galois group are equal to* 1.
4. *Is representable by exponentials of integrals and by algebraic functions or by exponentials of integrals over the field K if and only if the Galois group is diagonal. (These kinds of solvability are different in general if the Galois group is not triangular.)*
5. *Is representable by algebraic functions or by radicals over the field K if and only if the Galois group is diagonal and all eigenvalues of all operators from the Galois group are roots of unity. (These kinds of solvability are different in general if the Galois group is not triangular.)*
6. *Lies in the field K if and only if the Galois group is trivial.*

□

Chapter 4
Coverings and Galois Theory

This chapter is devoted to the geometry of coverings and its relation to Galois theory. There is a surprising analogy between the classification of coverings over a connected, locally connected, and locally simply connected topological space and the fundamental theorem of Galois theory. We state the classification results for coverings so that this analogy becomes evident.

There is a whole series of closely related problems on classification of coverings. Apart from the usual classification, there is a classification of coverings with marked points. One can fix a normal covering and classify coverings (and coverings with marked points) that are subordinate to this normal covering. For our purposes, it is necessary to consider ramified coverings over Riemann surfaces and to solve similar classification problems for ramified coverings, etc.

In Sect. 4.1, we consider coverings over topological spaces. We discuss in detail the classification of coverings with marked points over a connected, locally connected, and locally simply connected topological space. Other classification problems reduce easily to this classification.

In Sect. 4.2, we consider finite ramified coverings over Riemann surfaces. Ramified coverings are first defined as those proper maps of real manifolds to a Riemann surface whose singularities are similar to the singularities of complex analytic maps. We then show that ramified coverings have a natural complex analytic structure. We discuss the operation of completion for coverings over a Riemann surface X with a removed discrete set O. This operation can be applied equally well to coverings and to coverings with marked points. It transforms a finite covering over $X \setminus O$ to a finite ramified covering over X.

The classification of finite ramified coverings with a fixed ramification set almost repeats the analogous classification of unramified coverings. Therefore, we allow ourselves to formulate results without proofs.

To compare the fundamental theorem of Galois theory and the classification of ramified coverings, we use the following fact. The set of orbits under a finite group action on a one-dimensional complex analytic manifold has a natural structure of a

A. Khovanskii, *Topological Galois Theory*, Springer Monographs in Mathematics,
DOI 10.1007/978-3-642-38871-2_4

complex analytic manifold. The proof uses Lagrange resolvents (in Galois theory, Lagrange resolvents are used to prove solvability by radicals of equations with a solvable Galois group).

In Sect. 4.2.4, the operation of completion for coverings is used to define the Riemann surface of an irreducible algebraic equation over the field $K(X)$ of meromorphic functions on a manifold X.

Section 4.3 is based on Galois theory and the Riemann existence theorem (which we accept without proof) and is devoted to the relation between finite ramified coverings over a manifold X and algebraic extensions of the field $K(X)$. For a covering space M over X, we show that the field $K(M)$ of meromorphic functions on M is an algebraic extension of the field $K(X)$ of meromorphic functions on X, and that every algebraic extension of the field $K(X)$ can be obtained in this way.

The following construction plays a key role. Fix a discrete subset O of a manifold X and a point $a \in X \setminus O$. Consider the field $P_a(X)$ of those meromorphic germs at the point a that admit a meromorphic continuation to finite-valued functions on $X \setminus O$ with algebraic singularities at points of the set O. The operation of meromorphic continuation of a germ along a closed path defines the action of the fundamental group $\pi_1(X \setminus O, a)$ on the field $P_a(O)$. The results of Galois theory are applied to this action of the fundamental group by automorphisms of the field $P_a(O)$. We describe a correspondence between the subfields of the field $P_a(O)$ that are algebraic extensions of the field $K(X)$ and the subgroups of finite index in the fundamental group $\pi_1(X \setminus O, a)$. We prove that this correspondence is bijective. Apart from Galois theory, the proof uses the Riemann existence theorem.

Normal ramified coverings over a connected complex manifold X are related to Galois extensions of the field $K(X)$. The fundamental theorem of Galois theory for such fields has a transparent geometric interpretation.

A local version of the relation between ramified coverings and algebraic extensions allows one to describe algebraic extensions of the field of convergent Laurent series. Extensions of this field are similar to algebraic extensions of the finite field $\mathbb{Z}/p\mathbb{Z}$ (under this analogy, the Frobenius automorphism corresponds to the monodromy map associated with a closed path in the plane that goes around the point 0).

At the end of this chapter, we consider compact one-dimensional complex manifolds. On the one hand, considerations of Galois theory show that the field of meromorphic functions on a compact manifold is a finitely generated extension of the field of complex numbers and has transcendence degree 1 (the proof uses the Riemann existence theorem). On the other hand, ramified coverings allow us to describe rather explicitly all algebraic extensions of the field of rational functions in one variable. The Galois group of an algebraic extension generated by roots of some algebraic equation has a geometric meaning: it coincides with the monodromy group of the Riemann surface of the algebraic function defined by this equation. Hence, Galois theory produces a topological obstruction to the representability of algebraic functions by radicals.

4.1 Coverings over Topological Spaces

This section is devoted to coverings over a connected, locally connected, and locally simply connected topological space. There is a series of closely related problems on classification of coverings. In Sect. 4.1.1, we discuss in detail the classification of coverings with marked points. Other classification problems (see Sect. 4.1.3) can be easily reduced to this classification. In Sect. 4.1.2, we discuss the correspondence between subgroups of the fundamental group and coverings with marked points. In Sect. 4.1.4, we describe a surprising formal analogy between classification of coverings and Galois theory.

4.1.1 Classification of Coverings with Marked Points

Continuous maps f_1 and f_2 from topological spaces Y_1 and Y_2, respectively, to a topological space X are called *right equivalent* if there exists a homeomorphism $h : Y_1 \to Y_2$ such that $f_1 = f_2 \circ h$. A topological space Y together with a projection $f : Y \to X$ to a topological space X is called a *covering with the fiber* D over X (where D is a discrete set) if the following holds: for each point $c \in X$, there exists an open neighborhood U such that the projection map of $U \times D$ onto the first factor is right equivalent to the map $f : Y_U \to U$, where $Y_U = f^{-1}(U)$. We formulate without proof the covering homotopy theorem (see [34]) that holds for coverings. We will use this theorem when the complex W_k is a point or the interval $[0, 1]$.

Theorem 4.1 (Covering homotopy theorem) *Let $f : Y \to X$ be a covering, W_k a k-dimensional CW-complex, and $F : W_k \to X$, $\tilde{F} : W_k \to Y$ mappings of W_k to X and Y, respectively, such that $f \circ \tilde{F} = F$. Then for every homotopy $F_t : W_k \times [0, 1] \to X$ of the map F, $F_0 = F$, there exists a unique lifting homotopy $\tilde{F} : W_k \times [0, 1] \to Y$, $f(\tilde{F}_t) = F_t$, of the map $\tilde{F}$ such that $\tilde{F}_0 = F$.*

Consider a covering $f : Y \to X$. A homeomorphism $h : Y \to Y$ is called a *deck transformation* of this covering if the equality $f = f \circ h$ holds. Deck transformations form a group. A covering is called *normal* if its group of deck transformations acts transitively on each fiber $f^{-1}(a)$, $a \in X$, of the covering and the following topological conditions on the spaces X and Y are satisfied: the space Y is connected, and the space X is locally connected and locally simply connected.

A triple $f : (Y, b) \to (X, a)$ consisting of spaces with marked points (X, a), (Y, b) and a map f is called a *covering with marked points* if $f : Y \to X$ is a covering and $f(b) = a$. Coverings with marked points are equivalent if there exists a homeomorphism between their covering spaces that commutes with projections and maps one marked point to the other marked point. It is usually clear from notation whether we mean coverings or coverings with marked points. In such cases, we will omit the words "with marked points" for brevity in talking about coverings.

A covering with marked points $f : (Y, b) \to (X, a)$ defines the homomorphism $f_* : \pi_1(Y, b) \to \pi_1(X, a)$ of the fundamental group $\pi_1(Y, b)$ of the space Y with the marked point b to the fundamental group $\pi_1(X, a)$ of the space X with the marked point a.

Lemma 4.2 *For a covering with marked points, the induced homomorphism of the fundamental groups has trivial kernel.*

Proof Let a closed path $\gamma : [0, 1] \to X$, $\gamma(0) = \gamma(1) = a$, in the space X be the image $f \circ \tilde{\gamma}$ of the closed path $\tilde{\gamma} : [0, 1] \to Y$, $\tilde{\gamma}(0) = \tilde{\gamma}(1) = b$, in the space Y. Let the path γ be homotopic to the identity path in the space of paths in X with fixed endpoints. Then the path $\tilde{\gamma}$ is homotopic to the identity path in the space of paths in Y with fixed endpoints. For the proof, it is enough to lift the homotopy with fixed endpoints to Y. □

The following theorem holds for every connected, locally connected, and locally simply connected topological space X with a marked point a.

Theorem 4.3 (On classification of coverings with marked points) *The following statements hold:*

1. *For every subgroup G of the fundamental group of the space X, there exist a connected space (Y, b) and a covering over (X, a) with the covering space (Y, b) such that the image of the fundamental group of the space (Y, b) coincides with the subgroup G.*
2. *Two coverings over (X, a) with connected covering spaces (Y, b_1) and (Y, b_2) are equivalent if the images of the fundamental groups of these spaces in the fundamental group of (X, a) coincide.*

Proof 1. Consider the space $\hat{\Omega}(X, a)$ of the paths $\gamma : [0, 1] \to X$ in X that originate at the point a, $\gamma(0) = a$, and its subspace $\hat{\Omega}(X, a, a_1)$ consisting of paths that terminate at a point a_1. On the spaces $\hat{\Omega}(X, a)$, $\hat{\Omega}(X, a, a_1)$, consider the topology of uniform convergence and the following equivalence relation. Say that paths γ_1 and γ_2 are equivalent if they terminate at the same point a_1 and if the path γ_1 is homotopic to the path γ_2 in the space $\hat{\Omega}(X, a, a_1)$ of paths with fixed endpoints. Let $\Omega(X, a)$ and $\Omega(X, a, a_1)$ denote the quotient spaces of $\hat{\Omega}(X, a)$ and $\hat{\Omega}(X, a, a_1)$ by this equivalence relation. The fundamental group $\pi_1(X, a)$ acts on the space $\Omega(X, a)$ by right multiplication (composition). For a fixed subgroup $G \subset \pi_1(X, a)$, denote by $\Omega_G(X, a)$ the space of orbits under the action of G on $\Omega(X, a)$. Points in $\Omega_G(X, a)$ are elements of the space $\hat{\Omega}(X, a)$ defined up to homotopy with fixed endpoints and up to right multiplication by elements of the subgroup G. There is a marked point $\tilde{a}$ in this space, namely, the equivalence class of the constant path $\gamma(t) = a$. The map $f : (\Omega_G(X, a), \tilde{a}) \to (X, a)$ that assigns to each path its right endpoint has the required properties. We omit a proof of this fact. Note, however, that the assumptions on the space X are necessary for the theorem to be true: if X is disconnected, then the map f has no preimages over the connected components of X disjoint from the point a, and if X is not locally connected and locally simply connected, then the map $f : (\Omega_G(X,a), \tilde{a}) \to (X,a)$ may not be a local homeomorphism.

2. We now show that a covering $f : (Y,b) \to (X,a)$ such that $f_*\pi_1(Y,b) = G \subset \pi_1(X,a)$ is right equivalent to the covering constructed using the subgroup G in the first part of the proof. To a point $y \in Y$, assign any element from the space of paths $\Omega(Y,b,y)$ in Y that originate at the point b and terminate at the point y, which are defined up to homotopy with fixed endpoints. Let $\tilde{\gamma}_1$, $\tilde{\gamma}_2$ be two paths from the space $\Omega(Y,b,y)$, and let $\tilde{\gamma} = (\tilde{\gamma}_1)^{-1} \circ \tilde{\gamma}_2$ denote the path comprising the path $\tilde{\gamma}_2$ and the path $\tilde{\gamma}_1$ traversed in the opposite direction. The path $\tilde{\gamma}$ originates and terminates at the point b, whence the path $f \circ \tilde{\gamma}$ lies in the group G. It follows that the image $f \circ \tilde{\gamma}$ of an arbitrary path $\tilde{\gamma}$ in the space $\Omega(Y,b,y)$ under the projection f is the same point of the space $\Omega_G(X,a)$ (that is, the same path in the space $\hat{\Omega}_G(X,a)$ up to homotopy with fixed endpoints and up to right multiplication by elements of the group G). In this way, we have assigned to each point $y \in Y$ a point of the space $\Omega_G(X,a)$. It is easy to check that this correspondence defines the right equivalence between the covering $f : (Y,b) \to (X,a)$ and the standard covering constructed using the subgroup $G = f_*\pi_1(Y,b)$.

□

4.1.2 Coverings with Marked Points and Subgroups of the Fundamental Group

Theorem 4.3 shows that coverings with marked points over a space X with a marked point a considered up to right equivalence are classified by subgroups G of the fundamental group $\pi_1(X,a)$. Let us discuss the correspondence between coverings with marked points and subgroups of the fundamental group.

Let $f : (Y,b) \to (X,a)$ be a covering that corresponds to the subgroup $G \subset \pi_1(X,a)$, and let $F = f^{-1}(a)$ denote the fiber over the point a. We have, then, the following lemma.

Lemma 4.4 *The fiber F is in bijective correspondence with the right cosets of the group $\pi_1(X,a)$ modulo the subgroup G. If a right coset h corresponds to a point c of the fiber F, then the group hGh^{-1} corresponds to the covering $f : (Y,c) \to (X,a)$ with marked point c.*

Proof The group G acts by right multiplication on the space $\Omega(X,a,a)$ of closed paths that originate and terminate at the point a and are defined up to homotopy with fixed endpoints. According to the description of the covering corresponding to the group G (see the first part of the proof of Theorem 4.3), the preimages of the point a with respect to this covering are orbits of the action of the group G on the space $\Omega(X,a,a)$, i.e., the right cosets of the group $\pi_1(X,a)$ modulo the subgroup G.

Let $h : [0, 1] \to X$, $h(0) = a$, be a loop in the space X, and $\tilde{h} : [0, 1] \to Y$, $f\tilde{h} = h$, the lift of this loop to Y that originates at the point b, $\tilde{h}(0) = b$, and terminates at the point c, $\tilde{h}(1) = c$. Let $G_1 \subset \pi_1(X, a)$ be the subgroup consisting of paths whose lifts to Y starting at the point c terminate at the same point c. It is easy to verify the inclusions $hGh^{-1} \subseteq G_1$, $h^{-1}G_1h \subseteq G$, which imply that $G_1 = hGh^{-1}$. □

Let us say that a covering $f_2 : (Y_2, b_2) \to (X, a)$ is *subordinate* to the covering $f_1 : (Y, b_1) \to (X, a)$ if there exists a continuous map $h : (Y_1, b_1) \to (Y_2, b_2)$ compatible with the projections f_1 and f_2, i.e., such that $f_1 = f_2 \circ h$.

Lemma 4.5 *The covering corresponding to a subgroup G_2 is subordinate to the covering corresponding to a subgroup G_1 if and only if the inclusion $G_2 \supseteq G_1$ holds.*

Proof Suppose that $\pi_1(X, a) \supseteq G_2 \supseteq G_1$, and let $f_2 : (Y_2, b_2) \to X$ be the covering corresponding to the subgroup G_2 of $\pi_1(X, a)$. By Lemma 4.2, the group G_2 coincides with the image $f_{2*}\pi_1(Y_2, b_2)$ of the fundamental group of the space Y_2 in $\pi_1(X, a)$. Let $g : (Y_1, b_1) \to (Y_2, b_2)$ be the covering corresponding to the subgroup $f_{2*}^{-1}G_1$ of the fundamental group $\pi_1(Y_2, b_2) = f_{2*}^{-1}G_2$. The map $f_2 \circ g : (Y_1, b_1) \to (X, a)$ defines the covering over (X, a) corresponding to the subgroup $G_1 \subset \pi_1(X, a)$. Hence, the covering $f_2 \circ g : (Y_2, b_2) \to (X, a)$ is right equivalent to the covering $f_1 : (Y_1, b_1) \to (X, a)$. We have proved the lemma in one direction. The proof in the opposite direction is similar. □

Consider a covering $f : Y \to X$ such that Y is connected and X is locally connected and locally simply connected. Suppose that for a point $a \in X$, the covering has the following properties: for all choices of preimages b and c of the point a, the coverings with marked points $f : (Y, b) \to (X, a)$ and $f : (Y, c) \to (X, a)$ are equivalent. Then:

1. The covering has this property for every point $a \in X$.
2. The covering $f : Y \to X$ is normal.

Conversely, if the covering is normal, then it has this property for every point $a \in X$. This statement follows immediately from the definition of a normal covering.

Lemma 4.6 *A covering is normal if and only if it corresponds to a normal subgroup H of the fundamental group $\pi_1(X, a)$. For this normal subgroup, the group of deck transformations is isomorphic to the quotient group $\pi_1(X, a)/H$.*

Proof Suppose that the covering $f : (Y, b) \to (X, a)$ corresponding to a subgroup $G \subset \pi_1(X, a)$ is normal. Then for every preimage c of the point a, this covering is right equivalent to the covering $f : (Y, c) \to (X, a)$. By Lemma 4.4, this means that the subgroup G coincides with each of its conjugate subgroups. It follows that the group G is a normal subgroup of the fundamental group. Similarly, one can show that if G is a normal subgroup of the fundamental group, then the covering corresponding to this subgroup is normal.

A deck homeomorphism that takes the point b to the point c is unique. Indeed, the set on which two such homeomorphisms coincide is open (since f is a local homeomorphism), and moreover, is closed (since homeomorphisms are continuous) and nonempty (since it contains the point b). Since the space Y is connected, this set must coincide with Y.

The fundamental group $\pi_1(X, a)$ acts by right multiplication on the space $\Omega(X, a)$. For every normal subgroup H, this action gives rise to an action on the equivalence classes in $\Omega_H(X, a)$. (Under multiplication by an element $g \in \pi_1(X, a)$, the equivalence class xH is mapped to the equivalence class $xHg = xgH$.) The action of the fundamental group on $\Omega_H(X, a)$ is compatible with the projection $f : \Omega_H(X, a) \to (X, a)$ that assigns to each path the point where it terminates. Hence, the fundamental group $\pi_1(X, a)$ acts on the space Y of the normal covering $f : (Y, b) \to (X, a)$ by deck homeomorphisms.

For the covering that corresponds to the normal subgroup H, the kernel of this action is the group H, i.e., there is an effective action of the quotient group $\pi_1(X, a)/H$ on the space of this covering. The quotient group action can map the point b to any other preimage c of the point a. Hence there are no other deck homeomorphisms $h : Y \to Y$ apart from the homeomorphisms of the action of the quotient group $\pi_1(X, a)/H$. The lemma is proved. □

The fundamental group $\pi_1(X, a)$ acts on the fiber $F = f^{-1}(a)$ of the covering $f : (Y, b) \to (X, a)$. We now define this action. Let γ be a path in the space X that originates and terminates at the point a. For every point $c \in F$, let $\tilde{\gamma}_c$ denote a lift of the path γ to Y such that $\tilde{\gamma}_c(0) = c$. The map $S_\gamma : F \to F$ that takes the point c to the point $\tilde{\gamma}_c(1) \in F$ belongs to the group $S(F)$ of bijections from the set F to itself. The map S_γ depends only on the homotopy class of the path γ, that is, on the element of the fundamental group $\pi_1(X, a)$ represented by the path γ. The homomorphism $S_\gamma : \pi_1(X, a) \to S(F)$ is called the *monodromy homomorphism*, and the image of the fundamental group in the group $S(F)$ is called the *monodromy group* of the covering $f : (Y, b) \to (X, a)$.

Let $f : (Y, b) \to (X, a)$ be the covering corresponding to a subgroup $G \subset \pi_1(X, a)$, $F = f^{-1}(a)$ the fiber of this covering over the point a, and $S(F)$ the permutation group of the fiber F. We have the following lemma.

Lemma 4.7 *The monodromy group of the of the above-mentioned covering is a transitive subgroup of the group $S(F)$ and is equal to the quotient group of $\pi_1(X, a)$ by the largest normal subgroup H that is contained in the group G, i.e.,*

$$H = \bigcap_{h \in \pi_1(X,a)} hGh^{-1}.$$

Proof The monodromy group is transitive. For the proof, we have to construct, for every point $c \in F$, a path γ such that $S_\gamma(b) = c$. Take an arbitrary path $\tilde{\gamma}$ in the connected space Y such that $\tilde{\gamma}$ connects the point b with the point c. To obtain the path γ, it is enough to take the image of the path $\tilde{\gamma}$ under the projection f.

It can be immediately seen from the definitions that the stabilizer of the point b under the action of the fundamental group on the fiber F coincides with the group $G \subset \pi_1(X, a)$. Let $h \in \pi_1(X, a)$ be an element in the fundamental group that takes the point b to the point $c \in F$. Then the stabilizer of the point c is equal to hGh^{-1}. The kernel H of the monodromy homomorphism is the intersection of stabilizers of all points in the fiber, that is, $H = \bigcap_{h\in\pi_1(X,a)} hGh^{-1}$. The intersection of all groups hGh^{-1} is the largest normal subgroup contained in the group G. □

4.1.3 Other Classifications of Coverings

In this subsection, we discuss the usual classification of coverings (without marked points). Then we prove the classification theorem for coverings and coverings with marked points subordinate to a given normal covering. At the end of the subsection, we give a description of intermediate coverings that directly relates such coverings to the subgroups of the deck transformation group acting on the normal covering.

We now pass to coverings without marked points. We will classify coverings with the connected covering space over a connected, locally connected, and locally simply connected space. This classification reduces to the analogous classification of coverings with marked points.

Two coverings $f_1 : Y_1 \to X$ and $f_2 : Y_2 \to X$ are called equivalent if there exists a homeomorphism $h : Y_1 \to Y_2$ compatible with the projections f_1 and f_2, i.e., such that $f_1 = f_2 \circ h$.

Lemma 4.8 *Coverings with marked points are equivalent as coverings (rather than as coverings with marked points) if and only if the subgroups corresponding to these coverings are conjugate in the fundamental group of the space X.*

Proof Let coverings $f_1 : (Y_1, b_1) \to (X, a)$ and $f_2 : (Y_2, b_2) \to (X, a)$ be equivalent as coverings. A homeomorphism h should map the fiber $f_1^{-1}(a)$ to the fiber $f_2^{-1}(a)$. Hence, the covering $f_1 : (Y_1, b_1) \to (X, a)$ is equivalent, as a covering with marked points, to the covering $f_2 : (Y_2, h(b_1)) \to (X, a)$, where $f_2(h(b_1)) = f_2(b_2)$. This means that the subgroups corresponding to the original coverings with marked points are conjugate. □

Therefore, coverings $f : Y \to X$, where Y is connected and X is locally connected and locally simply connected, are classified by subgroups of the fundamental group $\pi_1(X)$ defined up to conjugation in the group $\pi_1(X)$. Note that the group $\pi_1(X)$, in contrast to the group $\pi_1(X, a)$, is also defined up to conjugation. More precisely, for every choice of marked point a, the group $\pi_1(X)$ can be identified with the group $\pi_1(X, a)$. However, the identification is not uniquely defined; it is defined only up to postcomposition with a conjugation.

In classifying coverings and coverings with marked points, one can confine oneself to considering coverings subordinate to a given normal covering. The definition of the subordinacy relation for coverings with marked points has been given above. One can also define an analogous relation for coverings, at least when one of the coverings is normal.

We say that a *covering* $f : Y \to X$ *is subordinate to the normal covering* $g : M \to X$ if there exists a map $h : M \to Y$ compatible with the projections g and f, i.e. such that $g = f \circ h$. It is clear that a covering is subordinate to a normal covering if and only if every subgroup from the corresponding class of conjugate subgroups in the fundamental group of X contains the normal subgroup corresponding to the normal covering.

Fix a marked point a in the space X. Let $g : (M, b) \to (X, a)$ be the normal covering corresponding to a normal subgroup H of the group $\pi_1(X, a)$, and $N = \pi_1(X, a)/H$ the deck transformation group of this normal covering. Consider all possible coverings and coverings with marked points subordinate to this normal covering. We can apply all classification theorems to these coverings. The role of the fundamental group $\pi_1(X, a)$ will be played by the deck transformation group N of the normal covering.

Let $f : (Y, b) \to (X, a)$ be a subordinate covering with marked points, and G the corresponding subgroup of the fundamental group. We associate to this subordinate covering the subgroup of the deck transformation group N equal to the image of the subgroup G under the quotient projection $\pi(X, a) \to N$. The following theorem holds for this correspondence.

Theorem 4.9 *The correspondence between coverings with marked points subordinate to a given normal covering and subgroups of the deck transformation group of this normal covering is bijective.*

Subordinate coverings with marked points are equivalent as coverings if and only if the corresponding subgroups are conjugate in the deck transformation group.

A subordinate covering is normal if and only if it corresponds to a normal subgroup M of the deck transformation group N. The deck transformation group of the subordinate normal covering is isomorphic to the quotient group N/M.

Proof For the proof, it is enough to apply the already proved "absolute" classification results and the following evident properties of the group quotients.

The quotient projection is a bijection between all subgroups of the original group that contain the kernel of the projection and all subgroups of the quotient group. This bijection has the following properties:

1. It preserves the partial order on the set of subgroups defined by inclusion.
2. It takes a class of conjugate subgroups of the original group to a class of conjugate subgroups of the quotient group.
3. It establishes a one-to-one correspondence between all normal subgroups of the original group that contain the kernel of the projection and all normal subgroups of the quotient group.

Under the correspondence of normal subgroups described in point 3, the quotient of the original group by a normal subgroup is isomorphic to the quotient of the quotient group by the corresponding normal subgroup. □

Let $f : M \to X$ be a normal covering (as usual, we assume that the space M is connected and that the space X is locally connected and locally simply connected). An *intermediate covering* between M and X is a space Y together with a surjective continuous map $h_Y : M \to Y$ and a projection $f_Y : Y \to X$ satisfying the condition $f = f_Y \circ h_Y$.

Let us introduce two different notions of equivalence for intermediate coverings. We say that two intermediate coverings

$$M \xrightarrow{h_1} Y_1 \xrightarrow{f_1} X \quad \text{and} \quad M \xrightarrow{h_2} Y_2 \xrightarrow{f_2} X$$

are equivalent as *subcoverings of the covering* $f : M \to X$ if there exists a homeomorphism $h : Y_1 \to Y_2$ that makes the diagram

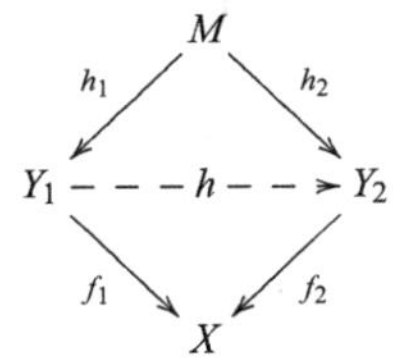

commutative, i.e., such that $h_2 = h \circ h_1$ and $f_1 = f_2 \circ h$. We say that two subcoverings are equivalent as *coverings over* X if there exists a homeomorphism $h : Y_1 \to Y_2$ such that $f_1 = h \circ f_2$ (the homeomorphism h is not required to make the upper part of the diagram commutative).

The classification of intermediate coverings regarded as subcoverings is equivalent to the classification of subordinate coverings with marked points. Indeed, if we mark a point b in the space M that lies over the point a, then we obtain a canonically defined marked point $h_Y(a)$ in the space Y.

The following statement is a reformulation of Theorem 4.9.

Proposition 4.10 *Intermediate coverings for a normal covering with the deck transformation group N regarded as subcoverings are classified by subgroups of the group N. Intermediate coverings for a normal covering with the deck transformation group N regarded as coverings over X are classified by the conjugacy classes of subgroups in the group N. A subordinate covering is normal if and only if it corresponds to a normal subgroup M of the deck transformation group N. The deck transformation group of the subordinate normal covering is isomorphic to the quotient group N/M.*

Let us give yet another description of intermediate coverings for a normal covering $f : M \to X$ with a deck transformation group N. The group N is a group of homeomorphisms of the space M with the following discreteness property: each point of the space M has a neighborhood such that its images under the action of different elements of the group N do not intersect. To construct such a neighborhood, take a connected component of the preimage under the projection

$f : M \to X$ of a connected and locally connected neighborhood of the point $f(z) \in X$.

For every subgroup G of the group N, consider the quotient space M_G of the space M under the action of the group G. A point in M_G is an orbit of the action of the group G on the space M. The topology in M_G is induced by the topology in the space M. A neighborhood of an orbit consists of all orbits that lie in an invariant open subset U of the space M with the following properties: the set U contains the original orbit, and a connected component of the set U intersects each orbit in at most one point. The space M_N can be identified with the space X. To do so, we identify a point $x \in X$ with the preimage $f^{-1}(x) \subset M$, which is an orbit of the deck transformation group N acting on M. Under this identification, the quotient projection $f_{e,N} : M \to M_N$ coincides with the original covering $f : M \to X$.

Let G_1, G_2 be two subgroups in N such that $G_1 \subseteq G_2$. Define the map $f_{G_1,G_2} : M_{G_1} \to M_{G_2}$ by assigning to each orbit of the group G_1 the orbit of the group G_2 that contains it. It is easy to see that the following hold:

1. The map f_{G_1,G_2} is a covering.
2. If $G_1 \subseteq G_2 \subseteq G_3$, then $f_{G_1,G_2} = f_{G_2,G_3} \circ f_{G_1,G_2}$.
3. Under the identification of M_N with X, the map $f_{G,N} : M_G \to M_N$ corresponds to a covering subordinate to the original covering $f_{e,N} : M \to M_N$ (since $f_{e,N} = f_{G,N} \circ f_{e,G}$).
4. If G is a normal subgroup of N, then the covering $f_{G,N} : M_G \to M_N$ is normal, and its deck transformation group is equal to N/G.

One can associate to an intermediate covering $f_{G,N} : M_G \to M_N$, either the triple of spaces

$$M \xrightarrow{f_{e,G}} M_G \xrightarrow{f_{G,N}} M_N$$

with the maps $f_{e,G}$ and $f_{G,N}$, or the pair of spaces $M_G \xrightarrow{f_{G,N}} M_N$ with the map $f_{G,N}$. These two possibilities correspond to two viewpoints on an intermediate covering, regarding it either as a subcovering or as a covering over M_N.

4.1.4 A Similarity Between Galois Theory and the Classification of Coverings

The fundamental theorem of Galois theory describes algebraic extensions that are intermediate between a base field and a fixed Galois extension of the base field (rather than all field extensions of the base field).

In the theory of coverings, one can also consider coverings that are intermediate between the base and a given normal covering of the base (rather than all coverings of the base simultaneously).

Classification of intermediate coverings regarded as subcoverings corresponds to the Galois-theoretic classification of intermediate extensions regarded as subfields of the field P. To see this, replace the words "normal covering," "deck transformation group," "subordinate covering" with the words "Galois extension," "Galois group," "intermediate field."

In Sect. 4.2, we consider finite ramified coverings over one-dimensional complex manifolds. Ramified coverings (with marked points or without marked points) over a manifold X whose ramification points lie over a given discrete set O are classified in the same way as coverings (with marked points or without marked points) over $X \setminus O$ (see Sect. 4.2.2). Finite ramified coverings correspond to algebraic extensions of the field of meromorphic functions on X. The fundamental theorem of Galois theory for these fields and the classification of intermediate coverings are not only formally similar but also very closely related to each other.

Note that the classification of intermediate coverings regarded as coverings over a base also have a formal analogue in Galois theory. It is similar to the classification of the algebraic extensions of a base field that can be embedded in a given Galois extension (this classification does not take into account how an algebraic extension embeds into the given Galois extension).

4.2 Completion of Ramified Coverings and Riemann Surfaces of Algebraic Functions

In this section, we consider finite ramified coverings over one-dimensional complex manifolds. We describe the operation of completion for coverings over a one-dimensional complex manifold X with a removed discrete set O. This operation can be applied equally well to coverings and to coverings with marked points. It transforms a finite covering over $X \setminus O$ to a finite ramified covering over X.

In Sect. 4.2.1, we consider the local case in which coverings of an open punctured disk are completed. In the local case, the operation of completion allows us to prove the existence of Puiseux expansions for multivalued functions with algebraic singularities.

In Sect. 4.2.2, we consider the global case. We first define the real operation of filling holes. Then we show that the ramified covering obtained using the real operation of filling holes has a natural structure of a complex manifold.

In Sect. 4.2.3, we classify finite ramified coverings with a fixed ramification set. The classification literally repeats the analogous classification of unramified coverings. Therefore, we allow ourselves to formulate results without proofs. We prove that the set of orbits under a finite group action on a one-dimensional complex analytic manifold has a natural structure of a complex analytic manifold.

In Sect. 4.2.4, we apply the operation of completion of coverings to define the Riemann surface of an irreducible algebraic equation over the field $K(X)$ of meromorphic functions over a manifold X.

Section 4.2.2 relies on the results of Sect. 4.2.1.

4.2.1 Filling Holes and Puiseux Expansions

Let D_r be an open disk of radius r on the complex line with center at the point 0, and $D_r^* = D_r \setminus \{0\}$ the punctured disk. For every positive integer k, consider the punctured disk D_q^*, where $q = r^{1/k}$, together with the map $f : D_q^* \to D_r^*$ given by the formula $f(z) = z^k$.

Lemma 4.11 *There exists a unique (up to right equivalence) connected k-fold covering $\pi : V^* \to D_r^*$ over the punctured disk D_r^*. This covering is normal. It is equivalent to the covering $f : D_q^* \to D_r^*$, where the map f is given by the formula $x = f(z) = z^k$.*

Proof The fundamental group of the domain D_r^* is isomorphic to the additive group $\mathbb{Z}$ of integers. The only subgroup in $\mathbb{Z}$ of index k is the subgroup $k\mathbb{Z}$. The subgroup $k\mathbb{Z}$ is a normal subgroup of $\mathbb{Z}$. The covering $z \to z^k$ of the punctured disk D_q^* over the punctured disk D_r^* is normal and corresponds to the subgroup $k\mathbb{Z}$. □

Let $\pi : V^* \to D_r^*$ be a connected k-fold covering over a punctured disk D_r^*. Let V denote the set consisting of the domain V^* and a point A. We can extend the map π to a map of the set V onto the disk D_r by setting $\pi(A) = 0$. Introduce the coarsest topology on the set V such that the following conditions are satisfied:

1. Identification of the set $V \setminus \{A\}$ with the domain V^* is a homeomorphism.
2. The map $\pi : V \to D$ is continuous.

Lemma 4.12 *The map $\pi : V \to D_r$ is right equivalent to the map $f : D_q \to D_r$ defined by the formula $x = f(z) = z^k$. In particular, V is homeomorphic to the open disk D_q.*

Proof Let $h : D_q^* \to V^*$ be the homeomorphism that establishes an equivalence of the covering $\pi : V^* \to D_r^*$ and the standard covering $f : D_q^* \to D_r^*$. Extend h to the map of the disk D_q to the set V by setting $h(0) = A$. We have to check that the extended map h is a homeomorphism. Let us check, for example, that h is a continuous map. By definition of the topology on V, every neighborhood of the point A contains a neighborhood V_0 of the form $V_0 = \pi^{-1}(U_0)$, where U_0 is a neighborhood of the point 0 on the complex line. Let $W_0 \subset D_q$ be the open set defined by the formula $W_0 = f^{-1}(U_0)$. We have $h^{-1}(V_0) = W_0$, which proves the continuity of the map h at the point 0. The continuity of the map h^{-1} can be proved similarly. □

We will use the notation of the preceding lemma.

Lemma 4.13 *The manifold V has a unique structure of an analytic manifold such that the map $\pi : V \to D_r$ is analytic. This structure is induced from the analytic structure on the disk D_q by the homeomorphism $h : D_q \to V$.*

Proof The homeomorphism h transforms the map π into the analytic map $f(z) \to z^k$. Hence, the analytic structure on V induced by the homeomorphism satisfies the condition of the lemma. Consider another analytic structure on V. The map $h : D \to V$ outside the point 0 can be locally represented as $h(z) = \pi^{-1}z^k$ and is therefore analytic. Thus the map $h : D \to V$ is continuous and analytic everywhere except at the point 0. By the removable singularity theorem, it is also analytic at the point 0, and therefore, there is a unique analytic structure on V such that the projection π is analytic. □

The transition from the real manifold V^* to the real manifold V and the transition from the covering $\pi : V^* \to D_r^*$ to the map $\pi : V \to D_r$ will be called the *real operation of filling a hole*. Lemma 4.13 shows that after a hole has been filled, the manifold V has a unique structure of a complex analytic manifold such that the map $\pi : V \to D_r$ is analytic. The transition from the complex manifold V^* to the complex manifold V and the transition from the analytic covering $\pi : V^* \to D_r^*$ to the analytic map $\pi : V \to D_r$ will be called the *operation of filling a hole*. In what follows, we will use precisely this operation.

The operation of filling a hole is intimately related to the definition of an algebraic singular point and to Puiseux series. Let us discuss this in more detail.

Definition 4.14 We say that an analytic germ φ_a at a point $a \in D$ defines a multivalued function on the disk D_r with an *algebraic singularity* at the point 0 if the following are satisfied:

1. The germ φ_a can be extended along any path that originates at the point a and lies in the punctured disk D_r^*.
2. The multivalued function φ in the punctured disk D_r^* obtained by extending the germ φ_a along paths in D_r^* takes a finite number k of values.
3. In approaching the point 0, the multivalued function φ grows no faster than a power function, i.e., there exist positive real numbers C, N such that each of the values of the multivalued function φ satisfies the inequality $|\varphi(x)| < C|x|^{-N}$.

Lemma 4.15 *A multivalued function φ with an algebraic singularity in the punctured disk D_r^* can be represented in that disk by the Puiseux series*

$$\varphi(x) = \sum_{m > -m_0} c_m x^{m/k}.$$

Proof If the function φ can be extended analytically along all paths in the punctured disk D_r^* and has k different values, then the germ $g_b = \varphi_a \circ z_b^k$, where $b^k = a$, defines a single-valued function in the punctured disk D_q^*, where $q = r^{1/k}$. Indeed, a simple loop around 0 in D_q^* is mapped by $z \mapsto z^k$ to a loop in D_r^* that goes k times around 0, and the function φ returns to the same value along such a loop. By the hypothesis, the function g grows no faster than a power function when approaching the point 0; hence in the punctured disk D_q^*, it can be represented by the Laurent series

$$g(z) = \sum_{m>-m_0} c_m z^m.$$

Substituting $x^{1/k}$ for z in the series for the function g, we obtain the Puiseux series for the function φ. □

4.2.2 Analytic-Type Maps and the Real Operation of Filling Holes

In this subsection, we define the real operation of filling holes. We show that the ramified covering resulting from the real operation of filling holes has a natural complex analytic structure.

Let X be a one-dimensional complex analytic manifold, M a two-dimensional real manifold, and $\pi : M \to X$ a continuous map. We say that the map π at a point $y \in M$ has an *analytic-type singularity*[1] *of multiplicity* $k > 0$ if there exist a connected punctured neighborhood $U^* \subset X$ of the point $x = \pi(y)$ and a connected component of the open set $\pi^{-1}(U^*)$ that is a punctured neighborhood $V^* \subset M$ of the point y such that the triple $\pi : V^* \to U^*$ is a k-fold covering. It is natural to regard the singular point y as a *multiplicity-k* preimage of the point x: the number of preimages (counted with multiplicity) of π in a neighborhood of an analytic-type singular point of multiplicity k is constant and equal to k.

A map $f : X \to M$ is an *analytic-type map*[2] if it has an analytic-type singularity at every point. Clearly, a complex analytic map $f : M \to X$ of a complex one-dimensional manifold M to a complex one-dimensional manifold X is an analytic-type map (when considered as a continuous map of a real manifold M to a complex manifold X). For an analytic-type map, a point y is called *regular* if its multiplicity is equal to 1, and *singular* if its multiplicity is greater than 1.[3] The set of all regular points of an analytic-type map is open. The map considered near a regular point is a local homeomorphism. The set O of singular points of an analytic-type map is a discrete subset of M.

Proposition 4.16 *Let M be a two-dimensional real manifold, and $f : M \to X$ an analytic-type map to a one-dimensional complex analytic manifold X. Then M has a unique structure of a complex analytic manifold such that the map f is analytic.*

---

[1] A point y that is an analytic-type singularity is also called a topological branch point.

[2] An analytic-type map is also called a topological branched covering.

[3] A singular point is also called a critical point.

Proof The map f is a local homeomorphism at the points of $M \setminus O$. This local homeomorphism to the analytic manifold X makes $M \setminus O$ into an analytic manifold. Near the points of the set O, one can define an analytic structure in the same way as near the points added by the operation of filling holes. We now prove that there are no other analytic structures such that f is analytic. Let M_1 and M_2 be two copies of the manifold M with two different analytic structures. Let O_1 and O_2 be distinguished discrete subsets of M_1 and M_2, and $h : M_1 \to M_2$ a homeomorphism identifying these two copies. It is clear from the hypothesis that the homeomorphism h is analytic everywhere except at the discrete set $O_1 \subset M_1$. By the removable singularities theorem, h is a biholomorphic map. Hence the two analytic structures on M coincide. □

We now return to the operation of filling holes. Let M be a real two-dimensional manifold, and $f : M \to X$ an analytic-type map of the manifold M to a complex one-dimensional manifold X.

Fix a local coordinate u near a point $a \in X$, $u(a) = 0$, that gives an invertible map of a small neighborhood of the point $a \in X$ to a small neighborhood of the origin on the complex line. Let U^* be the preimage of a small punctured disk D_r^* with center at 0 under the map u. Suppose that among all connected components of the preimage $\pi^{-1}(U^*)$, there exists a component V^* such that the restriction of the map π to V^* is a k-fold covering. In that case, one can apply the *real operation of filling a hole*. The operation does the following. Cut a neighborhood V^* out of the manifold M. The covering $\pi : V^* \to U^*$ is replaced by the map $\pi : V \to U$ by the operation of filling a hole described above. The manifold V^* lies in V and differs from V at one point. The real operation of filling a hole attaches the neighborhood V to the manifold $M \setminus V^*$ together with the map $\pi : V \to X$.

The *real operation of filling holes* consists in real operations of filling a hole applied to all holes simultaneously. It is well defined: if V^* is a connected component of the preimage $\pi^{-1}(U^*)$, where U^* is a punctured neighborhood of the point $o \in X$, and the map $\pi : V^* \to U^*$ is a finite covering, then the operation of filling all holes adds to the closure of the domain V^* exactly one point lying over the point o. The topology near this new point is defined in the same way as under the operation of filling one hole.

The *operation of filling holes* is the complexification of the real operation of filling holes. The operation of filling holes can be applied to a one-dimensional complex analytic manifold M endowed with an analytic map $f : M \to X$. Namely, the triple $f : M \to X$ should be regarded as an analytic-type map from a real manifold M to X. Then the real operation of filling holes should be applied to this triple. The result is a real manifold $\tilde{M}$ together with an analytic-type map $\pi : \tilde{M} \to X$. The manifold $\tilde{M}$ has a unique structure of a complex one-dimensional manifold such that the analytic-type map π is analytic. This complex manifold $\tilde{M}$ together with the analytic map π is the result of the operation of filling holes applied to the initial triple $f : M \to X$. In what follows, we will need only the operation of filling holes and not its real version.

Let X and M be one-dimensional complex manifolds, O a discrete subset of X, and $\pi : M \to U$, where $U = X \setminus O$, an analytic map that is a finite covering. Let X be connected (the covering space M may be disconnected).

Near every point $o \in O$, one can take a small punctured neighborhood U^* that does not contain other points of the set O. Over the punctured neighborhood U^*, there is a covering $f : V^* \to U^*$, where $V^* = f^{-1}(U^*)$. The manifold V^* splits into connected components V_i^*. Let us apply the operation of filling holes. Over the point $o \in O$, we attach a finite number of points. The number of points is equal to the number of connected components of V^*.

Lemma 4.17 *If the operation of filling holes is applied to a k-fold covering $\pi : M \to U$, then the result is a complex manifold $\tilde{M}$ endowed with a proper analytic map $\tilde{\pi} : \tilde{M} \to X$ of degree k.*

Proof We should check the properness of the map $\tilde{\pi}$. First of all, this map is analytic; hence the image of every open subset under this map is open. Next, the number of preimages of every point $x_0 \in X$ under the map $\tilde{\pi}$, counted with multiplicity, is equal to k. Hence the map $\tilde{\pi}$ is proper. □

4.2.3 Finite Ramified Coverings with a Fixed Ramification Set

In this subsection, we classify finite ramified coverings with a fixed ramification set.

Let X be a connected complex one-dimensional manifold with a distinguished discrete subset O and a marked point $a \notin O$. A triple consisting of complex manifolds M and X and a proper analytic map $\pi : (M, b) \to (X, a)$ whose critical values are all contained in the set O is called a *ramified covering over X with ramification over O*. We consider ramified coverings up to right equivalence. In other words, two triples $\pi_1 : M_1 \to X_1$ and $\pi_2 : M_2 \to X_2$ are considered the same if there exists a homeomorphism $h : M_1 \to M_2$ compatible with the projections π_1 and π_2, i.e., $\pi_1 = \pi_2 \circ h$. The homeomorphism h that establishes the equivalence of ramified coverings is automatically an analytic map from the manifold M_1 to the manifold M_2. This is proved in the same way as Proposition 4.16.

The following operation will be called the *ramification puncture*. To every connected ramified (over O) covering $\pi : M \to X$, the operation assigns the nonramified covering $\pi : M \setminus \tilde{O} \to X \setminus O$ over $X \setminus O$, where $\tilde{O}$ is the full preimage of the set O under the map π. The following lemma is a direct consequence of definitions.

Lemma 4.18 *The operation of ramification puncture and the operation of filling holes are inverse to each other. They establish an isomorphism between the category of ramified coverings over X with ramifications over the set O and the category of finite coverings over $X \setminus O$.*

All definitions and statements about coverings can be extended to ramified coverings. This is done automatically: it is enough to apply arguments used in the proof of Proposition 4.16. Thus we formulate definitions and propositions only about ramified coverings. Let us begin with definitions concerning ramified coverings.

A homeomorphism $h : M \to M$ is called a *deck transformation of a ramified covering* $\pi : M \to X$ with ramification over O if the equality $\pi = \pi \circ h$ holds. (The deck transformation h is automatically analytic.)

For a connected manifold M, a ramified covering $\pi : M \to X$ with ramification over O is called *normal* if its group of deck transformations acts transitively on every fiber of the map π. The group of deck transformations is automatically a group of analytic transformations of M.

A ramified covering $f_2 : M_2 \to X$ with ramification over O is said to be *subordinate* to a normal ramified covering $f_1 : M_1 \to X$ with ramification over O if there exists a ramified covering $h : M_1 \to M_2$ with ramification over $f_2^{-1}(O)$ such that $f_1 = f_2 \circ h$. (The map h is automatically analytic.)

We now proceed with definitions concerning coverings with marked points. A triple $\pi : (M, b) \to (X, a)$, where $\pi : M \to X$ is a ramified covering with ramification over O, and $a \in X$, $b \in M$ are marked points such that $a \notin O$ and $\pi(b) = a$, is called a *ramified covering over* X *with marked points with ramification over* O.

A ramified covering $f_2 : (M_2, b_2) \to (X, a)$ with ramification over O is said to be *subordinate* to a ramified covering $f_1 : (M_1, b_1) \to (X, a)$ with ramification over O if there exists a ramified covering $h : (M_1, b_1) \to (M_2, b_2)$ with ramification over $f_2^{-1}(O)$ such that $f = f_2 \circ h$. (The map h is automatically analytic.) In particular, such coverings are called *equivalent* if the map h is a homeomorphism. (The homeomorphism h is automatically a bianalytic bijection between M_1 and M_2.)

To a ramified covering $f : (Y, b) \to (X, a)$ with marked points and to a covering $\pi : M \to X$ with ramification over O, the operation of ramification puncture assigns the covering with marked points $f : (Y \setminus f^{-1}(O), b) \to (X \setminus O, a)$ and the covering $\pi : M \setminus \pi^{-1}(O) \to X \setminus O$. To these coverings over $X \setminus O$, one associates respectively a subgroup of finite index in the group $\pi_1(X \setminus O, a)$ and the class of conjugate subgroups of finite index in this group. We say that this subgroup corresponds to the ramified covering $f : (Y, b) \to (X, a)$ with marked points and that this class of conjugate subgroups corresponds to the ramified covering $\pi : M \to X$.

Consider all possible ramified coverings with marked points with a connected covering space over a manifold X with a marked point a that have ramification over a set O, $a \notin O$. Transferring the statements proved for coverings with marked points to ramified coverings, we obtain the following:

1. Such coverings are classified by subgroups of finite index in $\pi_1(X \setminus O, a)$.
2. Such a covering corresponding to the group G_2 is subordinate to the covering corresponding to the group G_1 if and only if the inclusion $G_2 \supseteq G_1$ holds.
3. Such a covering is normal if and only if the corresponding subgroup of the fundamental group $\pi_1(X \setminus O, a)$ is a normal subgroup H. The group of

deck transformations of the normal ramified covering is isomorphic to $\pi_1(X \setminus O, a)/H$.

Consider all possible ramified coverings over a manifold X with a connected covering space that have ramification over a set O, $a \notin O$. Transferring the statements proved for coverings to ramified coverings, we obtain the following:

4. Such coverings are classified by classes of conjugate subgroups of finite index in the group $\pi_1(X \setminus O, a)$.

One can literally translate the description of ramified coverings subordinate to a given normal covering with the deck transformation group N to ramified coverings.

To a ramified covering with a marked point, assign the subgroup of the deck transformation group N that is equal to the image under the quotient projection $\pi_1(X, a) \to N$ of the subgroup of the fundamental group corresponding to the ramified covering. For this correspondence, we have the following theorem.

Theorem 4.19 *The correspondence between ramified coverings with marked points subordinate to a given normal covering and subgroups of the deck transformation group of this normal covering is bijective.*

Subordinate ramified coverings with marked points are equivalent as coverings if and only if the corresponding subgroups are conjugate in the deck transformation group. A subordinate ramified covering is normal if and only if it corresponds to a normal subgroup M of the deck transformation group N. The deck transformation group of the subordinate normal covering is isomorphic to the quotient group N/M.

The notion of subcovering extends to ramified coverings. Let $f : M \to X$ be a normal ramified covering (as usual, we assume that the complex one-dimensional manifold M is connected). An *intermediate ramified covering* between M and X is a one-dimensional complex manifold Y together with a surjective continuous map $h_Y : M \to Y$ and a projection $f_Y : Y \to X$ satisfying the condition $f = f_Y \circ h_Y$ (it follows that the map h_Y is complex analytic).

We say that two intermediate ramified coverings

$$M \xrightarrow{h_1} Y_1 \xrightarrow{f_1} X \quad \text{and} \quad M \xrightarrow{h_2} Y_2 \xrightarrow{f_2} X$$

are equivalent as *ramified subcoverings of the covering* $f : M \to X$ if there exists a homeomorphism $h : Y_1 \to Y_2$ that makes the diagram

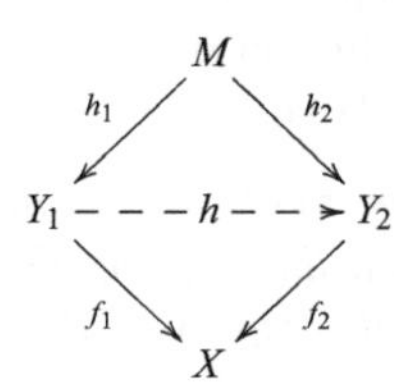

commutative, i.e., such that $h_2 = h \circ h_1$ and $f_1 = f_2 \circ h$. (It follows that h is bianalytic.)

We say that two ramified subcoverings are equivalent as *ramified coverings over* X if there exists an analytic map $h : Y_1 \to Y_2$ such that $f_1 = h \circ f_2$ (the map h is not required to make the upper part of the diagram commutative).

The classification of intermediate ramified coverings regarded as ramified subcoverings is equivalent to the classification of subordinate coverings with marked points. Indeed, if we mark a point b in the manifold M that lies over the point a, then we obtain a canonically defined marked point $h_Y(a)$ in the space Y.

Let us reformulate Proposition 4.10.

Proposition 4.20 *Intermediate ramified coverings for a normal covering with the deck transformation group N regarded as ramified subcoverings are classified by subgroups of the group N. Those regarded as ramified coverings over X are classified by the classes of conjugate subgroups in the group N.*

A subordinate ramified covering is normal if and only if it corresponds to a normal subgroup H of the deck transformation group N. The deck transformation group of a subordinate ramified normal covering is isomorphic to the quotient group N/H.

Let us give one more description of ramified coverings subordinate to a given normal ramified covering. Let $\pi : M \to X$ be a normal finite ramified covering with deck transformation group N.

The deck transformation group N is a group of analytic transformations of the manifold M commuting with the projection π. It induces a transitive transformation group of the fiber of π. Transformations in the group N can have isolated fixed points among the critical points of the map π.

Lemma 4.21 *The set M_N of orbits under the action of the deck transformation group N on a ramified normal covering M is in one-to-one correspondence with the manifold X.*

Proof By definition, deck transformations act transitively on the fiber of the map $\pi : M \to X$ over every point $x_0 \notin O$. Let $o \in O$ be a point in the ramification set. Let U^* be a small punctured coordinate disk around the point o not containing points of the set O. The preimage $\pi^{-1}(U^*)$ of the domain U^* splits into connected components V_i^* that are punctured neighborhoods of preimages b_i of the point o. The deck transformation group gives rise to a transitive permutation of the domains V_i^*. Indeed, each of these domains intersects the fiber $\pi^{-1}(c)$, where c is any point in the domain U^*, and the group N acts transitively on the fiber $\pi^{-1}(c)$. The transitivity of the action of N on the set of components V_i^* implies the transitivity of the action of N on the fiber $\pi^{-1}(o)$. □

Theorem 4.22 *The set of orbits M/G of a one-dimensional complex analytic manifold M under the action of a finite group G of analytic transformations has the structure of a one-dimensional complex analytic manifold.*

Proof 1. The stabilizer G_{x_0} of every point $x_0 \in M$ under the action of the group G is cyclic. Indeed, consider the homomorphism of the group G_{x_0} to the group of linear transformations of a one-dimensional complex vector space that assigns to a transformation its differential at the point x_0. This map cannot have a nontrivial kernel: if the first few terms of the Taylor series of the transformation f have the form $f(x_0+h) = x_0+h+ch^k+\cdots$, then the Taylor series of the ℓth iteration $f^{\circ\ell}$ of f has its first few terms of the form $f^{\circ\ell} = x_0+h+\ell ch^k+\cdots$. Hence, none of the iterations of the transformation f is the identity map, which contradicts the finiteness of the group G_{x_0}. A finite group of linear transformations of the space $\mathbb{C}^1$ is a cyclic group generated by multiplication by one of the primitive mth roots of unity ξ_m, where m is the order of the group G_{x_0}.

2. The stabilizer G_{x_0} of the point x_0 can be linearized, i.e., one can introduce a local coordinate u near x_0 such that the transformations in the group G_{x_0} written in this coordinate system are linear. Let f be a generator of the group G_{x_0}. Then the equality $f^{\circ m} = \mathrm{Id}$ holds, where Id is the identity transformation.

   The differential of the function f at the point x_0 is equivalent to multiplication by ξ_m, where ξ_m is one of the mth primitive roots of unity. Consider any function φ whose differential is not equal to zero at the point x_0. To the map f, one associates the linear operator f^* on the space of functions. Let us write the Lagrange resolvent $R_{\xi_m}(\varphi)$ of the function φ for the action of the operator f^*:

$$R_{\xi_m}(\varphi) = \frac{1}{m}\sum \xi_m^{-k}(f^*)^k(\varphi).$$

   The function $u = R_{\xi_m}(\varphi)$ is the eigenvector of the transformation f^* with eigenvalue ξ_m. The differentials at the point x_0 of the functions u and φ coincide (this can be verified by a simple calculation). The map f at the coordinate u becomes linear, since $f^*u = \xi_m u$.

3. We now introduce an analytic structure on the space of orbits. Consider any orbit. Suppose first that the stabilizers of the points in the orbit are trivial. Then a small neighborhood of the point in the orbit intersects each orbit at most once. A local coordinate near this point parameterizes neighboring orbits. If a point in the orbit has a nontrivial stabilizer, then we choose a local coordinate u near the point such that in this coordinate system, the stabilizer acts linearly, multiplying u by the powers of the root ξ_m. The neighboring orbits are parameterized by the function $t = u^m$. The theorem is proved.

□

To every subgroup G of the group N, we associate the analytic manifold M_G, which is the space of orbits under the action of the group G. Identify the manifold M_N with the manifold X. Under this identification, the quotient map $f_{e,N} : M \to M_N$ coincides with the original covering $f : M \to X$.

Let G_1, G_2 be two subgroups of N such that $G_1 \subseteq G_2$. Define the map $f_{G_1,G_2} : M_{G_1} \to M_{G_2}$ by assigning to each orbit of the group G_1 the orbit of the group G_2 that contains it. It is easy to see that the following statements hold:

1. The map f_{G_1,G_2} is a ramified covering.
2. If $G_1 \subseteq G_2 \subseteq G_3$, then $f_{G_1,G_2} = f_{G_2,G_3} \circ f_{G_1,G_2}$.

3. Under the identification of M_N with X, the map $f_{G,N} : M_G \to M_N$ corresponds to a ramified covering subordinate to the original covering $f_{e,N} : M \to M_N$ (since $f_{e,N} = f_{G,N} \circ f_{e,G}$).
4. If G is a normal subgroup of N, then the ramified covering $f_{G,N} : M_G \to M_N$ is normal, and its deck transformation group is equal to N/G.

To an intermediate ramified covering $f_{G,N} : M_G \to M_N$, one can associate either the triple of spaces

$$M \xrightarrow{f_{e,G}} M_G \xrightarrow{f_{G,N}} M_N$$

with the maps $f_{e,G}$ and $f_{G,N}$, or the pair of spaces $M_G \xrightarrow{f_{G,N}} M_N$ with the map $f_{G,N}$. These two possibilities correspond to two viewpoints, whereby an intermediate covering can be regarded either as a ramified subcovering or as a ramified covering over M_N.

4.2.4 The Riemann Surface of an Algebraic Equation over the Field of Meromorphic Functions

Our goal is a geometric description of algebraic extensions of the field $K(X)$ of meromorphic functions on a connected one-dimensional complex manifold X. In this subsection, we construct the Riemann surface of an algebraic equation over the field $K(X)$.

Let $T = y^n + a_1 y^{n-1} + \cdots + a_n$ be a polynomial in the variable y over the field $K(X)$ of meromorphic functions on X. We will assume that in the factorization of T, every irreducible factor occurs with multiplicity 1. In this case, the discriminant D of the polynomial T is a nonzero element of the field $K(X)$. Let O denote the discrete subset in X containing all poles of the coefficients a_i and all zeros of the discriminant D. For every point $x_0 \in X \setminus O$, the polynomial $T_{x_0} = y^n + a_1(x_0)y^{n-1} + \cdots + a_n(x_0)$ has exactly n distinct roots.

The *Riemann surface* of the equation $T = 0$ is an n-fold ramified covering $\pi : M \to X$ together with a meromorphic function $y : M \to \mathbb{C}P^1$ such that for every point $x_0 \in X \setminus O$, the set of roots of the polynomial T_{x_0} coincides with the set of values of the function y on the preimage $\pi^{-1}(x_0)$ of the point x_0 under the projection π. Let us show that there exists a unique Riemann surface of the equation (up to an analytic homeomorphism compatible with the projection to X and the function y).

We will consider the polynomial T as a function of two variables $x \in X$ and $y \in \mathbb{C}$ such that $T(x, y) = T_x(y)$ is the polynomial T whose coefficients are evaluated at the point x. Define the projection π of the Cartesian product $(X \setminus O) \times \mathbb{C}^1$ onto the first factor, and the function y on this Cartesian product as the projection onto the second factor. Consider the hypersurface M_O in the Cartesian product given by the equation $T(\pi(a), y(a)) = 0$. The partial derivative of T

with respect to the second argument is nonzero at every point of the hypersurface M_O, since the polynomial $T_{\pi(a)}$ has no multiple roots. By the implicit function theorem, the hypersurface M_O is nonsingular, and its projection onto $X \setminus O$ is a local homeomorphism. The projection $\pi : M_O \to X \setminus O$ and the function $y : M_O \to \mathbb{C}^1$ are defined on the manifold M_O. Applying the operation of filling holes to the covering $\pi : M_O \to X \setminus O$, we obtain an n-fold ramified covering $\pi : M \to X$.

Theorem 4.23 *The function $y : M_O \to \mathbb{C}^1$ can be extended to a meromorphic function $y : M \to \mathbb{C}P^1$. The ramified covering $\pi : M \to X$ endowed with the meromorphic function $y : M \to \mathbb{C}P^1$ is the Riemann surface of the equation $T = 0$. There are no other Riemann surfaces of the equation $T = 0$.*

Proof We need the following lemma.

Lemma 4.24 (From high-school mathematics) *Every root y_0 of the equation $y^n + a_1 y^{n-1} + \cdots + a_n = 0$ satisfies the inequality $|y_0| \leq \max(1, \sum |a_i|)$.*

Proof If $|y_0| > 1$ and $y_0 = -a_1 - \cdots - a_n y_0^{1-n}$, then $|y_0| \leq \max(1, \sum |a_i|)$. □

Let us now prove the theorem. The functions $\pi^* a_i$ are meromorphic on M. In the punctured neighborhood of every point, the function y satisfies the inequality $|y| \leq \max(1, \sum |\pi^* a_i|)$ and therefore has a pole or removable singularity at every added point.

By construction, the triple $\pi : M_O \to X \setminus O$ is an n-fold covering, and for every $x_0 \in X \setminus O$, the set of roots of the polynomial T_{x_0} coincides with the image of the set $\pi^{-1}(x_0)$ under the map $y : M_O \to \mathbb{C}P^1$. Therefore, the ramified covering $\pi : M \to X$ endowed with the meromorphic function $y : M \to \mathbb{C}P^1$ is the Riemann surface of the equation $T = 0$.

Let a ramified covering $\pi_1 : M_1 \to X_1$ and a function $y : M_1 \to \mathbb{C}P^1$ be another Riemann surface of this equation. Let O_1 denote the set $\pi_1^{-1} O$. There exists a natural bijective map $h_1 : M_O \to M_1 \setminus O_1$ such that $\pi_1 \circ h_1 = \pi$ and $y_1 \circ h_1 = y$. Indeed, by definition of the Riemann surface, the sets of numbers

$$\{y \circ \pi^{-1}(x)\} \quad \text{and} \quad \{y_1 \circ \pi_1^{-1}(x)\}$$

coincide with the set of roots of the polynomial $T_{\pi(x)}$. It is easy to see that the map h_1 is continuous and that it can be extended by continuity to an analytic homeomorphism $h : M \to M$ such that $\pi_1 \circ h = \pi$ and $y_1 \circ h = y$. The theorem is proved. □

Remark 4.25 Sometimes the manifold M in the definition of the Riemann surface of an equation is by itself called the Riemann surface of the equation. The same manifold is called the *Riemann surface of the function* y satisfying the equation. We will use this slightly ambiguous terminology whenever it will not lead to confusion.

The set $\tilde{O}$ of critical values of the ramified covering $\pi : M \to X$ associated with the Riemann surface of the equation $T = 0$ can be a proper subset of the set O used in the construction (the inclusion $\tilde{O} \subset O$ always holds). The set $\tilde{O}$ is called the *ramification set of the equation* $T = 0$. Over a point $a \in X \setminus \tilde{O}$, the equation $T_a = 0$ might have multiple roots. However, in the field of germs of meromorphic functions at the point $a \in X \setminus \tilde{O}$, the equation $T = 0$ has only simple roots, and their number is equal to the degree of the equation $T = 0$. Each of the meromorphic germs at the point a satisfying the equation $T = 0$ corresponds to a point over a in the Riemann surface of the equation.

4.3 Finite Ramified Coverings and Algebraic Extensions of Fields of Meromorphic Functions

Let $\pi : M \to X$ be a finite ramified covering of complex one-dimensional manifolds. Galois theory and the Riemann existence theorem allow us to describe a relationship between the field $K(M)$ of meromorphic functions on M and the field $K(X)$ of meromorphic functions on X. The field $K(M)$ is an algebraic extension of the field $K(X)$, and every algebraic extension of the field $K(X)$ is obtained in this way. This section is devoted to the relation between finite ramified coverings over a complex one-dimensional manifold X and algebraic extensions of the field $K(X)$.

In Sect. 4.3.1, we define the field $P_a(O)$ consisting of the meromorphic germs at the point $a \in X$ that can be meromorphically continued to multivalued functions on $X \setminus O$ with finitely many branches and with algebraic singularities at the points of the set O.

In Sect. 4.3.2, the action of the fundamental group $\pi_1(X \setminus O)$ on the field $P_a(O)$ is considered, and the results of Galois theory are applied to the action of this group of automorphisms. We describe the correspondence between subfields of the field $P_a(O)$ that are algebraic extensions of the field $K(X)$ and the subgroups of finite index in the fundamental group $\pi_1(X \setminus O)$. We prove that this correspondence is bijective (apart from Galois theory, the proof uses the Riemann existence theorem). Consider the Riemann surface of an equation ramified over O. We show that this Riemann surface is connected if and only if the equation is irreducible. The field of meromorphic functions on the Riemann surface of an irreducible equation coincides with the algebraic extension of the field $K(X)$ obtained by adjoining a root of the equation.

In Sect. 4.3.3, we show that the field of meromorphic functions on every connected ramified finite covering of X is an algebraic extension of the field $K(X)$, and different extensions correspond to different coverings.

4.3.1 *The Field $P_a(O)$ of Germs at the Point $a \in X$ of Algebraic Functions with Ramification over O*

Let X be a connected complex manifold of dimension 1, O a discrete subset of X, and a a marked point in X not belonging to the set O.

Let $P_a(O)$ denote the collection of germs φ_a of meromorphic functions at the point a that satisfy the following properties:

1. The germ φ_a can be extended meromorphically along every path that originates at the point a and lies in $X \setminus O$.
2. For the germ φ_a, there exists a subgroup $G_0 \subset \pi_1(X \setminus O, a)$ of finite index in the group $\pi_1(X \setminus O, a)$ such that under the continuation of the germ φ_a along a path in the subgroup G_0, one obtains the initial germ φ_a.
3. The multivalued analytic function on $X \setminus O$ obtained by analytic continuation of the germ φ_a has algebraic singularities at the points of the set O.

Let us discuss property 3 in more detail. Let $\gamma : [0, 1] \to X$ be any path that goes from the point a to a singular point $o \in O$, $\gamma(0) = a$, $\gamma(1) = o$, inside the domain $X \setminus O$; that is, $\gamma(t) \in X \setminus O$ if $t < 1$. Property 3 means the following: For all values of the parameter t sufficiently close to 1 ($t_0 < t < 1$), consider the germs obtained by analytic continuation of φ_a along the path γ up to the point $\gamma(t)$. These germs are analytic, and they define a k-valued analytic function φ_γ in a small punctured neighborhood V_o^* of the point o. The restriction of the function φ_γ to a small punctured coordinate disk $D^*_{|u|<r}$ with center at the point o, where u is a local coordinate near the point o such that $u(o) = 0$, must have an algebraic singularity in the sense of Definition 4.14. The last condition does not depend on the choice of a coordinate function u. This means that the function φ_γ can be expanded into a Puiseux series in u (or equivalently, the function grows no faster than a power of u as it approaches the point o).

Lemma 4.26 *The set of germs $P_a(O)$ is a field. The fundamental group G of the domain $X \setminus O$ acts on the field $P_a(O)$ by analytic continuation. The invariant subfield of this action is the field of meromorphic functions on the manifold X.*

Proof Suppose that the germs $\varphi_{1,a}$ and $\varphi_{2,a}$ lie in the field $P_a(O)$ and do not change under continuations along the subgroups G_1 and G_2 of finite index in the group $G = \pi_1(X \setminus O, a)$. Then the germs $\varphi_{1,a} \pm \varphi_{2,a}$, $\varphi_{1,a}\varphi_{2,a}$ and $\varphi_{1,a}/\varphi_{2,a}$ (the germ $\varphi_{1,a}/\varphi_{2,a}$ is well defined, provided that the germ $\varphi_{2,a}$ is not identically equal to zero) can be extended meromorphically along every path that originates at the point a and lies in the domain $X \setminus O$. These germs do not change under continuation along the subgroup $G_1 \cap G_2$ of finite index in the group $G = \pi_1(X \setminus O, a)$.

Multivalued functions defined by these germs have algebraic singularities at the points of the set O, since the germs of functions representable by Puiseux series form a field. (Of course, one needs to be careful in applying arithmetic operations to multivalued functions. However, for a fixed path passing through the point o, one can apply arithmetic operations to the fixed branches of the functions representable by Puiseux series. As a result, one obtains a branch of the function representable by a Puiseux series.)

Thus we have shown that $P_a(O)$ is a field. Meromorphic continuation preserves arithmetic operations. Therefore, the fundamental group G acts on $P_a(O)$ by automorphisms. The invariant subfield consists of the germs in the field $P_a(O)$ that are the germs of meromorphic functions in the domain $X \setminus O$. At the points of the set O, these single-valued functions have algebraic singularities and therefore are meromorphic functions on the manifold X. The lemma is proved. □

4.3.2 Galois Theory for the Action of the Fundamental Group on the Field $P_a(O)$

In this subsection, we apply Galois theory to the action of the fundamental group $G = \pi_1(X \setminus O, a)$ on the field $P_a(O)$.

Theorem 4.27 *The following properties hold:*

1. *Every element φ_a of the field $P_a(O)$ is algebraic over the field $K(X)$.*
2. *The set of germs at the point a satisfying the same irreducible equation as the germ φ_a coincides with the orbit of the germ φ_a under the action of the group G.*
3. *The germ φ_a lies in the field obtained by adjoining an element f_a of the field $P_a(O)$ to the field $K(X)$ if and only if the stabilizer of the germ φ_a under the action of the group G contains the stabilizer of the germ f_a.*

Proof The proof of parts 1 and 2 follows from Theorem 2.16, and the proof of part 3 follows from Theorem 2.21. □

Part 1 of the theorem can be reformulated as follows.

Proposition 4.28 *A meromorphic germ at the point a lies in the field $P_a(O)$ if and only if it satisfies an irreducible equation $T = 0$ whose set of ramification points is contained in the set O.*

Part 2 of Theorem 4.27 is equivalent to the following statement.

Proposition 4.29 *Consider an equation $T = 0$ whose set of ramification points is contained in the set O. The equation T is irreducible if and only if the Riemann surface of the equation is connected.*

Proof Let $f : M \to X$ be a Riemann surface of an equation whose set of ramification points is contained in the set O. By part 2 of Theorem 4.27, the equation is irreducible if and only if the manifold $M \setminus f^{-1}(O)$ is connected. Indeed, the connectedness of the covering space is equivalent to the fact that the fiber $F = f^{-1}(a)$ lies in a single connected component of the covering space. This, in turn, implies the transitivity of the action of the monodromy group on the fiber F. It remains to note that the manifold M is connected if and only if the manifold $M \setminus f^{-1}(O)$ obtained by removing a discrete subset from M is also connected. □

Proposition 4.30 *A subfield of the field $P_a(O)$ is a normal extension of the field $K(X)$ if and only if it is obtained by adjoining all germs at the point a of a multivalued function on X satisfying an irreducible algebraic equation $T = 0$ over X whose ramification lies over O. The Galois group of this normal extension is isomorphic to the monodromy group of the Riemann surface of the equation $T = 0$.*

Proof A normal extension is always obtained by adjoining all roots of an irreducible equation. In the setting of the proposition, the ramification set of this equation must be contained in O. Both the Galois group of the normal covering and the monodromy group of the equation $T = 0$ are isomorphic to the image of the fundamental group $\pi_1(X \setminus O, a)$ under its action on the orbit in the field $P_a(O)$ consisting of the germs at the point a that satisfy the equation $T = 0$. □

Consider the Riemann surface of the equation $T = 0$ whose root is a germ $\varphi_a \in P_a(O)$. The points of this Riemann surface lying over the point a correspond to the roots of the equation $T = 0$ in the field $P_a(O)$. The germ φ_a is one of these roots. In this way, we assign to each germ φ_a of the field $P_a(O)$, first, the ramified covering $\pi_{\varphi_a} : M_{\varphi_a} \to X$, whose set of critical values is contained in O, and second, the marked point $\varphi_a \in M_{\varphi_a}$ lying over the point a (the symbol φ_a denotes the point of the Riemann surface corresponding to the germ φ_a). Part 3 of Theorem 4.27 can be reformulated as follows.

Proposition 4.31 *A germ φ_a lies in the field obtained by adjoining an element f_a of the field $P_a(O)$ to the field $K(X)$ if and only if the ramified covering $\pi_{\varphi_a} : (M_{\varphi_a}, \varphi_a) \to (X, a)$ is subordinate to the ramified covering $\pi_{f_a} : (M_{f_a}, f_a) \to (X, a)$.*

Indeed, according to the classification of ramified coverings with marked points, the covering corresponding to the germ φ_a is subordinate to the covering corresponding to the germ f_a if and only if the stabilizer of the germ φ_a under the action of the fundamental group $\pi_1(X \setminus O)$ contains the stabilizer of the germ f_a.

Corollary 4.32 *The fields obtained by adjoining elements φ_a and f_a of the field $P_a(O)$ to the field $K(X)$ coincide if and only if the ramified coverings with marked points $\pi_{\varphi_a} : (M_{\varphi_a}, \varphi_a) \to (X, a)$ and $\pi_{f_a} : (M_{f_a}, f_a) \to (X, a)$ are equivalent.*

Is it true that for every subgroup H of finite index in the fundamental group $\pi_1(X \setminus O, a)$, there exists a germ $f_a \in P_a(O)$ whose stabilizer is equal to H?

The answer to this question is positive. Galois theory alone does not suffice to prove this fact: in order to apply algebraic arguments, we need to have plenty of meromorphic functions on the manifold.[4] It will be sufficient for us to use the fact formulated below, which we will call the Riemann existence theorem and apply without proof. (The proof uses functional analysis and is not algebraic. Note that there exist two-dimensional compact complex analytic manifolds such that the only meromorphic functions on these manifolds are constants.)

---

[4]Galois theory allows one to obtain the following result. Suppose that the answer for a subgroup H is positive, and let $f_a \in P_a(O)$ be a germ whose stabilizer is equal to H. Let $\tilde{H}$ denote the largest normal subgroup lying in H. Then for every subgroup containing the group $\tilde{H}$, the answer is also positive. For the proof, it suffices to apply the fundamental theorem of Galois theory to the minimal Galois extension of the field $K(X)$ containing the germ f_1.

Theorem 4.33 (The Riemann existence theorem) *For every finite subset of a one-dimensional analytic manifold, there exists a meromorphic function on that manifold, analytic in a neighborhood of the subset and taking different values at different points of the subset.*

Theorem 4.34 *For every subgroup H of finite index in the fundamental group $\pi_1(X \setminus O, a)$, there exists a germ $f_a \in P_a(O)$ whose stabilizer is equal to H.*

Proof Let $\pi : (M, b) \to (X, a)$ be a finite ramified covering over X whose critical points lie over O. Let us assume that the covering corresponds to a subgroup $H \subset \pi_1(X \setminus O)$. Let $F = \pi^{-1}(a)$ denote the fiber of the covering over the point a. By the Riemann existence theorem, there exists a meromorphic function on the manifold M that takes different values at different points of the set F. Let $\pi_{b,a}^{-1}$ be a germ of the inverse map to the projection π that takes the point a to a point b. The germ of the function $f \circ \pi_{b,a}^{-1}$ lies in the field $P_a(O)$ by construction, and its stabilizer under the action of the fundamental group $\pi_1(X \setminus O)$ is equal to H. □

Thus we have shown that the classification of algebraic extensions of the meromorphic function field $K(X)$ that are contained in the field $P_a(O)$ is equivalent to the classification of ramified finite coverings $\pi : (M, b) \to (X, a)$ whose critical values lie in the set O. Both types of objects are classified by subgroups of finite index in the fundamental group $\pi_1(X \setminus O, a)$. In particular, the following theorem holds. (In this theorem, $K_a(X)$ denotes the field of meromorphic germs at $a \in X$ that are germs of globally defined meromorphic functions on X.)

Theorem 4.35 *There is a bijective correspondence between subgroups of finite index in the fundamental group and algebraic extensions of the field $K_a(X)$ that are contained in the field $P_a(O)$. If a subgroup G_1 lies in the subgroup G_2, then the field corresponding to the subgroup G_2 lies in the field corresponding to the subgroup G_1. A subfield of $P_a(O)$ is a Galois extension of the field $K_a(X)$ if it corresponds to a normal subgroup H of the fundamental group. The Galois group of this extension is isomorphic to the quotient group $\pi_1(X \setminus O, a)/H$.*

4.3.3 Field of Functions on a Ramified Covering

Here we show that irreducible algebraic equations over the field $K(X)$ give rise to isomorphic extensions of this field if and only if the Riemann surfaces of these equations provide equivalent ramified coverings over the manifold X.

Proposition 4.29 implies the following corollary.

Corollary 4.36 *An algebraic equation over the field $K(X)$ is irreducible if and only if its Riemann surface is connected.*

Let $\pi : (M, b) \to (X, a)$ be a finite ramified covering with marked points such that the one-dimensional complex manifold M is connected, and the point a does not belong set of critical values of the map π. We can apply the results about the field $P_a(O)$ and its subfields to describe the field of meromorphic functions on M. The following construction is useful.

Let $\pi_{b,a}^{-1}$ denote a germ of the inverse map to the projection π that takes the point a to a point b. Let $K_b(M)$ be the field of germs at the point b of meromorphic functions on the manifold M. This field is isomorphic to the field $K(M)$. The map $(\pi_{b,a}^{-1})^*$ embeds the field $K_b(M)$ in the field $P_a(O)$. Taking different preimages b of the point a, we obtain different embeddings of the field $K_b(M)$ in the field $P_a(O)$.

Suppose that an equation $T = 0$ is irreducible over the field $K(X)$. Then its Riemann surface is connected, and the meromorphic functions on this surface form the field $K(M)$. The field $K(M)$ contains the subfield $\pi^*(K(X))$ isomorphic to the field of meromorphic functions on the manifold X. Let $y : M \to \mathbb{C}P^1$ be a meromorphic function that appears in the definition of the Riemann surface. We have the following proposition.

Proposition 4.37 *The field $K(M)$ of meromorphic functions on the surface M is generated by the function y over the subfield $\pi^*(K(X))$. The function y satisfies the irreducible algebraic equation $T = 0$ over the subfield $\pi^*(K(X))$.*

Proof Let $b \in M$ be a point of the manifold M that is projected to the point a, $\pi(b) = a$, and $\pi_{b,a}^{-1}$ a germ of the inverse map to the projection π that takes the point a to a point b. Let $K_b(M)$ denote the field of germs at the point b of meromorphic functions on the manifold M. This field is isomorphic to the field $K(M)$. The map $(\pi_{b,a}^{-1})^*$ embeds the field $K_b(M)$ into the field $P_a(O)$.

For every meromorphic function $g : M \to \mathbb{C}P^1$, the germ $g_b \circ \pi_{b,a}^{-1}$ lies in the field $P_a(O)$. The stabilizer of this germ under the action of the group $\pi_1(X \setminus O, a)$ contains the stabilizer of the point b under the action of the monodromy group. For the germ $y_b \circ \pi_{b,a}^{-1}$, the stabilizer is equal to the stabilizer of the point b under the action of the monodromy group, since the function y by definition takes distinct values at the points of the fiber $\pi^{-1}(a)$. The proposition now follows from part 2 of Theorem 4.27. □

Theorem 4.38 *Irreducible equations $T_1 = 0$ and $T_2 = 0$ over the field $K(X)$ give rise to isomorphic extensions of this field if and only if the ramified coverings $\pi_1 : M_1 \to X$ and $\pi_2 : M_2 \to X$ that occur in the definition of the Riemann surfaces of these equations are equivalent.*

Proof Consider the points of the Riemann surfaces of the equations $T_1 = 0$ and $T_2 = 0$ that lie over a point x of the manifold X. For almost all x, these points are uniquely defined by the values of the roots y_1 and y_2 of the equations $T_1 = 0$ and $T_2 = 0$ over the point x. If the equations $T_1 = 0$ and $T_2 = 0$ define the same extensions of the field $K(X)$, then $y_1 = Q_1(y_2)$ and $y_2 = Q_2(y_1)$, where Q_1 and Q_2 are polynomials with coefficients in the field $K(X)$. These polynomials define almost everywhere an invertible map of one Riemann surface to the other that is compatible with projections of these surfaces to X. By continuity, it extends to an isomorphism of coverings.

If the Riemann surfaces of the equations give rise to equivalent coverings, and a map $h : M_1 \to M_2$ establishes the equivalence, then h is compatible with the projections and hence is analytic. The map $h^* : K(M_2) \to K(M_1)$ establishes the isomorphism of the fields $K(M_1)$ and $K(M_2)$ and takes the subfield $\pi_2^*(K(X))$ to the subfield $\pi_1^*(K(X))$, since $\pi_1 = \pi_2 \circ h$. □

4.4 Geometry of Galois Theory for Extensions of the Field of Meromorphic Functions

In this section, we summarize the previous results. In Sect. 4.4.1, we discuss the relationship between normal ramified coverings over a connected complex one-dimensional manifold X and Galois extensions of the field $K(X)$. In Sect. 4.4.2, this relation is used to describe extensions of the field of convergent Laurent series.

In Sect. 4.4.3, we talk about complex one-dimensional manifolds. Galois theory helps to describe the field of meromorphic functions on a compact manifold, and the geometry of ramified coverings allows us to describe explicitly enough all algebraic extensions of the field of rational functions in one variable.

The Galois group of an extension of the field of rational functions coincides with the monodromy group of the Riemann surface of an algebraic function defining this extension. Therefore, Galois theory gives a topological obstruction to the representability of algebraic functions by radicals.

4.4.1 Galois Extensions of the Field $K(X)$

By Theorem 4.38, algebraic extensions of the field of rational functions on a connected complex one-dimensional manifold X have a transparent geometric classification that coincides with the classification of connected finite ramified coverings over the manifold X. By this classification, Galois extensions of the field $K(X)$ correspond to normal ramified coverings over the manifold X. Let us describe all intermediate extensions for such Galois extensions.

Let X be a connected complex analytic one-dimensional manifold, $\pi : M \to X$ a normal ramified finite covering over X; also, let O be a finite subset in X containing all critical values of the map π, and $a \in X$ any point not in O. We have the field $K(X)$ of meromorphic functions on the manifold X and a Galois extension of this field, namely, the field $K(M)$ of meromorphic functions on the manifold M.

By Proposition 4.20, intermediate ramified coverings $M \xrightarrow{x_1} Y_1 \xrightarrow{f_1} X$ are in one-to-one correspondence with the subgroups of the deck transformation group N of the normal covering $\pi : M \to X$. To every ramified covering $M \xrightarrow{x_1} Y_1 \xrightarrow{f_1} X$, one can associate the subfield $x_1^*(K(Y_1))$ of the field $K(M)$ of meromorphic functions on the manifold M. As follows from the fundamental theorem of Galois theory, every intermediate field between $K(M)$ and $\pi^* K(X)$ is of this form, i.e., it is the field $x^*(K(Y))$ for an intermediate ramified covering $M \xrightarrow{x} Y \xrightarrow{f} X$. By this classification, intermediate Galois extensions of the field $K(X)$ correspond to intermediate normal coverings $M \xrightarrow{x} Y \xrightarrow{f} X$, and the Galois groups of intermediate Galois extensions are equal to the deck transformation groups of intermediate normal coverings.

Here is a slightly different description of the same Galois extension. The finite deck transformation group N acts on a normal ramified covering M. To each subgroup G of the group N, one can associate the subfield $K_G(M)$ of meromorphic functions on M invariant under the action of the group G.

Proposition 4.39 *The field $K(M)$ is a Galois extension of the field $K_N(M) = \pi^*(K(X))$. The Galois group of this Galois extension is equal to N. Under the Galois correspondence, a subgroup $G \subset N$ corresponds to the field $K_G(M)$.*

4.4.2 Algebraic Extensions of the Field of Germs of Meromorphic Functions

In this subsection, the relation between normal coverings and Galois extensions is used to describe extensions of the field of convergent Laurent series.

Let L_0 be the field of germs of meromorphic functions at the point $0 \in \mathbb{C}^1$. This field can be identified with the field of convergent Laurent series $\sum_{m>m_0} c_m x^m$.

Theorem 4.40 *For every k, there exists a unique extension of the field L_0 of degree k. It is generated by the element $z = x^{1/k}$. This extension is normal, and its Galois group is equal to $\mathbb{Z}/k\mathbb{Z}$.*

Proof Let $y^k + a_{k-1}y^{k-1} + \cdots + a_0 = 0$ be an irreducible equation over the field L_0. The irreducibility of the equation implies the existence of a small open disk D_r with center at the point 0 satisfying the following conditions:

1. All Laurent series a_i, $i = 1, \ldots, k$, converge in the punctured disk D_r^*.
2. The equation is irreducible over the field $K(D_r)$ of meromorphic functions on the disk D_r.
3. The discriminant of the equation does not vanish at all points of the punctured disk D_r^*.

Let $\pi : M \to D_r$ be the Riemann surface of the irreducible equation over the disk D_r. By the assumption, the point 0 is the only critical value of the map π. The fundamental group of the punctured disk D_r^* is isomorphic to the additive group of integers $\mathbb{Z}$. The group $k\mathbb{Z}$ is the only subgroup of index k in the group $\mathbb{Z}$. This subgroup is a normal subgroup, and the quotient group $\mathbb{Z}/k\mathbb{Z}$ is the cyclic group of order k. Hence, there exists a unique extension of degree k. It corresponds to the germ of a k-fold covering $f : (\mathbb{C}^1, 0) \to (\mathbb{C}^1, 0)$, where $f = z^k$. The extension is normal, and its Galois group is equal to $\mathbb{Z}/k\mathbb{Z}$. Next, the function $z : D_q \to \mathbb{C}^1$, where $q = r^{1/k}$, takes distinct values on all preimages of the point $a \in D_r^*$ under the map $x = z^k$. Hence, the function $z = x^{1/k}$ generates the field $K(D_q)$ over the field $K(D_r)$. The theorem is proved. □

By the theorem, the function z and its powers $1, z = x^{1/k}, \dots, z^{k-1} = x^{(k-1)/k}$ form a basis in the extension L of degree k of the field L_0 regarded as a vector space over the field L_0. Functions $y \in L$ can be regarded as multivalued functions of x. The expansion $y = f_0 + f_1 z + \dots + f_{k-1} z^{k-1}$, $f_0, \dots, f_{k-1} \in L_0$, of the element $y \in L$ in the given basis is equivalent to the expansion of the multivalued function $y(x)$ into the Puiseux series

$$y(x) = f_0(x) + f_1(x)x^{1/k} + \dots + f_{k-1}(x)x^{(k-1)/k}.$$

Note that the elements $1, z, \dots, z^{k-1}$ are the eigenvectors of the isomorphism of the field L over the field L_0 defined by analytic continuation along the loop around the point 0. It generates the Galois group. The eigenvalues of the given eigenvectors are equal to $1, \xi, \dots, \xi^{k-1}$, where ξ is a primitive kth root of unity. The existence of such a basis of eigenvectors is proved in Galois theory (see Proposition 2.1).

Remark 4.41 The field L_0 is in many respects similar to the finite field $\mathbb{Z}/p\mathbb{Z}$. Continuation along the loop around the point 0 is similar to the Frobenius isomorphism. Indeed, each of these fields has a unique extension of degree k for every positive integer k. All these extensions are normal, and their Galois groups are isomorphic to the cyclic group of k elements. The generator of the Galois group of the first field corresponds to a loop around the point 0, and the generator of the Galois group of the second field is the Frobenius isomorphism. Every finite field has similar properties. For the field F_q consisting of $q = p^n$ elements, the role of a loop around the point 0 is played by the nth iterate of the Frobenius automorphism.

4.4.3 Algebraic Extensions of the Field of Rational Functions

Let us now consider the case of connected compact complex one-dimensional manifolds. Using Galois theory, we show that the field of meromorphic functions on such a manifold is a finite extension of the transcendence degree 1 of the field of complex numbers. On the other hand, the geometry of ramified coverings over the Riemann sphere provides a clear description of all finite algebraic extensions of the field of rational functions.

The Riemann sphere $\mathbb{C}P^1$ is the simplest of all compact complex manifolds. It is isomorphic to the projective line $\mathbb{C}P^1$, on which we fix the point ∞ at infinity, $\mathbb{C}^1 \cap \{\infty\} = \mathbb{C}P^1$, and a holomorphic coordinate function $x : \mathbb{C}P^1 \to \mathbb{C}P^1$ that has a pole of order 1 at the point ∞. Every meromorphic function on $\mathbb{C}P^1$ is a rational function of x.

We say that a pair of meromorphic functions f, g on a manifold M *separates almost all points* of the manifold M if there exists a finite set $A \subset M$ such that the vector function (f, g) is defined on the set $M \setminus A$ and takes distinct values at all points of $M \setminus A$.

Theorem 4.42 *Let M be a connected compact one-dimensional complex manifold.*

1. *Then every pair of meromorphic functions f, g on M are related by a polynomial relation (i.e., there exists a polynomial Q in two variables such that the identity $Q(f, g) = 0$ holds).*
2. *Let functions f, g separate almost all points of the manifold M. Then every meromorphic function φ on the manifold M is the composition of a rational function R in two variables with the functions f and g, that is, $\varphi = R(f, g)$.*

Proof 1. If the function f is identically equal to a constant C, then one can take the relation $f \equiv C$ as a polynomial relation. Otherwise, the map $f : M \to \mathbb{C}P^1$ is a ramified covering with a certain subset O of ramification points. It remains to use part 1 of Theorem 4.27.

2. If the function f is identically equal to a constant C, then the function g takes distinct values at the points of the set $M \setminus A$. Therefore, the ramified covering $g : M \to \mathbb{C}P^1$ is a bijective map of the manifold M to the Riemann sphere $\mathbb{C}P^1$. In this case, every meromorphic function φ on M is the composition of a rational function R in one variable with the function g; that is, $\varphi = R(g)$.

If the function f is not constant, then it gives rise to the ramified covering $f : M \to \mathbb{C}P^1$ over the Riemann sphere $\mathbb{C}P^1$. Let O be the union of the set $f(A)$ and the critical value set of the map f. Let a be a point of the Riemann sphere not lying in O, and let F be the fiber of the ramified covering $f : M \to \mathbb{C}P^1$ over the point a. By our assumption, the function g must separate the points of the set F. It remains to use part 3 of Theorem 4.27. □

Let

$$y^n + a_{n-1}y^{n-1} + \cdots + a_0 \tag{4.1}$$

be an irreducible equation over the field of rational functions. The Riemann surface $\pi : M \to \mathbb{C}P^1$ of this equation is also called the *Riemann surface of an algebraic function* defined by this equation. The monodromy group of the ramified covering $\pi : M \to \mathbb{C}P^1$ is also called the *monodromy group of this algebraic function.* By Proposition 4.30, the Galois group of (4.1) coincides with the monodromy group.

Hence the Galois group of the irreducible equation (4.1) over the field of rational functions has a topological meaning: it is equal to the monodromy group of the Riemann surface of the algebraic function defined by (4.1). This fact was known to Frobenius, but it was probably discovered even earlier.

The results of Galois theory yield a topological obstruction to the solvability of (4.1) by radicals and k-radicals. Galois theory implies the following theorems.

Theorem 4.43 *An algebraic function y defined by* (4.1) *is representable by radicals over the field of rational functions if and only if its monodromy group is solvable.*

Theorem 4.44 *An algebraic function y defined by* (4.1) *is representable by k-radicals over the field of rational functions if and only if its monodromy group is k-solvable.*

Connected ramified coverings over the Riemann sphere $\mathbb{C}P^1$ whose critical values lie in a fixed finite set O admit a complete and finite description. Connected k-fold ramified coverings with marked points $\pi : (M, b) \to (\mathbb{C}P^1 \setminus O, a)$ are classified by subgroups of index k in the fundamental group $\pi_1(\mathbb{C}P^1 \setminus O)$. For every group G, the following lemma holds.

Lemma 4.45 *The classification of index-k subgroups of the group G is equivalent to the classification of transitive actions of the group G on a k-point set with a marked point.*

Proof Indeed, to a subgroup G_0 of index k in the group G one can associate the transitive action of the group G on the set of right cosets of the group G by the subgroup G_0. This set consists of k points, and the right coset of the identity element is a marked point. In the other direction, to every transitive action of the group G, one can assign the stabilizer G_0 of the marked point. This subgroup has index k in the group G. □

The fundamental group $\pi_1(\mathbb{C}P^1 \setminus O, a)$ is a free group with a finite number of generators. It has finitely many different transitive actions on the set of k elements. All these actions can be described as follows.

Let us number the points of the set O. Suppose that this set contains $m + 1$ points. The fundamental group $\pi_1(\mathbb{C}P^1 \setminus O, a)$ is a free group generated by paths $\gamma_1, \dots, \gamma_m$, where γ_i is a path going around the ith point of the set O. Take a set of k elements with one marked element. In the group $S(k)$ of permutations of this set, choose m arbitrary elements $\sigma_1, \dots, \sigma_m$. We are interested in ordered collections $\sigma_1, \dots, \sigma_m$ satisfying a single relation: the group of permutations generated by these elements must be transitive. There is a finite number of collections $\sigma_1, \dots, \sigma_m$. One can check each of them, and choose all collections generating transitive groups. To every such collection, one can associate a unique ramified covering $\pi : (M, b) \to (\mathbb{C}P^1, a)$ with a marked point. It corresponds to the stabilizer of the marked element under the homomorphism $F : \pi_1(\mathbb{C}P^1 \setminus O, a) \to S(k)$ that maps the generator γ_i to the element σ_i. Hence in a finite number of steps, one can list all transitive actions

$$F : \pi_1\left(\mathbb{C}P^1 \setminus O, a\right) \to S(k)$$

of the fundamental group $\pi_1(\mathbb{C}P^1 \setminus O, a)$ on the set of k elements.

Conjugations of the group $S(k)$ act on the finite set of homomorphisms $F : \pi_1(\mathbb{C}P^1 \setminus O, a) \to S(k)$ with transitive images. The orbits of a finite group action on a finite set can in principle be enumerated. Hence, conjugacy classes of the subgroups of index k in the fundamental group can also be listed in a finite number of steps.

Therefore, we obtain a complete geometric description of all possible Galois extensions of the field of rational functions in one variable. Note that in this description, we used the Riemann existence theorem. The Riemann existence theorem does not help to describe algebraic extensions of other fields, such as the field of rational numbers. The problem of describing algebraic extensions of the field of rational numbers is open. For instance, it is unknown in general whether there exists an extension of the field of rational numbers whose Galois group is a given finite group.

Chapter 5
One-Dimensional Topological Galois Theory

The monodromy group of an algebraic function is isomorphic to the Galois group of the associated extension of the field of rational functions. Therefore, the monodromy group is responsible for the representability of an algebraic function by radicals. However, not only algebraic functions have a monodromy group. It is defined for the logarithm, arctangent, and many other functions for which the Galois group does not make sense. It is thus natural to try using the monodromy group for these functions instead of the Galois group to prove that they do not belong to a certain Liouville class. This particular approach is implemented in one-dimensional topological Galois theory [45–48, 50, 54, 56].

In the one-dimensional version of topological Galois theory, we consider functions representable by quadratures as multivalued analytic functions of one complex variable. It turns out that there exist topological restrictions on the way the Riemann surface of a function representable by quadratures can be positioned over the complex plane. If a function does not satisfy these restrictions, then it cannot be expressed by quadratures.

Besides its geometric appeal, this approach has the following advantage. Topological obstructions relate to branching. These obstructions persist not only for functions representable by quadratures but also for a much wider class of functions. This wider class of functions can be obtained if we add all meromorphic functions to functions representable by quadratures and allow them to enter all formulas. For this reason, topological results on nonrepresentability by quadratures turn out to be stronger than the corresponding algebraic results. Indeed, the composition of functions is not an algebraic operation. In differential algebra, this operation is replaced with a differential equation describing it. However, the Euler Γ-function, for example, does not satisfy an algebraic differential equation. Hence it is pointless to look for an equation satisfied, say, by the function $\Gamma(\exp x)$. The only known results on nonrepresentability of functions by quadratures and functions like the Euler Γ-function are obtained by our method.

A. Khovanskii, *Topological Galois Theory*, Springer Monographs in Mathematics,
DOI 10.1007/978-3-642-38871-2_5

On the other hand, this method cannot be used to prove that a particular single-valued meromorphic function is not representable by quadratures.

Using differential Galois theory (to be more precise, only its linear-algebraic part dealing with linear algebraic groups and their differential invariants), one can prove that the only reasons for unsolvability of linear Fuchsian systems of differential equations are topological (see Sect. 6.1). In other words, if there are no topological obstructions to solvability of a Fuchsian system by quadratures, then that system is solvable by quadratures.

There are topological obstructions to representability of functions by quadratures, generalized quadratures, and k-quadratures.

Firstly, the functions representable by generalized quadratures, and in particular, the functions representable by quadratures and k-quadratures, may have no more than countably many singular points in the complex plane (see Sect. 5.4). (However, even for the simplest functions representable by quadratures, the set of singular points can be everywhere dense.)

Secondly, the monodromy group of a function representable by quadratures is necessarily solvable (see Sect. 5.7.2). (However, even for the simplest functions representable by quadratures, the monodromy group can have the cardinality of the continuum.)

There are similar restrictions on the positioning of the Riemann surface for functions representable by generalized quadratures and k-quadratures. However, these restrictions are more involved. To state them, we should regard the monodromy group not as an abstract group but rather as a subgroup of the permutation group of the branches. In other words, these restrictions make use not only of the monodromy group but of the *monodromy pair* of the function consisting of the monodromy group and the stabilizer of some germ of the function (see Sect. 5.5.3).

In Sect. 5.1, we discuss the notion of topological unsolvability due to Arnold and prove (following Arnold) that elliptic functions are topologically nonelementary. In Sect. 5.2, we give a criterion for representability of functions by radicals, the proof of which contains an idea of topological Galois theory.

5.1 On Topological Unsolvability

Arnold proved that a number of classical mathematical problems are topologically unsolvable [2–10]. The problem of solvability of algebraic equations by quadratures is among these problems (see Sect. 5.2).

As was shown by Jordan, the Galois group of an algebraic equation over the field of rational functions has a topological interpretation. Consider an irreducible algebraic equation

$$y^n + r_1 y^{n-1} + \cdots + r_n = 0 \tag{5.1}$$

over the field of rational functions on the Riemann sphere. The Galois group of Eq. (5.1) over the field of rational functions is isomorphic to the monodromy group of the (multivalued) algebraic function y defined by Eq. (5.1). The following is a consequence of Galois theory.

Corollary 5.1 *The following statements hold:*

1. *An algebraic function is representable by radicals if and only if its monodromy group is solvable.*
2. *An algebraic function is representable by k-radicals if and only of its monodromy group is k-solvable.*

We now give a definition due to Arnold.

Definition 5.2 (Arnold) A map $f : X \to Y$ is said to be *topologically bad* (for example, a topologically nonelementary function) if among the maps left–right topologically equivalent to it,[1] there are no good ones (elementary, for example).

To each multivalued analytic function f of a complex variable, one associates its Riemann surface M_f and the projection $\pi_f : M_f \to S^2$ of that surface onto the Riemann sphere S^2.

Corollary 5.3 *Suppose that the projections π_f and π_g of the Riemann surfaces M_f and M_g of functions f and g onto the Riemann sphere are topologically equivalent. Then the functions f and g are either both representable or both unrepresentable by radicals (k-radicals) (i.e., the topological type of the projection from the Riemann surface of a function onto the Riemann sphere is responsible for the representability of the function by radicals and by k-radicals).*

Proof The statement follows immediately from Corollary 5.1. Indeed, the algebraicity of the function is related to the compactness of its Riemann surface, its representability by radicals is related to the solvability of its monodromy group, and its representability by k-radicals is related to the k-solvability of its monodromy group. All these properties are topological. □

For algebraic functions, the Galois group is isomorphic to the monodromy group. This gives the possibility of proving results of a Galois-theoretic nature for algebraic functions separately, without introducing new notions or proving general theorems. Classical authors used this trick often. Thus, in the book [26], one can read the following in the section devoted to the monodromy group:

> The following theorems will make [that the monodromy group is the Galois group] convincing to the reader who knows Galois theory, and the reader who does not know it will have an introduction to Galois theory adapted to some special fields.

---

[1] A map $g : X \to Y$ between topological spaces is said to be *left–right topologically equivalent* (or just *topologically equivalent*) to a map $f : X \to Y$ if there are homeomorphisms $h_X : X \to X$ and $h_Y : Y \to Y$ such that $g = h_Y \circ f \circ h_X$.

In the 1960s, during his time as a teacher in Kolmogorov's boarding school for gifted high-school students, Arnold found a new proof that a generic algebraic function of degree 5 is not representable by radicals. He proved in a purely topological way (without using Galois theory) that a function whose monodromy group is not solvable cannot be represented by radicals (see Sect. 5.2). Arnold gave a series of lectures on this proof in Kolmogorov's boarding school. This lecture course was later revised and published by B.V. Alekseev [1]. According to Arnold, a topological proof of the unsolvability of a problem usually implies new consequences. For example, from the topological proof of the nonrepresentability by radicals of an algebraic function with an unsolvable monodromy group it follows easily that such a function is not representable by a formula including not only radicals but also arbitrary entire functions; see [46].

During that same period, Arnold proved the topological nonelementarity of elliptic functions and integrals and also obtained a number of similar results, but he published nothing on the subject. In December 2003, he wrote a letter to me about this. The following theorem, as well as the definition given above, are taken from that letter.

Theorem 5.4 (Arnold) *If a meromorphic function $g : U \to \mathbb{C}P^1$ defined in a complex domain $U \subset \mathbb{C}$ is topologically equivalent to an elliptic function $f : \mathbb{C} \to \mathbb{C}P^1$, then g is also an elliptic function (perhaps with different periods from those of f).*

Proof The elliptic function f is invariant under a group Λ of translations

$$z \mapsto z + k_1 w_1 + k_2 w_2, \quad k_1, k_2 \in \mathbb{Z}^2.$$

The group Λ is isomorphic to the group $\mathbb{Z}^2$. Thus the function g is invariant under a group, isomorphic to $\mathbb{Z}^2$, of certain homeomorphisms of the domain U. Each homeomorphism h from this group is in fact a biholomorphic map of the domain U to itself. Indeed, by the inverse function theorem, the identity $g(z) = g(h(z))$ implies that the map h is holomorphic in a neighborhood of every point outside the preimage of the critical-point set of g under the map h. At the points of this preimage, the map h is holomorphic by the removable singularity theorem. By the assumption, the region U is homeomorphic to $\mathbb{C}$. Therefore, by Riemann's theorem, the domain U either coincides with $\mathbb{C}$ or is biholomorphically equivalent to the interior of the unit disk. The domain U coincides with $\mathbb{C}$, since the group of biholomorphic transformations of the unit disk does not contain $\mathbb{Z}^2$ as a subgroup. In the group of biholomorphic transformations of $\mathbb{C}$, a subgroup isomorphic to $\mathbb{Z}^2$ with no fixed points is a group of translations

$$z \mapsto k_1 \tau_1 + k_2 \tau_2, \quad k_1, k_2 \in \mathbb{Z}^2.$$

Therefore, g is an elliptic function. □

As is well known, elliptic functions are not elementary.[2] From this classical result and from the theorem proved above, the topological nonelementarity of elliptic functions follows.

Below is a quotation from the letter of Arnold.

> As far as I remember, these arguments proved the topological nonelementarity not only of elliptic functions f, but also of elliptic integrals f^{-1}, and of many other things. Besides, all this generalizes to curves of other genera (with other coverings, or at least with the universal coverings). I do not remember whether I proved this rigorously, but I think that I had reasons for the analogous *multidimensional* statements to be false: as far as I remember, *the conservation of the topological type in the multidimensional case does not guarantee the conservation of algebraicity*. By itself, this does not obstruct the topological nonelementarity (badness), but it obstructs my proof of it, the proof by reduction to the nonelementarity of a classical (algebraic) object like an elliptic function.

5.2 Topological Nonrepresentability of Functions by Radicals

In this section, we give a version of Arnold's topological proof of the nonrepresentability of functions by radicals based on the note [46] and containing a germ of one-dimensional topological Galois theory.

The class of functions of one variable representable by radicals can be defined by the following two lists.

List of basic functions:

- Complex constants $y = C$
- The independent variable $y = x$
- The functions $y = x^{1/n}$, where $n > 1$ is any positive integer

List of admissible operations:

- Arithmetic operations
- Composition.

Definition 5.5 The *class of functions representable by radicals* is the set of all functions that can be obtained from basic functions by admissible operations.

[2] A differential field of elliptic functions is generated over $\mathbb{C}$ by the corresponding Weierstrass $\wp$-function, which satisfies a certain nonlinear first-order differential equation (see, for example, [36,37]). The nonelementarity of elliptic functions follows from a generalization of Picard–Vessiot theory due to Kolchin [65]. Kolchin's generalization is applicable not only to linear differential equations, but also to some that are nonlinear, for example, to the equation of the $\wp$-function. The Galois group of the differential field of elliptic functions over the field $\mathbb{C}$ obviously contains the group $\mathbb{C}/\Lambda$, which is the quotient group of the group $\mathbb{C}$ of translations $f(z) \mapsto f(z + a)$ by the subgroup Λ of periods of elliptic functions. It is not hard to show that the Galois group coincides with this group. The nonrepresentability of elliptic functions by generalized quadratures, according to Kolchin, follows from the nonexistence of a normal tower in the group $\mathbb{C}/\Lambda$ such that each quotient of it is either a finite group, the additive group of complex numbers, or the multiplicative group of complex numbers.

In this section, we discuss the following criterion for representability by radicals.

Proposition 5.6 (Criterion for representability by radicals) *A function is representable by radicals if and only if it is an algebraic function and its monodromy group is solvable.* □

This criterion is easy to deduce from Galois theory. Indeed, the monodromy group of an algebraic function is isomorphic to the Galois group of the field extension obtained from the field of all rational functions by adjoining all branches of the algebraic function.

In this section, we give a simple proof of the criterion that is independent of Galois theory and the other parts of this book (except that in Sect. 5.2.4, we use an uncomplicated linear-algebraic argument from Chap. 2). Sect. 5.2.1 provides an (almost obvious) verification of the fact that the monodromy groups of basic functions representable by radicals are solvable. In Sect. 5.2.2, we state the properties of solvable groups that we need. In Sect. 5.2.4, we prove that algebraic functions with solvable monodromy groups are representable by radicals.

Remark 5.7 The class of functions representable by radicals was defined in Sect. 1.2 slightly differently. It is easy to see, however, that this definition is equivalent to the definition given above.

Remark 5.8 In this section, as in all other parts of this book (except Chap. 7), the operations on multivalued functions are understood in the sense of Definition 1.1.

5.2.1 Monodromy Groups of Basic Functions

It is easy to compute the monodromy groups of basic functions representable by radicals. The monodromy groups of constants and of the independent variable are trivial (i.e., they contain the identity element only), since those functions are single-valued.

Proposition 5.9 *The monodromy group of the function $y = x^{1/n}$ is the cyclic group $\mathbb{Z}/n\mathbb{Z}$.*

Proof The function $x^{1/n}$ has no ramification points in the domain $\mathbb{C}^* = \overline{\mathbb{C}}\setminus\{0, \infty\}$, whose fundamental group is isomorphic to the group of integers. A generator of the group $\pi_1(\mathbb{C}^*, 1)$ is represented by the loop $\gamma(t) = \exp(2\pi i t)$, $0 \le t \le 1$. The function $y = x^{1/n}$ has n germs y_k, $0 \le k < n$, at the point 1, whose values at this point are $y_k(1) = \exp 2k\pi i/n$. The analytic continuation of the germ y_k along the curve γ takes the value $\exp 2(k+t)\pi i/n$ at the point $\gamma(t)$. As the point $\gamma(t)$ makes a complete turn, the parameter t increases from 0 to 1. Hence, after a complete turn, the germ y_k is transformed into the germ y_m, where $m \equiv (k+1) \pmod n$. □

Corollary 5.10 *The monodromy groups of the basic functions representable by radicals are solvable.*

5.2.2 Solvable Groups

We now state some facts about solvable groups that we need. A group G is called a *solvable group of depth* $\ell \geq 0$ if it satisfies the following properties:

1. There exists a nested chain of subgroups $G = G_0 \subset \cdots \subset G_l = e$ in which the group G_l coincides with the identity element e, and for $i = 1, \ldots, \ell$, the group G_i is a normal subgroup of the group G_{i-1}, the quotient group G_{i-1}/G_i being abelian.
2. There is no shorter nested chain of subgroups with the same properties.

A group is said to be *solvable* if it is a solvable group of some depth $\ell \geq 0$.

Set $[G]^0 = G$. Let $[G]^1$ denote the commutator of the group G. For every positive integer ℓ, we define the *ℓth commutator $[G]^\ell$ of the group G* as the commutator of the group $[G]^{\ell-1}$. The following proposition is straightforward:

Proposition 5.11 *The following properties hold:*

1. *For every positive integer ℓ, the group $[G]^\ell$ is a normal subgroup of the group G (and moreover, every automorphism of the group G maps the subgroup $[G]^\ell$ to itself).*
2. *A group G is a solvable group of depth $\ell > 0$ if and only if $[G]^\ell = e$ and $[G]^{\ell-1} \neq e$.*
3. *For every homomorphism $\pi : G_1 \to G_2$ of a group G_1 to a group G_2 and every positive integer ℓ, we have the inclusion $[\pi(G_1)]^\ell \subseteq [G_2]^\ell$.*
4. *For every pair of positive integers ℓ, m and every group G, the following inclusion holds: $[G]^\ell \supseteq [G]^{\ell+m}$.*

5.2.3 The Class of Algebraic Functions with Solvable Monodromy Groups Is Stable

It is easy to show that the class of algebraic functions is stable under composition and arithmetic operations. We will not spell out the proof of this well-known fact. We now show that the class of functions with solvable monodromy groups is also stable under these operations.

Let y be an algebraic function of one complex variable, A the set of ramification points of y, and $U_A = \overline{\mathbb{C}} \setminus A$ the complement of the set A in the Riemann sphere $\overline{\mathbb{C}}$. Let $\pi_1(U_A, *)$ be the fundamental group of the domain U_A with a marked point $* \in U_A$.

Proposition 5.12 *The following properties hold:*

1. *If the monodromy group of the algebraic function y is a solvable group of depth at most ℓ, then every germ of the function y is mapped to itself by analytic*

continuation along any element of the ℓth commutator $[\pi_1(U_A, *)]^\ell$ of the group $\pi_1(U_A, *)$.

2. *If at least one germ of the function y at the point $*$ is mapped to itself under analytic continuation along the ℓth commutator $[\pi_1(U_A, *)]^\ell$ of the group $\pi_1(U_A, *)$, then the monodromy group of the algebraic function y is a solvable group of depth at most ℓ.*

Proof The result follows immediately from Proposition 5.11. It suffices to make the following remark. The group $[G]^\ell$ is a normal subgroup of the fundamental group $G = \pi_1(U_A, *)$ (see part 1 of Proposition 5.11). Hence if at least one germ of the function y remains unchanged under analytic continuation along the paths from the subgroup $[G]^\ell$, then the analytic continuation along the paths from this subgroup changes no germ of the function y. □

Let B be any finite set containing the set A of ramification points of the function y, let $U_B = \overline{\mathbb{C}} \setminus B$ be the complement of the set B in the Riemann sphere $\overline{\mathbb{C}}$, and let $*$ be any point in the domain U_B. The following corollary differs from Proposition 5.12 only in that A is replaced by B.

Corollary 5.13 *The following properties hold:*

1. *If the monodromy group of an algebraic function y is a solvable group of depth at most ℓ, then every germ of the function y remains unchanged under analytic continuation along the ℓth commutator $[\pi_1(U_B, *)]^\ell$ of the group $\pi_1(U_B, *)$.*
2. *If at least one germ of the function y at the point $*$ remains unchanged under analytic continuation along the ℓth commutator $[\pi_1(U_B, *)]^\ell$ of the group $\pi_1(U_B, *)$, then the monodromy group of the algebraic function y is a solvable group of depth at most ℓ.*

Proof Every permutation of the germs of the function y corresponding to a loop from the group $\pi_1(U_A, *)$ can also be obtained by going along a loop from the group $\pi_1(U_B, *)$. To this end, it suffices to perturb the initial loop slightly, getting rid of its intersection points (if any) with the finite set $B \setminus A$. Therefore, Corollary 5.13 is essentially identical to Proposition 5.12. □

Theorem 5.14 *The class of algebraic functions with solvable monodromy groups is stable under the arithmetic operations.*

Proof Suppose that algebraic functions y and z have sets of ramification points A_1 and A_2 and solvable monodromy groups M_1 and M_2 of depths ℓ_1 and ℓ_2, respectively. Set $B = A_1 \cup A_2$, $U_B = \overline{\mathbb{C}} \setminus B$, and let $*$ be a point in the domain U_B. Let m denote the maximum of ℓ_1 and ℓ_2. By Corollary 5.13, meromorphic continuation along paths from the group $[\pi_1(U_B, *)]^m$ does not change the germs of the functions y and z. Therefore, as we go along these loops, the germs of the functions $y + z$, $y - z$, $y \times z$, and $y : z$ return to their initial values. Hence, by Corollary 5.13, the monodromy groups of these germs are solvable and are of depth not exceeding m. □

Theorem 5.15 *The class of algebraic functions with solvable monodromy groups is stable under composition.*

Before we proceed with the proof of Theorem 5.15, we make it a little more precise. Consider algebraic functions y and z. Let A_1 and A_2 be the sets of ramification points of the functions y and z, and M_1 and M_2 the monodromy groups of these functions that are solvable groups of depths ℓ_1 and ℓ_2, respectively. Let $y^{-1}(A_2)$ be the full preimage of the set A_2 under the multivalued map generated by the function y (i.e., $x \in y^{-1}(A_2)$ if there exists a germ of the function y at the point x whose value at x belongs to the set A_2).

Theorem 5.16 *Under the assumptions made above, the composition $z(y)$ of the algebraic functions y and z has a solvable monodromy group of depth at most $\ell_1 + \ell_2$. The set of ramification points of the function $z(y)$ is a subset of the set $B = A_1 \cup y^{-1}(A_2)$.*

Proof It is clear that the set $B = A_1 \cup y^{-1}(A_2)$ contains all ramification points of the function $z(y)$. Set $U_B = \overline{\mathbb{C}} \setminus B$, and let $*$ be a point in the domain U_B. Set $U_{A_2} = \overline{\mathbb{C}} \setminus A_2$. Let G denote the group $\pi_1(U_B, *)$, and $[G]^{\ell_1}$ its ℓ_1th commutator.

By Corollary 5.13, meromorphic continuation along the paths from the group $[G]^{\ell_1}$ does not change the germs of the function y. This means that for every germ y_i of the function y at the point $*$, we have a homomorphism $\tau_i : [G]^{\ell_1} \to \pi_1(U_{A_2}, \star)$, where $\star = y_i(*)$, which maps a path

$$\gamma : [0, 1] \to U_B, \quad \gamma(0) = \gamma(1) = *,$$

from the group $[G]^{\ell_1} = [\pi_1(U_B, *)]^{\ell_1}$ to the path

$$y_i(\gamma) : [0, 1] \to U_{A_2}, \quad y_i(\gamma(0)) = y_i(\gamma(1)) = y_i(*) = \star$$

obtained by meromorphic continuation of the germ y_i along the path γ.

By definition, meromorphic continuation along the paths from $[\pi_1(U_{A_2}, \star)]^{\ell_2}$ does not change the germ of z. Therefore, meromorphic continuation along the paths from the group $\tau_i^{-1}([\pi_1(U_{A_2}, \star]^{\ell_2})$ does not change the germ $z(y_i)$. By part 3 of Proposition 5.11, the group $\tau_i^{-1}([\pi_1(U_{A_2}, \star]^{\ell_2})$ contains the ℓ_2th commutator of the group $[G]^{\ell_1}$. Hence the meromorphic continuation along the paths from the group $[\pi_1(U_B, *)]^{\ell_1+\ell_2}$ does not change the germ of the composition $z(y_i)$. To conclude the proof of Theorem 5.16, it remains to use Corollary 5.13. □

5.2.4 An Algebraic Function with a Solvable Monodromy Group Is Representable by Radicals

Let y be an n-valued algebraic function of one variable, A its set of ramification points, $U_A = \overline{\mathbb{C}} \setminus A$, and $* \in U_A$ a marked point. Let V be an open disk centered at the point $*$ and disjoint from the set A. On the disk V, there are n holomorphic branches $y_1, \dots, y_n$ of the function y. Therefore, on this disk, all functions in the field $R\langle y\rangle = R\langle y_1, \dots, y_n\rangle$ obtained by adjoining elements $y_1, \dots, y_n$ to the field R of all rational functions are meromorphic. Every function z from the field $R\langle y\rangle$ admits a meromorphic continuation along any path γ from the fundamental group $\pi_1(U_A, *)$, since all rational functions and all branches of the function y admit meromorphic continuations along γ. The meromorphic continuation along the path γ preserves all arithmetic operations. Hence it defines an automorphism τ_γ of the field $R\langle y\rangle$. The automorphism τ_γ is the identity if and only if the path γ gives rise to the identity monodromy transformation M of the function y. A function $z \in R\langle y\rangle$ that is stable under all automorphisms τ_γ is a rational function. Indeed, z is a single-valued algebraic function, and therefore, z is a rational function.

We have thus proved the following theorem:

Theorem 5.17 *The monodromy group M of the function y acts on the field $R\langle y\rangle$ as a group of automorphisms. The invariant subfield with respect to this action is the field R of all rational functions.*

Combining Theorem 5.17 with the linear-algebraic results of Proposition 2.1 and Theorem 2.3, we obtain the following corollary.

Corollary 5.18 *An algebraic function with a solvable monodromy group is representable by radicals.*

This concludes the proof of the criterion for representability by radicals.

5.3 On the One-Dimensional Version of Topological Galois Theory

Not only algebraic functions have monodromy groups. Such groups are also defined for basic elementary functions and many more functions for which the Galois group does not make sense. For such functions, it is natural to use the monodromy group instead of the Galois group for proving that a function does not belong to a certain Liouville class. This approach is implemented in the topological version of Galois theory. Let us give an example that shows some difficulties that we need to overcome in this way. Consider the elementary function f defined by the following formula:

$$f(z) = \log\left(\sum_{j=1}^{n} \lambda_j \log(z - a_j)\right),$$

where a_j, $j = 1, \dots, n$, are distinct points on the complex line, and λ_j, $j = 1, \dots, n$, are complex constants. Let Λ denote the additive subgroup of complex numbers generated by the constants $\lambda_1, \dots, \lambda_n$. It is clear that if $n > 2$, then for almost every collection of constants $\lambda_1, \ldots, \lambda_n$, the group Λ is everywhere dense on the complex line.

Proposition 5.19 *If the group Λ is dense on the complex line, then the elementary function f has a dense set of logarithmic ramification points.*

Proof Let g_a be one of the germs of the function g defined by the formula

$$g(z) = \sum_{j=1}^{n} \lambda_j \log(z - a_j)$$

at a point $a \neq a_j$, $j = 1, \ldots, n$. A loop around the points $a_1, \ldots, a_n$ adds the number $2\pi i \lambda$ to the germ g_a, where λ is an element of the group Λ. Conversely, every germ $g_a + 2\pi i \lambda$, where $\lambda \in \Lambda$, can be obtained from the germ g_a by analytic continuation along some loop. Let U be a small neighborhood of the point a, and $G : U \to \mathbb{C}$ an analytic function whose germ at the point a is g_a. The image V of the domain U under the map $G : U \to \mathbb{C}$ is open. Therefore, in the domain V, there is a point of the form $2\pi i \lambda$, where $\lambda \in \Lambda$. The function $G - 2\pi i \lambda$ is one of the branches of the function g over the domain U, and the zero set of this branch in the domain U is nonempty. Hence, one of the branches of the function $f = \log g$ has a logarithmic ramification point in U. □

It is not hard to verify that under the assumptions of the proposition, the monodromy group of the function f has the cardinality of the continuum (this is not surprising: the fundamental group $\pi_1(S^2 \setminus A)$ obviously has the cardinality of the continuum, provided that A is a countable dense set in the Riemann sphere S^2).

One can also prove that the image of the fundamental group $\pi_1(S^2 \setminus \{A \cup b\})$ of the complement of the set $A \cup b$, where $b \notin A$, in the permutation group of branches of the function f is a proper subgroup of the monodromy group of f. (The fact that the removal of one extra point can change the monodromy group makes all proofs more complicated.)

Thus even the simplest elementary functions can have dense singular sets and monodromy groups of cardinality of the continuum. Nevertheless, the set of singular points of an elementary function is at most countable, and its monodromy group is solvable. If a function does not satisfy these restrictions, then it cannot be elementary. There exist similar topological obstructions to the membership of a function of one complex variable in other Liouvillian classes of functions.

We now proceed with a detailed description of this geometric approach to the problem of solvability.

5.4 Functions with at Most Countable Singular Sets

In this section, we define a broad class of functions of one complex variable needed in the construction of the topological version of Galois theory.

5.4.1 Forbidden Sets

We begin by defining the class of functions that we well be dealing with in the sequel. A multivalued analytic function of one complex variable is called an $\mathscr{S}$-*function* if the set of its singular points is at most countable. Let us make this definition more precise. Two regular germs f_a and g_b defined at points a and b of the Riemann sphere S^2 are called *equivalent* if the germ g_b is obtained from the germ f_a by regular (analytic) continuation along some path. Each germ g_b equivalent to the germ f_a is also called a regular germ of the multivalued analytic function f generated by the germ f_a.

A point $b \in S^2$ is said to be *singular* for the germ f_a if there exists a path $\gamma : [0, 1] \to S^2$, $\gamma(0) = a$, $\gamma(1) = b$ such that the germ has no regular continuation along this path but for every t, $0 \leq t < 1$, it admits a regular continuation along the truncated path $\gamma : [0, t] \to S^2$. It is easy to see that equivalent germs have the same set of singular points.

A regular germ is called an $\mathscr{S}$*-germ* if the set of its singular points is at most countable. A multivalued analytic function is called an $\mathscr{S}$-function if each of its regular germs is an $\mathscr{S}$-germ. We will need a lemma that allows us to free a plane path from a countable set by a small deformation.

Lemma 5.20 (On freeing a path from a countable set) *Let A be an at most countable subset of the plane of complex numbers, $\gamma : [0, 1] \to \mathbb{C}$ a continuous path, and φ a continuous positive function on the interval $0 < t < 1$. Then there exists a path $\hat{\gamma} : [0, 1] \to \mathbb{C}$ such that for $0 < t < 1$, we have $\hat{\gamma}(t) \notin A$ and $|\gamma(t) - \hat{\gamma}(t)| < \varphi(t)$.*

A high-tech proof of the lemma is as follows. In the functional space of paths $\overline{\gamma}$ close to the path γ, say satisfying the inequality $|\gamma(t) - \overline{\gamma}(t)| \leq \varphi(t)/2$, the paths avoiding one particular point of A form an open dense set. The intersection of countably many open dense sets in such functional spaces is nonempty (it is easy to see that the space is complete).

Let us give an elementary proof of the lemma (it carries over almost verbatim to a more general case in which the set A is uncountable but has zero Hausdorff length; cf. Sect. 5.8). Let us first construct a continuous broken line $\overline{\gamma}$ with infinitely many edges such that its vertices do not belong to A and $|\gamma(t) - \overline{\gamma}(t)| < \frac{1}{2}\varphi(t)$. Such a broken line can indeed be constructed, since the complement of the set A is dense. Let us show how to change each edge $[p, q]$ of the broken line $\overline{\gamma}$ to make it avoid the set A. Take an interval $[p, q]$. Let m be its perpendicular bisector. Consider broken lines with two edges $[p, b]$, $[b, q]$, where $b \in m$ and the point b is sufficiently close to the interval. These broken lines intersect at the endpoints p, q only, and their cardinality is that of the continuum. Therefore, there exists a broken line among them that does not intersect the set A. Changing each edge of the initial broken line in this way, we obtain the desired curve.

Besides the set of singular points, it is also convenient to consider other sets such that the function admits an analytic continuation everywhere in the complement. An at most countable set A is called a *forbidden set* for a regular germ f_a if the germ f_a admits a regular continuation along every path $\gamma(t)$, $\gamma(0) = a$, that never intersects the set A except possibly at the initial moment.

Theorem 5.21 (On forbidden sets) *An at most countable set is a forbidden set of a germ if and only if it contains the set of its singular points. In particular, a germ has a forbidden set if and only if it is a germ of an $\mathcal{S}$-function.*

Proof Suppose that there exists a singular point b of a germ f_a that does not lie in a forbidden set A of this germ. By definition, there must be a path $\gamma : [0, 1] \to S^2$, $\gamma(0) = a$, $\gamma(1) = b$, such that there is no regular continuation of the germ f_a along it, but the germ can be continued up to every $t < 1$. Without loss of generality, we can assume that the points a, b and the path $\gamma(t)$ lie in the finite part of the Riemann sphere, i.e., $\gamma(t) \neq \infty$ for $0 \leq t \leq 1$. Let $R(t)$ denote the radius of convergence of the series $f_{\gamma(t)}$ obtained by continuation of the germ f_a along the path $\gamma : [0, t] \to S^2$. The function $R(t)$ is continuous on the half-open interval $[0, 1)$. By the lemma, there exists a path $\hat{\gamma}(t)$, $\hat{\gamma}(0) = a$, $\hat{\gamma}(1) = b$, such that $|\gamma(t) - \hat{\gamma}(t)| < \frac{1}{3}R(t)$ and $\hat{\gamma}(t) \notin A$ for $t > 0$. By the assumption, the germ f_a admits a continuation along the path $\hat{\gamma}$ up to the point 1. But it follows easily that the germ f_a admits a continuation along the path γ. This contradiction proves that the singular set of the germ f_a is contained in every forbidden set of this germ. The converse statement (that a countable set containing the singular set of the germ is forbidden for the germ) is obvious. □

5.4.2 *The Class of $\mathcal{S}$-Functions Is Stable*

We now prove that the class of functions introduced above is stable under all natural operations.

Theorem 5.22 (On the stability of the class of $\mathcal{S}$-functions) *The class $\mathcal{S}$ of all $\mathcal{S}$-functions is stable under the following operations:*

1. *Differentiation: if $f \in \mathcal{S}$, then $f' \in \mathcal{S}$.*
2. *Integration: if $f \in \mathcal{S}$ and $g' = f$, then $g \in \mathcal{S}$.*
3. *Composition: if $g, f \in \mathcal{S}$, then $g \circ f \in \mathcal{S}$.*
4. *Meromorphic operations: if $f_i \in \mathcal{S}$, $i = 1, \dots, n$, the function $F(x_1, \dots, x_n)$ is a meromorphic function of n variables, and $f = F(f_1, \dots, f_n)$, then $f \in \mathcal{S}$.*
5. *Solving algebraic equations: if $f_i \in \mathcal{S}$, $i = 1, \dots, n$, and $f^n + f_1 f^{n-1} + \cdots + f_n = 0$, then $f \in \mathcal{S}$.*
6. *Solving linear differential equations: if $f_i \in \mathcal{S}$, $i = 1, \dots, n$, and $f^{(n)} + f_1 f^{(n-1)} + \cdots + f_n = 0$, then $f \in \mathcal{S}$.*

Proof 1 and 2. Let f_a, $a \neq \infty$, be the germ of an $\mathcal{S}$-function with a singular set A. If the germ f_a admits a regular continuation along some path γ lying in the finite part of the Riemann sphere, then the integral and the derivative of this germ admit a regular continuation along the path γ. Hence it suffices to take the set $A \cup \{\infty\}$ as a forbidden set for the integral and for the derivative of the germ f_a.

3. Let f_a and g_b be the germs of $\mathcal{S}$-functions with singular sets A and B, and $f_a(a) = b$. Let $f^{-1}(B)$ denote the full preimage of the set B under the multivalued correspondence generated by the germ f_a. In other words, $x \in f^{-1}(B)$ if and only if there exists a germ ψ_x equivalent to the germ f_a such that $\psi_x(x) \in B$. The set $f^{-1}(B)$ is at most countable. It suffices to take the set $A \cup f^{-1}(B)$ as a forbidden set of the germ $g_b \circ f_a$.
4. Let $f_{i,a}$ be the germs of $\mathcal{S}$-functions f_i at a point a in the plane of complex numbers, A_i their singular sets, and F a meromorphic function of n variables. We are assuming that the germs $f_{i,a}$ and the function F are such that the germ $f_a = F(f_{1,a}, \dots, f_{n,a})$ is a well-defined meromorphic germ. Replacing the point a by a nearby point if necessary, we can assume that the germ f_a is regular. If a path $\gamma(t)$ does not intersect the set $A = \bigcup A_i$ for $t > 0$, then the germ f_a admits a meromorphic continuation along this path. Define the set B as the projection of the poles of the multivalued function f generated by f_a onto the Riemann sphere. It suffices to take the set $A \cup B$ as a forbidden set of the germ.
5. Let $f_{i,a}$ be germs of $\mathcal{S}$-functions f_i at a, A_i their singular sets, and f_a a regular germ satisfying the equality

$$f_a^n + f_{1,a} \cdot f_a^{n-1} + \dots + f_{n,a} = 0.$$

   If a path $\gamma(t)$ does not intersect the set $A = \bigcup A_i$ for $t > 0$, then there exists a continuation of the germ f_a along that path. This continuation contains, in general, meromorphic and algebraic elements. Let B be the projection of the poles of the function f and the ramification points of its Riemann surface onto the Riemann sphere S^2. It suffices to take the set $A \cup B$ as a forbidden set for the germ f_a.
6. If the coefficients of the equation

$$f_a^{(n)} + f_{1,a} \cdot f_a^{(n-1)} + \dots + f_{n,a} = 0$$

   admit regular continuations along some path γ lying in the finite part of the Riemann sphere, then every solution f_a of this equation also admits a regular continuation along the path γ. Therefore, it suffices to take the set $A = \bigcup A_i \cup \{\infty\}$ as a forbidden set of the germ f_a, where A_i are the singular sets for the germs $f_{i,a}$.

□

Remark 5.23 Arithmetic operations and exponentiation are examples of meromorphic operations; hence the class of $\mathcal{S}$-functions is stable under the arithmetic operations and exponentiation.

Corollary 5.24 *If a multivalued function f can be obtained from single-valued $\mathcal{S}$-functions by integration, differentiation, meromorphic operations, compositions,* *and solutions of algebraic and linear differential equations, then the function f has at most a countable number of singular points. In particular, functions having uncountably many singular points cannot be expressed by generalized quadratures.*

5.5 Monodromy Groups

In this section, we discuss different notions related to the monodromy group.

5.5.1 Monodromy Group with a Forbidden Set

The monodromy group of an $\mathscr{S}$-function f with a forbidden set A is the group of all permutations of branches of f that correspond to loops around the points of A. We now give a precise definition.

Let F_a be the set of all germs of an $\mathscr{S}$-function f at a point a not lying in some forbidden set A. Take a closed path γ in $S^2 \setminus A$ that originates at the point a. The continuation of every germ in the set F_a along the path γ is another germ in the set F_a.

Thus, to every path γ, we assign a map of the set F_a to itself; moreover, homotopic paths in $S^2 \setminus A$ give rise to the same map. The composition of paths gives rise to the product of maps. This defines a homomorphism τ from the fundamental group of the set $S^2 \setminus A$ to the group $S(F_a)$ of invertible transformations of the set F_a. This homomorphism will be called the homomorphism of *A-monodromy*. The monodromy group of an $\mathscr{S}$-function f with a forbidden set A or, for short, the *group of A-monodromy* is, by definition, the image of the fundamental group $\pi_1(S^2 \setminus A, a)$ in the group $S(F_a)$ under the homomorphism τ.

Proposition 5.25 *The following properties hold.*

1. *The A-monodromy group of an $\mathscr{S}$-function is independent of the choice of the point a.*
2. *The A-monodromy group of an $\mathscr{S}$-function f acts transitively on the branches of f.*

Both claims can be easily proved with the help of Lemma 5.20. Let us give a proof of the second claim.

Proof Let $f_{1,a}$ and $f_{2,a}$ be any two germs of the function f at the point a. Since the germs $f_{1,a}$ and $f_{2,a}$ are equivalent, there exists a path γ such that under continuation along this path, the germ $f_{1,a}$ is transformed into the germ $f_{2,a}$. By Lemma 5.20, there exists an arbitrarily close path $\hat{\gamma}$ avoiding the set A. If the path $\hat{\gamma}$ is sufficiently close to the path γ, then the corresponding permutation of branches takes the germ $f_{1,a}$ to the germ $f_{2,a}$.[3] □

[3]This last point is proved similarly to Theorem 5.21.

5.5.2 Closed Monodromy Groups

The dependence of the A-monodromy group on the choice of the set A (see Sect. 5.3) suggests that we should introduce something like the Tychonoff topology (i.e., the direct product topology) on the permutation group of the branches. It turns out that the closure of the A-monodromy group with respect to this topology is already independent of the set A.

The group $S(M)$ of invertible transformations of the set M is equipped with the following topology. For every finite set $L \subset M$, define a neighborhood U_L of the identity as the collection of transformations $p \in S(M)$ such that $p(\ell) = \ell$ for $\ell \in L$. The neighborhoods of the form U_L, where L runs over all finite subsets of M, form a base of neighborhoods of the identity.

Lemma 5.26 (On the closure of the monodromy group) *The closure of the monodromy group of an $\mathcal{S}$-function f with a forbidden set A in the group $S(F)$ of all permutations of the branches of f is independent of the choice of forbidden set A.*

Proof Let A_1 and A_2 be two forbidden sets of the function f, and F_a the collection of branches of f at a point $a \notin A_1 \cup A_2$. Let $\Gamma_1, \Gamma_2 \subseteq S(F_a)$ be the monodromy groups of f with these forbidden sets. It suffices to show that for every permutation $\mu_1 \in \Gamma_1$ and for every finite set $L \subseteq F_a$, there exists a permutation $\mu_2 \in \Gamma_2$ such that $\mu_1|_L = \mu_2|_L$. Suppose that a path $\gamma \in \pi_1(S^2 \setminus A_1, a)$ gives rise to the permutation μ_1. Since the set L is finite, every path $\hat{\gamma} \in \pi_1(S^2 \setminus A_1, a)$ sufficiently close to the path γ gives rise to a permutation $\hat{\mu}_1$ that coincides with μ_1 on the set L, $\mu_1|_L = \hat{\mu}_1|_L$.[4] By Lemma 5.20, such a path $\hat{\gamma}$ can be chosen so that it does not intersect the set A_2. In this case, the permutation $\hat{\mu}_1$ lies in the group Γ_2. □

The lemma justifies the following definition.

Definition 5.27 The *closed monodromy group* of an $\mathcal{S}$-function f is the closure in the group $S(F)$ of the monodromy group of the function with any forbidden set A.

5.5.3 Transitive Action of a Group on a Set and the Monodromy Pair of an $\mathcal{S}$-Function

The monodromy group of a function f is not only an abstract group, but also a transitive permutation group of the branches of this function. In this subsection, we recall an algebraic description of transitive group actions.

An *action* of a group Γ on a set M is a homomorphism τ of the group Γ to the group $S(M)$. Two actions $\tau_1 : \Gamma \to S(M_1)$ and $\tau_2 : \Gamma \to S(M_2)$ are said to

[4]This is proved by a similar argument to that used in the proof of Theorem 5.21.

be *equivalent* if there exists a one-to-one correspondence $q : M_1 \to M_2$ such that $\overline{q} \circ \tau_1 = \tau_2$, where $\overline{q} : S(M_1) \to S(M_2)$ is the isomorphism induced by the map q.

The *stabilizer* Γ_a of a point $a \in M$ under the action τ is the subgroup consisting of all elements $\mu \in \Gamma$ such that $\tau\mu(a) = a$. The action τ is called *transitive* if for every pair of points $a, b \in M$, there exists an element $\mu \in \Gamma$ such that $\tau\mu(a) = b$. The following proposition is obvious:

Proposition 5.28 *The following properties hold:*

1. *An action τ of a group Γ is transitive if and only if the stabilizers of every two points $a, b \in M$ are conjugate. The image of the group Γ under a transitive action τ is isomorphic to the quotient group $\Gamma / \bigcap_{\mu\in\Gamma} \mu\Gamma_a\mu^{-1}$.*
2. *There exists a transitive action of the group Γ with a given stabilizer of some point, and this transitive action is unique up to equivalence.*

Thus transitive actions of a group Γ are described by pairs of groups. The pair of groups $[\Gamma, \Gamma_a]$, where Γ_a is the stabilizer of some point a under a transitive action τ of the group Γ, is called the *monodromy pair of the point a* with respect to the action τ. The group $\tau(\Gamma) \sim \Gamma / \bigcap_{\mu\in\Gamma} \mu\Gamma_a\mu^{-1}$ is called the *monodromy group of the pair* $[\Gamma, \Gamma_a]$.

The A-monodromy homomorphism τ gives rise to a transitive action of the fundamental group $\pi_1(S^2 \setminus A)$ on the set F_a of branches of the function f over the point a.

The monodromy pair of the germ f_a with respect to the action τ is called the *monodromy pair of the germ f_a with forbidden set A*. The monodromy pair of the germ f_a with respect to the action of the closed monodromy group is called the *closed monodromy pair of the germ f_a*. Different germs of the $\mathcal{S}$-function f have isomorphic monodromy pairs with forbidden set A; hence it makes sense to speak of the monodromy pair with the forbidden set A and the closed monodromy pair of the $\mathcal{S}$-function f. The closed monodromy pair of the $\mathcal{S}$-function f will be denoted by $[f]$.

5.5.4 Almost Normal Functions

A pair of groups $[\Gamma, \Gamma_0]$, $\Gamma_0 \subseteq \Gamma$, is called an *almost normal pair* if there exists a finite set $P \subset \Gamma$ such that

$$\bigcap_{\mu\in\Gamma} \mu\Gamma_0\mu^{-1} = \bigcap_{\mu\in P} \mu\Gamma_0\mu^{-1}.$$

Proposition 5.29 (On discrete actions) *The image $\tau(\Gamma)$ of the group Γ under a transitive action $\tau : \Gamma \to S(M)$ is a discrete subgroup of $S(M)$ if and only if the monodromy pair $[\Gamma, \Gamma_0]$ of some element $x_0 \in M$ is almost normal.*

Proof Let the group $\tau(\Gamma)$ be discrete. Let $\overline{P}$ denote a finite subset of the set M such that the neighborhood $U_{\overline{P}}$ of the identity contains no transformations of the group $\tau(\Gamma)$ different from the identity. This means that the intersection $\bigcap_{x \in \overline{P}} \Gamma_x$ of the stabilizers of points $x \in \overline{P}$ acts trivially on the set M, i.e., $\bigcap_{x \in \overline{P}} \Gamma_x \subseteq \bigcap_{\mu \in \Gamma} \mu \Gamma_0 \mu^{-1}$. The groups Γ_x are conjugate to the group Γ_0, and hence we can choose a finite set $P \subset \Gamma$ such that $\bigcap_{\mu \in P} \mu \Gamma_0 \mu^{-1} = \bigcap_{\mu \in \Gamma} \mu \Gamma_0 \mu^{-1}$. The converse statement can be proved similarly. □

An $\mathcal{S}$-function f is called *almost normal* if its monodromy group is discrete. From the lemma, it follows that the function f is almost normal if and only if its closed monodromy pair $[f]$ is almost normal.

A *differential rational function* of several functions is a rational function of those functions and their derivatives.

Lemma 5.30 (On finitely generated functions) *Suppose that every germ of an $\mathcal{S}$-function f over the point a is a differential rational function of finitely many fixed germs of f over a. Then the function f is almost normal.*

Indeed, if under continuation along a closed path, the specified germs of the function are unchanged, then a differential rational function of them is also unchanged.

From Lemma 5.30, on finitely generated functions, it follows that every solution of a linear differential equation with rational coefficients is an almost normal function. The same is also true for many other functions appearing naturally in differential algebra.

5.5.5 Classes of Group Pairs

In Sect. 5.6, we will describe how closed monodromy pairs of functions are transformed under composition, integration, differentiation, etc. To this end, we will need to introduce some notions concerning pairs of groups (group pairs).

A *group pair* always means a pair consisting of a group and a subgroup of it. We will identify a group with the group pair consisting of the group and its trivial subgroup.

Definition 5.31 A collection $\mathcal{L}$ of group pairs will be referred to as an *almost complete class of group pairs* if the following conditions are satisfied:

1. For every group pair $[\Gamma, \Gamma_0] \in \mathcal{L}$, $\Gamma_0 \subseteq \Gamma$, and every homomorphism $\tau : \Gamma \to G$, where G is any group, the group pair $[\tau\Gamma, \tau\Gamma_0]$ is also contained in $\mathcal{L}$.
2. For every group pair $[\Gamma, \Gamma_0] \in \mathcal{L}$, $\Gamma_0 \subseteq \Gamma$, and every homomorphism $\tau : G \to \Gamma$, where G is any group, the group pair $[\tau^{-1}\Gamma, \tau^{-1}\Gamma_0]$ is also contained in $\mathcal{L}$,
3. For every group pair $[\Gamma, \Gamma_0] \in \mathcal{L}$, $\Gamma_0 \subseteq \Gamma$, and a group G equipped with a T_2-topology and containing the group $\Gamma \subseteq G$, the group pair $[\overline{\Gamma}, \overline{\Gamma_0}]$ is

also contained in $\mathscr{L}$, where $\overline{\Gamma}, \overline{\Gamma}_0$ are the closures of the groups Γ, Γ_0 in the group G.

Definition 5.32 An almost complete class $\mathscr{M}$ of group pairs will be called a *complete class of group pairs* if the following conditions are satisfied:

1. For every group pair $[\Gamma, \Gamma_0] \in \mathscr{M}$ and a group Γ_1, $\Gamma_0 \subseteq \Gamma_1 \subseteq \Gamma$, the group pair $[\Gamma, \Gamma_1]$ is also contained in $\mathscr{M}$.
2. For every two group pairs $[\Gamma, \Gamma_1]$, $[\Gamma_1, \Gamma_2] \in \mathscr{M}$, the group pair $[\Gamma, \Gamma_2]$ is also contained in $\mathscr{M}$.

The minimal almost complete and complete classes of group pairs containing a fixed set $\mathscr{B}$ of group pairs are denoted respectively by $\mathscr{L}\langle\mathscr{B}\rangle$ and $\mathscr{M}\langle\mathscr{B}\rangle$.

Lemma 5.33 *The following properties hold:*

1. *If the monodromy group of the pair $[\Gamma, \Gamma_0]$ is contained in some complete class $\mathscr{M}$ of pairs, then the pair $[\Gamma, \Gamma_0]$ is also contained in $\mathscr{M}$.*
2. *If an almost normal pair $[\Gamma, \Gamma_0]$ is contained in some complete class $\mathscr{M}$ of group pairs, then its monodromy group is also contained in $\mathscr{M}$.*

Let us give a proof of the second statement. Let Γ_i, $i = 1, \dots, n$, be a finite number of subgroups conjugate to Γ_0 such that $\bigcap_{i=1}^n \Gamma_i = \bigcap_{\mu\in\Gamma} \mu\Gamma_0\mu^{-1}$. The pair $[\Gamma, \Gamma_i]$ is isomorphic to the pair $[\Gamma, \Gamma_0]$, whence $[\Gamma, \Gamma_i] \in \mathscr{M}$. Let $\tau : \Gamma_2 \to \Gamma$ be the inclusion homomorphism. Then $\tau^{-1}(\Gamma_1) = \Gamma_2 \bigcap \Gamma_1$, whence $[\Gamma_2, \Gamma_2 \bigcap \Gamma_1] \in \mathscr{M}$. The class $\mathscr{M}$ contains the pairs $[\Gamma, \Gamma_2]$ and $[\Gamma_2, \Gamma_2 \bigcap \Gamma_1]$. Therefore, $[\Gamma, \Gamma_1 \bigcap \Gamma_2] \in \mathscr{M}$. Continuing this argument, we obtain that the class $\mathscr{M}$ contains the pair $[\Gamma, \bigcap_{i=1}^n \Gamma_i]$ and together with it, the group $\Gamma / \bigcap_{\mu\in\Gamma} \mu\Gamma_0\mu^{-1}$.

Proposition 5.34 *An almost complete class of pairs $\mathscr{L}$ contains the closed monodromy pair $[f]$ of an $\mathscr{S}$-function f if and only if this class contains the monodromy pair of the function f with a forbidden set A.*

Proof Let $[\Gamma, \Gamma_0]$ be the monodromy pair of the function f with forbidden set A. Then $[f] = [\overline{\Gamma}, \overline{\Gamma}_0]$. Hence every almost complete class $\mathscr{L}$ containing the pair $[\Gamma, \Gamma_0]$ also contains the pair $[f]$. Conversely, if $[\overline{\Gamma}, \overline{\Gamma}_0]$ is contained in the class $\mathscr{L}$, then $[\Gamma, \Gamma_0] \in \mathscr{L}$. Indeed, the topology on the permutation group is such that $\Gamma_0 = \Gamma \bigcap \overline{\Gamma}_0$. Hence, the pair $[\Gamma, \Gamma_0]$ is the preimage of the pair $[\overline{\Gamma}, \overline{\Gamma}_0]$ under the inclusion of the group Γ in its closure. □

5.6 The Main Theorem

Here we formulate and prove the main theorem of the topological version of Galois theory.

Theorem 5.35 *The class of $\mathscr{S}$-functions $\hat{\mathscr{M}}$ consisting of all $\mathscr{S}$-functions whose closed monodromy pairs lie in some complete class $\mathscr{M}$ of pairs is stable under* *differentiation, composition, and meromorphic operations. Furthermore, if the class $\mathscr{M}$ contains the additive group $\mathbb{C}$ of complex numbers, then the class $\hat{\mathscr{M}}$ is stable under integration. If it contains the permutation group $S(k)$ of k elements, then the class $\hat{\mathscr{M}}$ is stable under solving algebraic equations of degree at most k.*

The proof of the main theorem consists of the following lemmas.

Lemma 5.36 (On derivatives) *For every $\mathscr{S}$-function f, the following inclusion holds: $[f'] \in \mathscr{M}\langle[f]\rangle$.*

Proof Let A be the set of singular points of the $\mathscr{S}$-function f, and f_a a germ of the function f at a nonsingular point a. We let Γ denote the fundamental group $\pi_1(S^2 \setminus A, a)$, and Γ_1 and Γ_2 the stabilizers of the germs f_a and f'_a. The group Γ_1 is contained in the group Γ_2. Indeed, under continuation along a path $\gamma \in \Gamma_1$, the germ f_a remains unchanged, and hence its derivative is also unchanged. From the definition of a complete class of pairs, it follows that $[\Gamma, \Gamma_2] \in \mathscr{M}\langle[\Gamma, \Gamma_1]\rangle$. Using Proposition 5.34, we obtain that $[f'] \in \mathscr{M}\langle[f]\rangle$. □

Lemma 5.37 (On composition) *For every pair of $\mathscr{S}$-functions f and g, the following inclusion holds: $[g \circ f] \in \mathscr{M}\langle[f], [g]\rangle$.*

Proof Let A and B be the sets of singular points of the functions f and g. Let $f^{-1}(B)$ be the full preimage of the set B under the multivalued function f. Set $Q = A \bigcup f^{-1}(B)$. Let f_a be any germ of the function f at the point $a \notin Q$, and g_b any germ of the function g at the point $b = f(a)$. The set Q is forbidden for the germ $g_b \circ f_a$. Let Γ denote the fundamental group $\pi_1(S^2 \setminus Q, a)$; let Γ_1 and Γ_2 denote the stabilizers of the germs f_a and $g_b \circ f_a$. We will write G for the fundamental group $\pi_1(S^2 \setminus B, b)$ and G_0 for the stabilizer of the germ g_b.

We now define a homomorphism $\tau : \Gamma_1 \to G$. To each path γ whose homotopy class belongs to Γ_1 (abusing notation, we will sometimes write $\gamma \in \Gamma_1$), we assign the path $\tau(\gamma)(t) = f_{\gamma(t)}(\gamma(t))$, where $f_{\gamma(t)}$ is the germ obtained by continuation of the germ f_a along the path γ up to the point t. The paths $\tau(\gamma)$ are closed, since under continuation along γ, the germ f_a remains unchanged. A homotopy of the path γ in the set $S^2 \setminus Q$ gives rise to a homotopy of the path $\tau \circ \gamma$ in the set $S^2 \setminus B$, since $f^{-1}(B) \subseteq Q$. Therefore, the homomorphism is well defined.

The germ $g_b \circ f_a$ is unchanged under continuation along the paths from the group $\tau^{-1}(G_0)$, or in other words, $\tau^{-1}(G_0) \subseteq \Gamma_2$. The lemma now follows. Indeed, we obtain the inclusions $\Gamma \supseteq \Gamma_2 \supseteq \tau^{-1}(G_0) \subseteq \tau^{-1}(G) = \Gamma_1 \subseteq \Gamma$, which imply that $[\Gamma, \Gamma_2] \in \mathscr{M}\langle[G, G_0], [\Gamma, \Gamma_1]\rangle$. From Proposition 5.34, we obtain that $[g \circ f] \in \mathscr{M}\langle[f], [g]\rangle$. □

Lemma 5.38 (On integrals) *For every $\mathscr{S}$-function f, the following inclusion holds: $[\int f(x)dx] \in \mathscr{M}\langle[f], \mathbb{C}\rangle$, where $\mathbb{C}$ is the additive group of complex numbers.*

Proof Let A be the set of singular points of the function f, and let $Q = A \bigcup\{\infty\}$. Let f_a be any germ of the function f at a point $a \notin Q$, and g_a a germ of $\int f(x)\,dx$ at that point: $g'_a = f_a$. We can take the set Q as a forbidden set for the germs f_a

and g_a. Let Γ denote the fundamental group $\pi_1(S^2 \setminus Q, a)$; let Γ_1 and Γ_2 denote the stabilizers of the germs f_a and g_a.

We now define a homomorphism $\tau : \Gamma_1 \to \mathbb{C}$. To each path $\gamma \in \Gamma_1$, assign the number $\int_\gamma f_{\gamma(t)}(\gamma(t))\,dx$, where $f_{\gamma(t)}$ is the germ obtained by continuation of the germ f_a along the path γ up to the point t, and $x = \gamma(t)$. The stabilizer Γ_2 of the germ g_a coincides with the kernel of the homomorphism τ, which implies that $[\Gamma, \Gamma_2] \in \mathscr{M}\langle[\Gamma, \Gamma_1], \mathbb{C}\rangle$. From Proposition 5.34, we obtain that $[\int f(x)dx] \in \mathscr{M}\langle[f], \mathbb{C}\rangle$. □

In the sequel, it will be convenient to use vector functions. The definitions of a forbidden set, an $\mathscr{S}$-function, and the monodromy group carry over automatically to vector functions.

Lemma 5.39 (On vector functions) *For every vector $\mathscr{S}$-function $\mathbf{f} = (f_1, \dots, f_n)$, the following equality holds:*

$$\mathscr{M}\langle[\mathbf{f}]\rangle = \mathscr{M}\langle[f_1], \dots, [f_n]\rangle.$$

Proof Let A_i be the sets of singular points of the functions f_i. The set of singular points of the vector function $\mathbf{f}$ is the set $Q = \bigcup A_i$. Let $\mathbf{f}_a = (f_{1,a}, \dots, f_{n,a})$ be any germ of the vector function $\mathbf{f}$ at a point $a \notin Q$. Let Γ denote the fundamental group $\pi_1(S^2 \setminus Q, a)$, let Γ_i denote the stabilizer of the germ $f_{i,a}$, and let Γ_0 denote the stabilizer of the vector germ $\mathbf{f}_a$. The stabilizer Γ_0 is exactly $\bigcap_{i=1}^n \Gamma_i$, which implies that

$$\mathscr{M}\langle[\Gamma, \Gamma_0]\rangle = \mathscr{M}\langle[\Gamma, \Gamma_1], \dots, [\Gamma, \Gamma_n]\rangle.$$

From Proposition 5.34, we obtain that

$$\mathscr{M}\langle[\mathbf{f}]\rangle = \mathscr{M}\langle[f_1], \dots, [f_n]\rangle.$$

□

Lemma 5.40 (On meromorphic operations) *For every vector $\mathscr{S}$-function $\mathbf{f} = (f_1, \dots, f_n)$ and meromorphic function $F(x_1, \dots, x_n)$ such that the function $F \circ \mathbf{f}$ is defined, the following inclusion holds: $[F \circ \mathbf{f}] \in \mathscr{M}\langle[\mathbf{f}]\rangle$.*

Proof Let A be the set of singular points of the function $\mathbf{f}$, and B the projection of the set of singular points of the function $F \circ \mathbf{f}$ to the Riemann sphere. We can take the set $Q = A \cup B$ as a forbidden set for the functions $F \circ \mathbf{f}$ and $\mathbf{f}$. Let $\mathbf{f}_a$ be any germ of the function $\mathbf{f}$ at a point $a \notin Q$. Let Γ denote the fundamental group $\pi_1(S^2 \setminus Q, a)$; let Γ_1 and Γ_2 denote the stabilizers of the germs $\mathbf{f}_a$ and $F \circ \mathbf{f}_a$. The group Γ_2 is contained in the group Γ_1. Indeed, under continuation along any path $\gamma \in \Gamma_1$, the vector function $\mathbf{f}$ remains unchanged, and therefore the meromorphic function of $\mathbf{f}$ is also unchanged. From the inclusion $\Gamma_2 \subseteq \Gamma_1$, it follows that $[\Gamma, \Gamma_2] \in \mathscr{M}\langle[\Gamma, \Gamma_1]\rangle$. From Proposition 5.34, we obtain that

$$[F \circ \mathbf{f}] \in \mathscr{M}\langle[\mathbf{f}]\rangle.$$

□

Lemma 5.41 (On algebraic functions) *For every vector $\mathscr{S}$-function $\mathbf{f} = (f_1, \dots, f_n)$ and algebraic function y in $\mathbf{f}$ defined by the equation*

$$y^k + f_1 y^{k-1} + \cdots + f_k = 0, \tag{5.2}$$

the following inclusion holds: $[y] \in \mathscr{M}\langle[\mathbf{f}], S(k)\rangle$, where $S(k)$ is the group of permutations on k elements.

Proof Let A be the set of singular points of the function $\mathbf{f}$, and B the projection of the set of algebraic ramification points of the function y to the Riemann sphere. We can take the set $Q = A \cup B$ as a forbidden set for the functions y and $\mathbf{f}$. Let y_a and $\mathbf{f}_a$ be any germs of the functions y and $\mathbf{f}$ at a point $a \notin Q$ that are related by the equality

$$y_a^k + f_{1,a} y_a^{k-1} + \cdots + f_{k,a} = 0.$$

Let Γ denote the fundamental group $\pi_1(S^2 \setminus Q, a)$; let Γ_1 and Γ_2 denote the stabilizers of the germs $\mathbf{f}_a$ and y_a. The coefficients of (5.2) are unchanged under continuation along every path $\gamma \in \Gamma_1$. Therefore, under continuation along the path γ, the roots of (5.2) are permuted. Thus we have a homomorphism τ of the group Γ_1 to the group $S(k)$, that is, $\tau : \Gamma_1 \to S(k)$. The group Γ_2 is contained in the kernel of the homomorphism τ, which implies that $[\Gamma, \Gamma_2] \in \mathscr{M}\langle[\Gamma, \Gamma_1], S(k)\rangle$. From Proposition 5.34, we obtain that

$$[y] \in \mathscr{M}\langle[f], S(k)\rangle.$$

□

This concludes the proof of the main theorem.

5.7 Group-Theoretic Obstructions to Representability by Quadratures

In this section, we compute the classes of group pairs that appear in the main theorem and formulate a necessary condition for the representability of functions by quadratures, k-quadratures, and generalized quadratures.

5.7.1 Computation of Some Classes of Group Pairs

The main theorem motivates the following problems: describe the minimal class of group pairs containing the additive group $\mathbb{C}$ of complex numbers; describe the minimal classes of group pairs containing, respectively, the group $\mathbb{C}$ and all finite groups, or the group $\mathbb{C}$ and the group $S(k)$. In this subsection, we give solutions to these problems.

Proposition 5.42 *The minimal complete class of pairs $\mathscr{M}\langle\mathscr{L}_\alpha\rangle$ containing given almost complete classes of pairs $\mathscr{L}_\alpha$ consists of the group pairs $[\Gamma, \Gamma_0]$ that admit a chain of subgroups $\Gamma = \Gamma_1 \supseteq \cdots \supseteq \Gamma_m \subseteq \Gamma_0$ such that for every i, $1 \le i \le m-1$, the group pair $[\Gamma_i, \Gamma_{i+1}]$ is contained in some almost complete class $\mathscr{L}_{\alpha(i)}$.*

To prove this, it suffices to verify that the group pairs $[\Gamma, \Gamma_0]$ satisfying the conditions of the proposition belong to the complete class $\mathscr{M}\langle\mathscr{L}_\alpha\rangle$ and form a complete class of pairs. Both statements follow immediately from the definitions. It is also easy to verify the following propositions.

Proposition 5.43 *The collection of group pairs $[\Gamma, \Gamma_0]$ such that Γ_0 is a normal subgroup of the group Γ and the group Γ/Γ_0 is abelian is the minimal almost complete class of pairs $\mathscr{L}\langle\mathscr{A}\rangle$ containing the class $\mathscr{A}$ of all abelian groups.*

Proposition 5.44 *The collection of group pairs $[\Gamma, \Gamma_0]$ such that Γ_0 is a normal subgroup of the group Γ and the group Γ/Γ_0 is finite is the minimal almost complete class of pairs $\mathscr{L}\langle\mathscr{K}\rangle$ containing the class $\mathscr{K}$ of all finite groups.*

Proposition 5.45 *The collection of group pairs $[\Gamma, \Gamma_0]$ such that* $\mathrm{ind}(\Gamma, \Gamma_0) \le k$ *is an almost complete class of group pairs.*

The class of group pairs from Proposition 5.45 will be denoted by $\mathscr{L}\langle\mathrm{ind} \le k\rangle$. Proposition 5.45 is of interest to us in connection with the characteristic property of subgroups in the group $S(k)$, Lemma 2.52.

A chain of subgroups Γ_i, $i = 1, \ldots, m$, $\Gamma = \Gamma_1 \supseteq \cdots \supseteq \Gamma_m \subseteq \Gamma_0$ is called a *normal tower of the group pair* $[\Gamma, \Gamma_0]$ if the group Γ_{i+1} is a normal subgroup of the group Γ_i for every $i = 1, \ldots, m-1$. The collection of quotient groups Γ_i/Γ_{i+1} is called the *collection of divisors with respect to the normal tower.*

Theorem 5.46 (On the classes of pairs $\mathscr{M}\langle\mathscr{A}, \mathscr{K}\rangle$, $\mathscr{M}\langle\mathscr{A}, S(k)\rangle$, and $\mathscr{M}\langle\mathscr{A}\rangle$)

1. *A group pair $[\Gamma, \Gamma_0]$ belongs to the minimal complete class $\mathscr{M}\langle\mathscr{A}, \mathscr{K}\rangle$ containing all finite groups and abelian groups if and only if it has a normal tower such that each divisor in the tower is either a finite group or an abelian group.*
2. *A group pair $[\Gamma, \Gamma_0]$ belongs to the minimal complete class $\mathscr{M}\langle\mathscr{A}, S(k)\rangle$ containing the group $S(k)$ and all abelian groups if and only if it has a normal tower such that each divisor in this tower is either a subgroup of the group $S(k)$ or an abelian group.*
3. *A group pair $[\Gamma, \Gamma_0]$ belongs to the minimal complete class $\mathscr{M}\langle\mathscr{A}\rangle$ containing all abelian groups if and only if the monodromy group of this pair is solvable.*

Proof The first claim of the theorem follows from the description of the classes $\mathscr{L}\langle\mathscr{A}\rangle$ and $\mathscr{L}\langle\mathscr{K}\rangle$ given in Propositions 5.43 and 5.44 and from Proposition 5.42.

To prove the second claim, consider the minimal complete class of group pairs containing the classes $\mathscr{L}\langle\mathscr{A}\rangle$ and $\mathscr{L}\langle\text{ind} \le k\rangle$. This class consists of group pairs $[\Gamma, \Gamma_0]$ that admit a chain of subgroups $\Gamma = \Gamma_1 \supseteq \cdots \supseteq \Gamma_m \subseteq \Gamma_0$ such that for every i, $1 \le i \le m-1$, either the group Γ_i/Γ_{i+1} is abelian, or we have $\text{ind}(\Gamma_i, \Gamma_{i+1}) \le k$ (see Propositions 5.44, 5.45, and 5.43). The class of group pairs just described contains the group $S(k)$ (see Lemma 2.55) together with all abelian groups, and it is obviously the minimal complete class of pairs possessing these properties. It remains only to reformulate the answer.

We can gradually transform the chain of subgroups $\Gamma = \Gamma_1 \supseteq \cdots \supseteq \Gamma_m \subseteq \Gamma_0$ into a normal tower for the pair $[\Gamma, \Gamma_0]$. Suppose that for $j < i$, the group Γ_{j+1} is a normal subgroup of the group Γ_j, and $\text{ind}(\Gamma_i, \Gamma_{i+1}) \le k$. Let $\overline{\Gamma}_{i+1}$ denote the maximal normal subgroup of the group Γ_i contained in the group Γ_{i+1}. It is clear that the quotient group $\Gamma_i/\overline{\Gamma}_{i+1}$ is a subgroup of the group $S(k)$. Instead of the initial chain of subgroups, consider the chain $\Gamma = G_1 \supseteq \cdots \supseteq G_m = \Gamma_0$ such that $G_j = \Gamma_j$ for $j \le i$ and $G_j = \Gamma_j \bigcap \overline{\Gamma}_{i+1}$ for $j > i$. Continuing this process (for at most m steps), we will pass from the initial chain of subgroups to a normal tower, thus obtaining a description of the class $\mathscr{M}\langle\mathscr{A}, S(k)\rangle$ in the desired terms.

We now prove claim 3. According to Propositions 5.43 and 5.44, the group pair $[\Gamma, \Gamma_0]$ belongs to the class $\mathscr{M}\langle\mathscr{A}\rangle$ if and only if there exists a chain $\Gamma = \Gamma_1 \supseteq \cdots \supseteq \Gamma_m \subseteq \Gamma_0$ such that Γ_i/Γ_{i+1} are abelian groups. Consider a chain of groups $\Gamma = G^1 \supseteq \cdots \supseteq G^m$ such that the group G^{i+1} is the commutator of the group G^i for $i = 1, \ldots, m-1$. Every automorphism of the group Γ takes the chain of groups G^i to itself; hence each group G^i is a normal subgroup of the group Γ. Induction on i shows that $G^i \subseteq \Gamma_i$ and in particular, $G^m \subseteq \Gamma_m \subseteq \Gamma_0$. The group G^m is a normal subgroup of the group Γ, and since $G^m \subseteq \Gamma_0$, we have $G^m \subseteq \bigcap_{\mu\in\Gamma} \mu\Gamma_0\mu^{-1}$. By definition of the chain G^i, the group Γ/G^m is solvable. The group $\Gamma/\bigcap_{\mu\in\Gamma}\mu\Gamma_0\mu^{-1}$ is solvable as a quotient group of the group Γ/G^m. The converse statement (a pair of groups with a solvable monodromy group lies in the class $\mathscr{M}\langle\mathscr{A}\rangle$) is obvious. □

Proposition 5.47 *Every abelian group Γ whose cardinality is at most the cardinality of the continuum belongs to the class $\mathscr{L}\langle\mathbb{C}\rangle$.*

Proof The set of complex numbers $\mathbb{C}$ is a vector space over the rational numbers whose dimension is the cardinality of the continuum. Let $\{e_\alpha\}$ be a basis of this space. The subgroup $\tilde{\mathbb{C}}$ of the group $\mathbb{C}$ spanned by the numbers $\{e_\alpha\}$ is a free abelian group with the number of generators equal to the cardinality of the continuum. Every abelian group Γ whose cardinality is at most the cardinality of the continuum is a quotient group of the group $\tilde{\mathbb{C}}$, and therefore, $\Gamma \in \mathscr{L}\langle\mathbb{C}\rangle$. □

From Proposition 5.47 and from the computation of the classes $\mathscr{M}\langle\mathscr{A}, \mathscr{K}\rangle$, $\mathscr{M}\langle S(n)\rangle$, and $\mathscr{M}\langle\mathscr{A}\rangle$, it follows that a pair of groups $[\Gamma, \Gamma_0]$ for which the cardinality of Γ is at most the cardinality of the continuum belongs to the classes $\mathscr{M}\langle\mathbb{C}, \mathscr{K}\rangle$, $\mathscr{M}\langle\mathbb{C}, S(n)\rangle$, and $\mathscr{M}\langle\mathbb{C}\rangle$ if and only if it belongs to the classes $\mathscr{M}\langle\mathscr{A}, \mathscr{K}\rangle$, $\mathscr{M}\langle\mathscr{A}, S(n)\rangle$, and $\mathscr{M}\langle\mathscr{A}\rangle$.

This result suffices for our purposes, since the permutation group of the branches of a function has at most the cardinality of the continuum.

Lemma 5.48 *A free nonabelian group Λ does not belong to the class $\mathscr{M}\langle\mathscr{A},\mathscr{K}\rangle$.*

Proof Suppose that $\Lambda \in \mathscr{M}\langle\mathscr{A},\mathscr{K}\rangle$, i.e., Λ has a normal tower $\Lambda = \Gamma_1 \supseteq \cdots \supseteq \Gamma_m = e$ such that each divisor in this tower is a finite group or an abelian group. Each group Γ_i is free as a subgroup of a free group (see [68]). The group $\Gamma_m = e$ is abelian. Let Γ_{i+1} be the abelian group with the smallest index. For any elements $a, b \in \Gamma_i$, there exists a nontrivial relation: if Γ_i/Γ_{i+1} is abelian, then, for example, elements $aba^{-1}b^{-1}$ and $ab^2a^{-1}b^{-2}$ commute; if Γ_i/Γ_{i+1} is finite, then some powers a^p, b^p of the elements a, b commute. Therefore, the group Γ_i has at most one generator, and it is therefore abelian. This contradiction proves that $\Lambda \notin \mathscr{M}\langle\mathscr{A},\mathscr{K}\rangle$. □

Lemma 5.49 *For $k > 4$, the symmetric group $S(k)$ does not belong to the class $\mathscr{M}\langle\mathbb{C}, S(k-1)\rangle$.*

Proof For $k > 4$, the alternating group $A(k)$ is simple and nonabelian. For this group, the criterion of being in the class $\mathscr{M}\langle\mathbb{C}, S(k-1)\rangle$ obviously fails. Therefore, the symmetric group $S(k)$ for $k > 4$ does not belong to the class $\mathscr{M}\langle\mathbb{C}, S(k-1)\rangle$. □

Lemma 5.50 *The only transitive group of permutations on k elements generated by transpositions is the symmetric group $S(k)$.*

Proof Let a group Γ be a transitive permutation group generated by transpositions of the set M with k elements. A subset $M_0 \subseteq M$ is said to be *complete* if every permutation of the set M_0 extends to some permutation of the set M from the group Γ. Complete subsets exist. For example, two elements of the set M that are interchanged by a basis transposition form a complete subset. Take a complete subset M_0 of maximal cardinality. Suppose that $M_0 \neq M$. Then by the induction hypothesis, the restriction of Γ to M_0 coincides with $S(M_0)$. Since the group Γ is transitive, there exists a basis transposition μ interchanging some elements $a \notin M_0$ and $b \in M_0$. The permutation group generated by the transposition μ and the group $S(M_0)$ is the group $S(M_0 \cup \{a\})$. The set $M_0 \cup \{a\}$ is complete and contains the set M_0. This contradiction proves that the group Γ is the group $S(M)$. □

5.7.2 Necessary Conditions for Representability by Quadratures, k-Quadratures, and Generalized Quadratures

The main theorem (see Sect. 5.6) together with the computation of classes of group pairs provides topological obstructions to the representability of functions by generalized quadratures, by k-quadratures, and by quadratures. In this subsection, we collect in one place all the information obtained thus far. Let us start with the definition of the class of functions representable by single-valued $\mathscr{S}$-functions and quadratures (k-quadratures, generalized quadratures). As in Sect. 1.2, we will define these classes by providing a list of basic functions and a list of admissible operations.

Functions Representable by Single-Valued $\mathscr{S}$-Functions and Quadratures

List of basic functions:

- Single-valued $\mathscr{S}$-functions.

List of admissible operations:

- Composition
- Meromorphic operations
- Differentiation
- Integration

Functions Representable by Single-Valued $\mathscr{S}$-Functions and k-Quadratures

This class of functions is defined in the same way. We only need to add the solutions of algebraic equations of degree $\leq k$ to the list of admissible operations.

Functions Representable by Single-Valued $\mathscr{S}$-Functions and Generalized Quadratures

This class of functions is defined in the same way. We have only to add the operation of solving algebraic equations to the list of admissible operations.

It is readily seen from the definition that the class of functions representable by single-valued $\mathscr{S}$-functions and quadratures (k-quadratures, generalized quadratures) contains the class of functions representable by quadratures (k-quadratures, generalized quadratures). It is clear that the classes of functions just defined are much broader than their classical analogues. Therefore, for example, the claim that a function f does not belong to the class of functions representable by single-valued $\mathscr{S}$-functions and quadratures is considerably stronger than the claim that f is not representable by quadratures.

Proposition 5.51 *The class of functions representable by single-valued $\mathscr{S}$-functions and quadratures (k-quadratures, generalized quadratures) is contained in the class of $\mathscr{S}$-functions.*

This proposition follows immediately from the theorem on the stability of the class of $\mathscr{S}$-functions (see Sect. 5.4.2).

Proposition 5.52 (A result on generalized quadratures) *The closed monodromy pair $[f]$ of a function f representable by generalized quadratures has a normal tower such that each divisor of this tower is either a finite group or an abelian group. Furthermore, this condition is fulfilled for the closed monodromy pair $[f]$ of every function f representable by single-valued $\mathcal{S}$-functions and generalized quadratures. If the function f is almost normal, then the monodromy group of the function $[f]$ also satisfies this condition.*

Proposition 5.53 (A result on k-quadratures) *The closed monodromy pair $[f]$ of a function f representable by k-quadratures has a normal tower such that each divisor of this tower is either a subgroup of the group $S(k)$ or an abelian group. Furthermore, this condition is fulfilled for the closed monodromy pair $[f]$ of every function f representable by single-valued $\mathcal{S}$-functions and k-quadratures. If the function f is almost normal, then the monodromy group of the function f also satisfies this condition.*

Proposition 5.54 (A result on quadratures) *The closed monodromy group of a function f representable by quadratures is solvable. Furthermore, the closed monodromy group of every function f representable by single-valued $\mathcal{S}$-functions and quadratures is solvable.*

To prove these results, it suffices to apply the main theorem to the classes $\hat{\mathcal{M}}\langle\mathbb{C},\mathcal{K}\rangle$, $\hat{\mathcal{M}}\langle\mathbb{C},S(k)\rangle$, and $\hat{\mathcal{M}}\langle\mathbb{C}\rangle$ of $\mathcal{S}$-functions and to use the computation of the classes $\mathcal{M}\langle\mathbb{C},\mathcal{K}\rangle$, $\mathcal{M}\langle\mathbb{C},S(k)\rangle$, and $\mathcal{M}\langle\mathbb{C}\rangle$.

Let us now give examples of functions not representable by generalized quadratures. Let the Riemann surface of a function f be the universal covering of the region $S^2 \setminus A$, where S^2 is the Riemann sphere and A is a finite set containing at least three points. Then the function f is not representable by single-valued $\mathcal{S}$-functions and generalized quadratures. Indeed, the function f is an almost normal function. The closed monodromy group of the function f is free and nonabelian, since the fundamental group of the region $S^2 \setminus A$ is free and nonabelian.

Example 5.55 Consider the function f that maps the upper half-plane conformally onto a triangle with zero angles bounded by arcs of circles. The function f is the inverse of the Picard modular function. The Riemann surface of the function f is the universal covering of the sphere with three punctures; hence the function f is not representable by single-valued $\mathcal{S}$-functions and generalized quadratures.

Note that the function f is closely related to the elliptic integrals

$$K_1(k) = \int_0^1 \frac{dx}{\sqrt{(1-x^2)(1-k^2x^2)}} \quad \text{and} \quad K_2(k) = \int_0^{1/k} \frac{dx}{\sqrt{(1-x^2)(1-k^2x^2)}}.$$

Each of the three functions K_1, K_2, and f can be expressed by quadratures through each function (see [38]). Hence *each of the integrals K_1 and K_2 is not representable by $\mathcal{S}$-functions and generalized quadratures.*

Example 5.55 admits a substantial generalization: in Sect. 6.3, all polygons bounded by arcs of circles are listed that are the images of the upper half-plane under functions representable by generalized quadratures.

Example 5.56 Let f be a k-valued algebraic function with simple ramification points located in different geometric points of the Riemann sphere. For $k > 4$, the function f is not representable by single-valued $\mathscr{S}$-functions and $(k-1)$-quadratures, compositions, and meromorphic operations. In particular, the function f is not representable by $(k-1)$-quadratures.

Indeed, a loop around a simple ramification point of the function f gives rise to a transposition of the set of branches of this function. The monodromy group of the function f is a transitive permutation group generated by transpositions, i.e., the group $S(k)$. For $k > 4$, the group $S(k)$ does not lie in the class $\mathscr{M}\langle\mathbb{C}, S(k-1)\rangle$.

In Chap. 7, the topological results on nonrepresentability of functions by quadratures (k-quadratures, and generalized quadratures) are generalized to the case of functions of several complex variables.

5.8 Classes of Singular Sets and a Generalization of the Main Theorem

In Sect. 5.4, we considered $\mathscr{S}$-functions, i.e., multivalued analytic functions of a complex variable whose singular sets are at most countable. Let $\mathscr{S}$ be the class of all at most countable subsets of the Riemann sphere S^2. Let us list the properties of the class $\mathscr{S}$ that we have actively used:

1. If $A \in \mathscr{S}$, then the set $S^2 \setminus A$ is dense and locally path connected.
2. There exists a nonempty set A such that $A \in \mathscr{S}$.
3. If $A \in \mathscr{S}$ and $B \subseteq A$, then $B \in \mathscr{S}$.
4. If $A_i \in \mathscr{S}$, $i = 1, 2, \ldots$, then $\bigcup_1^\infty A_i \in \mathscr{S}$.
5. Let U_1 and U_2 be open subsets of the sphere and $f : U_1 \to U_2$ an invertible analytic map. Then if $A \subseteq U_1$ and $A \in \mathscr{S}$, then $f(A) \in \mathscr{S}$.

A *complete class of sets* is a class of subsets of the Riemann sphere satisfying properties 1–5 above. A multivalued analytic function will be called a *Q-function* if its set of singular points lies in some complete class Q of sets. All definitions and theorems from Sect. 5.5 carry over to Q-functions. Thus, for example, we have the following version of the main theorem.

Theorem 5.57 (A version of the main theorem) *For every complete class Q of sets and every complete class $\mathscr{M}$ of group pairs, the class $\hat{\mathscr{M}}$ consisting of all Q-functions f such that $[f] \in \mathscr{M}$ is stable under differentiation, composition, and meromorphic operations. If in addition, $\mathbb{C} \in \mathscr{M}$, then the class $\hat{\mathscr{M}}$ of Q-functions is stable under the integration. Alternatively, if in addition, $S(k) \in \mathscr{M}$, then the class*

$\hat{\mathcal{M}}$ of Q-functions is stable under the operation of solving algebraic equations of degrees at most k.

Let us give an example of a complete class of sets. Let X_α be the set of all subsets of the Riemann sphere with zero Hausdorff measure of weight α. It is not hard to show that for $\alpha \leq 1$, the set X_α is a complete class of subsets of the sphere. Note that the new formulation of the main theorem allows us to strengthen all negative results. Consider, for example, the result of nonrepresentability of functions by quadratures. (The results of nonrepresentability by k-quadratures and by generalized quadratures can be generalized in the same way.) Define the following class of functions.

5.8.1 Functions Representable by Single-Valued X_1-Functions and Quadratures

List of basic functions:

- Single-valued X_1-functions

List of admissible operations:

- Composition
- Meromorphic operations
- Differentiation
- Integration

By the new formulation of the main theorem, a $\mathcal{S}$-function having an unsolvable monodromy group is not only unrepresentable by quadratures, but also unrepresentable by single valued X_1-functions and quadratures.

Corollary 5.58 *If a polygon G bounded by circle arcs satisfies none of the three integrability conditions (see Sect. 6.3), then the function f_G cannot be expressed by generalized quadratures, compositions, and meromorphic operations through single-valued X_1-functions.*

Chapter 6
Solvability of Fuchsian Equations

In this chapter, we discuss the "permissive" part of topological Galois theory. It is based on the following classical results: a simple linear-algebraic part of Picard–Vessiot theory and Frobenius's theorem (Theorem 6.2). To prove that a Fuchsian equation with a k-solvable monodromy group is solvable by k-quadratures, we also need to use standard Galois theory.

In Sect. 6.1, we construct solutions of Fuchsian linear differential equations with a solvable (almost solvable, k-solvable) monodromy group. In Sect. 6.2, we give explicit criteria for different kinds of solvability of linear Fuchsian systems with sufficiently small coefficients. In the course of construction, we use the theory of Lappo-Danilevsky. In Sect. 6.3, we classify all planar polygons G bounded by circular arcs for which the function f_G realizing the Riemann map of the upper half-plane onto the polygon G can be represented by generalized quadratures.

6.1 Picard–Vessiot Theory for Fuchsian Equations

In this section, we show that the topology of the covering of the complex plane by the Riemann surface of a generic solution of a Fuchsian linear differential equation is directly responsible for solvability of the equation in finite terms.

6.1.1 The Monodromy Group of a Linear Differential Equation and Its Connection with the Galois Group

Consider a linear differential equation

$$y^{(n)} + r_1 y^{(n-1)} + \cdots + r_n y = 0, \tag{6.1}$$

A. Khovanskii, *Topological Galois Theory*, Springer Monographs in Mathematics,
DOI 10.1007/978-3-642-38871-2_6

where the r_i are rational functions of a complex variable x. The poles of the rational functions r_i and the point ∞ are called *singular points* of (6.1). In a neighborhood of a nonsingular point x_0, the solutions of the equation form an n-dimensional vector space V^n. Take an arbitrary path γ in the complex plane going from a point x_0 to a point x_1 and avoiding the singular points a_i. The analytic continuations along this path of the solutions of the equation remain solutions of the equation. Hence to each path γ connecting the point x_0 with a point x_1, one can assign a linear map M_γ of the solution space $V^n_{x_0}$ at the point x_0 to the solution space $V^n_{x_1}$ at the point x_1. If the path γ is deformed so as to avoid the singular points while the endpoints remain fixed, then the map M_γ remains unchanged. Closed paths (loops) correspond to linear transformations of the space V^n. The collection of all such linear transformations of the space V^n forms a group, which is called the *monodromy group of equation* (6.1). Thus the monodromy group of an equation is a group of linear transformations of solutions that correspond to loops around singular points. The monodromy group of an equation characterizes the multivaluedness of its solutions.

Lemma 6.1 *The following statements hold:*

1. *The monodromy group of almost every solution of* (6.1) *is isomorphic to the monodromy group of this equation.*
2. *The monodromy pair of every solution of* (6.1) *is almost normal.*

Proof The second statement of the lemma follows from Sect. 5.5.4. Let us give a proof of the first statement. The monodromy group of (6.1) is a linear group containing at most a countable number of elements. For every element of this group but the identity, its set of fixed points is a proper subspace of the finite-dimensional space of solutions of (6.1). The set of solutions that are fixed under at least one nontrivial transformation from the monodromy group has measure zero in the space of solutions (since the union of at most a countable number of proper subspaces in a finite-dimensional space has measure zero in that space). Each of the remaining solutions of (6.1) has monodromy group isomorphic to the monodromy group of the equation. □

There exist n linearly independent solutions $y_1, \dots, y_n$ of (6.1) in a neighborhood of every nonsingular point x_0. In this neighborhood, we can consider the field of functions $\mathscr{R}\langle y_1, \dots, y_n\rangle$ obtained by adjoining all solutions y_i and all their derivatives to the field $\mathscr{R}$ of rational functions.

Every transformation M_γ in the monodromy group extends to an automorphism of the field $\mathscr{R}\langle y_1, \dots, y_n\rangle$. Indeed, together with the functions $y_1, \dots, y_n$, every element of the field $\mathscr{R}\langle y_1, \dots, y_n\rangle$ admits a meromorphic continuation along the path γ. This continuation provides the desired automorphism, since the arithmetic operations and differentiation are preserved under the analytic continuation, and rational functions return to their initial values because they are single-valued.

Thus the monodromy group of (6.1) embeds into the Galois group of this equation over the field of rational functions.

The invariant subfield of the monodromy group is a subfield of $\mathscr{R}\langle y_1, \dots, y_n\rangle$, a field consisting of single-valued functions. In contrast to algebraic equations, for differential equations, the invariant subfield with respect to the action of the monodromy group can be larger than the field of rational functions.

For example, for the differential equation (6.1) all of whose coefficients $r_i(x)$ are polynomials, all solutions are entire functions. But certainly, the solutions of such equations are not always polynomials. The reason is that solutions of differential equations can grow exponentially fast as they approach singular points. A broad class of linear differential equations is known for which there is no such complication, i.e., for which the solutions grow no faster than a power as they approach every singular point (in every sector with its apex at that point). Differential equations having this property are called *Fuchsian differential equations* (see [13, 40]). For Fuchsian differential equations, we have the following theorem of Frobenius.

Theorem 6.2 (Frobenius) *For Fuchsian differential equations, the subfield of the differential field $\mathscr{R}\langle y_1, \dots, y_n\rangle$ consisting of single-valued functions coincides with the field of rational functions.*

Before we prove Frobenius's theorem, let us state some immediate corollaries.

Corollary 6.3 *The algebraic closure of the monodromy group M (i.e., the minimal algebraic group containing M) of a Fuchsian equation coincides with the Galois group of that equation over the field of rational functions.*

Proof The corollary follows from Frobenius's theorem and the fundamental theorem of differential Galois theory (see Sect. 3.3). □

Theorem 6.4 *A Fuchsian linear differential equation is solvable by quadratures, by k-quadratures, or by generalized quadratures if and only if its monodromy group is respectively solvable, k-solvable, or almost solvable.*

This theorem follows from the Picard–Vessiot theorem (see Sect. 3.5) and the preceding corollary. Therefore, differential Galois theory gives us the following two results.

Proposition 6.5 *If the monodromy group of a Fuchsian differential equation is solvable (k-solvable, almost solvable), then this equation is solvable by quadratures (k-quadratures, generalized quadratures).*

Proposition 6.6 *If the monodromy group of a Fuchsian differential equation is unsolvable (not k-solvable, not almost solvable), then this equation is unsolvable by quadratures (k-quadratures, generalized quadratures).*

The first of these results does not use the fundamental theorem of Galois theory and belongs essentially to linear algebra. The reason is that a group of automorphisms of the differential field $\mathscr{R}\langle y_1, \dots, y_n\rangle$ that fix the field of rational functions only does not need to be specially constructed. The monodromy group is such a group. Hence for a proof of solvability of Fuchsian equations with solvable or with almost solvable monodromy groups by quadratures or by generalized quadratures, it suffices to use linear-algebraic arguments from Sect. 3.7. For a proof of solvability of Fuchsian equations with k-solvable monodromy groups by

k-quadratures, these linear-algebraic arguments are not enough. We also need to apply Galois theory to algebraic extensions of the field of rational functions. On the other hand, for such extensions, Galois theory can be made very intuitive and geometric (see Chap. 4).

Our theorem allows us to strengthen the second (negative) result. This will be discussed in Sect. 6.2.4. Let us now proceed with the proof of Frobenius's theorem.

6.1.2 Proof of Frobenius's Theorem

We will show that every single-valued function in the differential field $\mathscr{R}\langle y_1, \dots, y_n\rangle$ is meromorphic on the Riemann sphere and is therefore rational. Let $p \in \overline{\mathbb{C}}$ be a singular point of the Fuchsian equation, and x a local parameter at that point such that $x(p) = 0$. According to Fuchsian theory, a solution is represented at the point p as the finite sum $y = \sum f_{\alpha k} x^\alpha \log^k x$, where $f_{\alpha k}$ are meromorphic functions in a neighborhood of the point p, α are complex numbers, and k are integers. It is clear that functions representable in the form $\sum f_{\alpha k} x^\alpha \log^k x$, where the functions $f_{\alpha k}$ are meromorphic at the point p, form a differential ring containing the field of functions meromorphic at the point p.

We need to show that the ratio of 2 functions from this differential ring is a single-valued function near the point p if and only if this function is meromorphic at p. The proof of this fact is based on Proposition 6.7, stated below. We need the following notation: $U(0, \varepsilon)$ is the ε-neighborhood of the point 0 on the complex line; $\hat{U}(0, \varepsilon)$ is the punctured ε-neighborhood of the point 0, i.e., we have $\hat{U}(0, \varepsilon) = U(0, \varepsilon) \setminus \{0\}$; $M(0, \varepsilon)$ and $\hat{M}(0, \varepsilon)$ are the fields of meromorphic functions on the regions $U(0, \varepsilon)$ and $\hat{U}(0, \varepsilon)$. Two meromorphic germs f_a and g_b are called *equivalent over a domain* U, $a, b \in U$, if the germ g_b is obtained from the germ f_a by continuation along some path in the domain U.

We now define the ring $K_a(0, \varepsilon)$. A meromorphic germ f_a defined at a point $a \in \hat{U}(0, \varepsilon)$ belongs to the ring $K_a(0, \varepsilon)$ if the following conditions are satisfied:

1. The germ f_a admits meromorphic continuations along all paths in $\hat{U}(0, \varepsilon)$.
2. The complex vector space spanned by all meromorphic germs at the point a equivalent to the germ f_a over the neighborhood $\hat{U}(0, \varepsilon)$ is finite-dimensional.

The ring $K_a(0, \varepsilon)$ contains the field $\hat{M}(0, \varepsilon)$; hence it is a vector space over this field.

Proposition 6.7 (On bases) *For every choice of branches of the functions* $\log x$ *and* x^α, *with* $\Re\alpha$, *the real part of* α, *equal to zero, the germs* $x_a^\alpha \log_a^k x$, $k = 0, 1, \dots$, *form a basis of the space* $K_a(0, \varepsilon)$ *over the field* $\hat{M}(0, \varepsilon)$.

Let us first prove the following lemma.

Lemma 6.8 *The germs* $1, \log_a x, \dots, \log_a^k x, \dots$ *are linearly independent over the field* $\hat{M}(0, \varepsilon)$.

Proof The existence of a nontrivial relation $\sum a_k \log_a^k x = 0$, $a_k \in \hat{M}(0, \varepsilon)$, implies that the function $\log x$ is finite-valued in a neighborhood of zero, which is false. □

The proof of the proposition is based on consideration of the monodromy operator $A : K_a(0, \varepsilon) \to K_a(0, \varepsilon)$ that takes each germ to its continuation along a loop around the point 0.

Lemma 6.9 *Fix a complex number α whose real part is between 0 and 1. The germs $x_a^\alpha \log_a^k x$, $k = 0, 1, \ldots, n-1$, form a basis of the vector space $\ker(A - \lambda E)^n$, where λ and α are subject to the relation $\lambda = e^{2\pi i \alpha}$.*

Proof Note that the space $\ker(A - \lambda E)$ is at most one-dimensional. Indeed, if $A f_a = \lambda f_a$ and $A g_a = \lambda g_a$, then $A(f_a/g_a) = f_a/g_a$. Therefore, the germ $\psi_a = f_a/g_a$ is the germ of some function ψ from the field $\hat{M}(0, \varepsilon)$, and $f_a = \psi g_a$. Hence the dimension of the space $\ker(A - \lambda E)^n$ is at most n. On the other hand, it is easy to verify that the germs $x_a^\alpha \log_a^k x$, $[\Re \alpha] = 0$, $k = 0, 1, \ldots, n-1$, lie in this space. By Lemma 6.8, these germs are linearly independent, and therefore, they form a basis of the space $\ker(A - \lambda E)^n$. □

The spaces $\ker(A - \lambda E)^n$ with different values of λ do not intersect. Hence all the germs $x_a^\alpha \log_a^k x$ are linearly independent. Let us show that every germ f_a in the space $K_a(0, \varepsilon)$ can be represented as a linear combination of these functions. By definition, the germ f_a lies in some finite-dimensional space V invariant under the monodromy operator. Let $\tilde{A}$ be the restriction of the operator A to the space V. By linear algebra, the space V splits into a direct sum of the root subspaces $\ker(\tilde{A} - \lambda E)^{n_\lambda}$, where λ are eigenvalues of the operator $\tilde{A}$, and n_λ are their multiplicities. From Lemma 6.9, it follows that every element of the space V can be represented as a linear combination of the vectors $x_a^\alpha \log_a^k x$.[1]

Remark 6.10 Different choices of branches of the functions $\log x$ and x^α lead to different bases in the space $K_a(0, \varepsilon)$. The entries of the transition matrix from one such basis to another are complex numbers.

Definition 6.11 1. A meromorphic germ f_a, $a \in \hat{U}(0, \varepsilon)$, has an *entire Fuchsian singularity* over the neighborhood $\hat{U}(0, \varepsilon)$ if $f_a \in K_a(0, \varepsilon)$, and the coordinates of the germ f_a in the basis $x_a^\alpha \log_a^k x$ are meromorphic, i.e., if

$$f_a = \sum f_{\alpha,k} \log_a^k x \cdot x_a^\alpha, \quad \text{where } f_{\alpha,k} \in M(0, \varepsilon).$$

2. A meromorphic germ f_a, $a \in \hat{U}(0, \varepsilon)$, has a *Fuchsian singularity* over the neighborhood $\hat{U}(0, \varepsilon)$ if it is representable as a ratio of 2 germs ψ_a, g_a having entire Fuchsian singularities over $\hat{U}(0, \varepsilon)$, so that $f_a = \psi_a / g_a$.

[1]Note that the functions f and xf are proportional over $\hat{M}(0, \varepsilon)$; thus we can arrange that the α's have real parts between 0 and 1.

Corollary 6.12 *A germ $f_a \in K_a(0,\varepsilon)$ has a Fuchsian singularity over the neighborhood $\hat{U}(0,\varepsilon)$ if and only if it has an entire Fuchsian singularity over that neighborhood.*

Proof The germ f_a is in $K_a(0,\varepsilon)$, and therefore, $f_a = \sum r_{\alpha,k} x_a^\alpha \log_a^k x$, where $r_{\alpha,k} \in \hat{M}(0,\varepsilon)$ are the coordinates of the germ f_a in the basis. The germ f_a has a Fuchsian singularity, and hence the following equality holds:

$$\frac{\sum p_{\alpha,k} x_a^\alpha \log_a^k x}{\sum q_{\alpha,k} x_a^\alpha \log_a^k x} - \sum r_{\alpha,k} x_a^\alpha \log_a^k x = 0,$$

where $p_{\alpha,k}, q_{\alpha,k}$ are elements of the field $M(0,\varepsilon)$. Let us multiply this equality by $\sum q_{\alpha,k} x_a^\alpha \log_a^k x$, remove the parentheses, and reduce, if necessary, the germs $x_a^\beta \log_a^k x$ to the form $x^n \cdot x_a^\alpha \log_a^k x$, where n is an integer and $\Re\alpha = 0$. Since the germs $x_a^\alpha \log_a^k x$ are linearly independent over the field $\hat{M}(0,\varepsilon)$, the equality is equivalent to the system of equations obtained by equating the coefficients with these functions to zero. The obtained system is a system of linear equations in the functions $r_{\alpha,k}$ with coefficients in the field $M(0,\varepsilon)$. The system has a unique solution, since the functions $r_{\alpha,k}$ are uniquely defined. Therefore, the functions $r_{\alpha,k}$ lie in the field $M(0,\varepsilon)$. □

Corollary 6.13 *If the germ f_a of a meromorphic function f on a neighborhood $\hat{U}(0,\varepsilon)$ has a Fuchsian singularity over that neighborhood, then the function f is meromorphic in the neighborhood $U(0,\varepsilon)$.*

Proof The germ f_a is in $K_a(0,\varepsilon)$, and its basis expansion has the form $f_a = f \cdot 1$. By Corollary 6.12, the germ f_a has an entire Fuchsian singularity, and hence $f \in M(0,\varepsilon)$. □

The proof of Corollary 6.13 concludes the proof of Frobenius's theorem.

6.1.3 The Monodromy Group of Systems of Linear Differential Equations and Its Connection with the Galois Group

The results of Sect. 6.1.1 carry over automatically to linear differential equations with regular singular points. Let us begin with a linear differential equation over the field of rational functions with singular points (not assuming that singular points are regular). Consider an equation

$$\mathbf{y}' = \mathbf{A}(x)\mathbf{y}, \tag{6.2}$$

where $\mathbf{y} = (y_1(x),\ldots,y_n(x))$, $\mathbf{A}(x) = (a_{i,j}(x))$, $1 \le i,j \le n$, are rational matrix-valued functions, and x is a complex variable. Let $a_1,\ldots,a_k$ be the poles of the matrix $\mathbf{A}(x)$. In a neighborhood of a nonsingular point x_0, $x_0 \neq \infty$, $x_0 \neq a_i$,

$i = 1, \dots, k$, the solutions of the equation form an n-dimensional vector space V^n. Now take an arbitrary path $\gamma(t)$ on the complex line going from the point x_0 to the point x_1 and not passing through the singular points a_i, $\gamma(0) = x_0$, $\gamma(1) = x_1$, $\gamma(t) \neq a_i$. The solutions of the equation can be analytically continued along the path while remaining solutions of the equation. Hence, to every path γ, we can associate a linear map M_γ of the space of solutions $V^n_{x_0}$ at the point x_0 to the space of solutions V_{x_1} at the point x_1. If the path γ is deformed so as to avoid the singular points while leaving the endpoints fixed, then the map M_γ is unchanged. Loops define linear transformations of the space V^n. The collection of all such linear transformations of the space V^n forms a group, which is called the *monodromy group of Eq.* (6.2). Thus the monodromy group of an equation is the group of linear transformations of solutions that correspond to loops around singular points. The monodromy group of an equation characterizes the multivaluedness of its solutions.

Lemma 6.14 *The following properties hold:*

1. *The monodromy group of almost every solution of system* (6.2) *coincides with the monodromy group of system* (6.2).
2. *The monodromy pair of every component (i.e., the monodromy pair of y_i for every i) of every solution of system* (6.2) *is almost normal.*
3. *If the monodromy group of system* (6.2) *does not belong to some complete class $\mathcal{M}$ of group pairs, then the monodromy pair of at least one component of almost every solution of this system is not in $\mathcal{M}$.*

Proof The first two statements of the lemma are proved in the same way as Lemma 6.1. The third statement follows from the first and from Lemma 5.39. □

In a neighborhood of a nonsingular point x_0, there exist n linearly independent solutions $\mathbf{y}_1, \dots, \mathbf{y}_n$ of (6.2). We can consider in this neighborhood the field of functions $\mathcal{R}\langle \mathbf{y}_1, \dots, \mathbf{y}_n \rangle$ obtained by adjoining all solutions $\mathbf{y}_i$ and all their derivatives to the field $\mathcal{R}$ of rational functions.

Each transformation M_γ from the monodromy group extends to an automorphism of the field $\mathcal{R}\langle \mathbf{y}_1, \dots, \mathbf{y}_n \rangle$. Indeed, together with the functions $\mathbf{y}_1, \dots, \mathbf{y}_n$, each element of the field $\mathcal{R}\langle \mathbf{y}_1, \dots, \mathbf{y}_n \rangle$ admits a meromorphic continuation along the path γ. This continuation provides the desired automorphism, since the arithmetic operations and differentiation are preserved under analytic continuation, and rational functions return to their initial values due to the single-valuedness.

A singular point of (6.2) is called *regular* if in every sector with apex at the singular point, all solutions grow no faster than a power as they approach this point (see [13, 40]). It is known that near a regular singular point, every component of every solution has an entire Fuchsian singularity (see Definition 6.11). Equation (6.2) is called regular if all its singular points (including ∞) are regular. For a regular Eq. (6.2), all single-valued functions from the field $\mathcal{R}\langle \mathbf{y}_1, \dots, \mathbf{y}_n \rangle$ are rational functions.

Theorem 6.15 *For a regular system of linear differential equations* (6.2), *the differential field $\mathcal{R}\langle \mathbf{y}_1, \dots, \mathbf{y}_n \rangle$ is a Picard–Vessiot extension of the field $\mathcal{R}$. The*

Galois group of this extension is the algebraic closure of the monodromy group of system (6.2).

Proof The monodromy group acts on the differential field $\mathscr{R}\langle \mathbf{y}_1, \dots, \mathbf{y}_n \rangle$ as a group of isomorphisms with invariant subfield $\mathscr{R}$. The field $\mathscr{R}\langle \mathbf{y}_1, \dots, \mathbf{y}_n \rangle$ is generated over $\mathscr{R}$ by a finite-dimensional $\mathbb{C}$-vector space invariant under the monodromy action, namely, the vector space spanned by all components of all solutions of (6.2). The theorem now follows from Corollary 3.11. □

Theorem 6.16 *Every component of every solution of a regular system of linear differential equations is representable by quadratures, by k-quadratures, or by generalized quadratures if and only if the monodromy group of the system is respectively solvable, k-solvable, or almost solvable.*

The proof follows from the Picard–Vessiot theorem (see Theorem 3.18) and from the preceding theorem. As in the case of a Fuchsian equation, the "permissive" part of the theorem concerning the solvability of the system, can be proved essentially by means of linear algebra (see Sects. 3.7 and 6.1.1). But the topological version of Galois theory makes the "prohibitive" part of the theorem considerably stronger (see Sect. 6.2.4).

6.2 Galois Theory for Fuchsian Systems of Linear Differential Equations with Small Coefficients

In this section, we give explicit criteria for different kinds of solvability of Fuchsian linear differential equations with sufficiently small coefficients [39]. The proof uses the theory of Kolchin (see Sect. 5.8) and that of Lappo-Danilevsky.

6.2.1 Fuchsian Systems of Equations

Among systems of regular linear differential equations, one distinguishes Fuchsian systems of linear differential equations. These are equations of the form $\mathbf{y}' = \mathbf{A}(x)\mathbf{y}$, where the matrix $\mathbf{A}(x)$ has no multiple poles and vanishes at infinity. In other words, these are equations of the form

$$\mathbf{y}' = \sum_{p=1}^{k} \frac{A_p}{(x - a_p)} \mathbf{y},$$

where A_p are complex $(n \times n)$ matrices, and $\mathbf{y} = (y_1, \dots, y_n)$ is a vector function taking values in $\mathbb{C}^n$. Points a_p are called *poles*, and the matrices A_p are called the *residue matrices* of the Fuchsian system of equations.

For Fuchsian systems of equations, as for other regular systems of differential equations, the algebraic closure of the monodromy group coincides with the Galois group of the Picard–Vessiot extension of the field of rational functions generated by the system of equations (see Sect. 1.3).

Ivan Lappo–Danilevsky developed the theory of analytic functions of matrices and applied it to differential equations [69]. We will need his results regarding Fuchsian systems of equations, which we will use in the form of Corollary 6.18 below.

Take any nonsingular point $x_0 \neq a_p$. Fix k paths $\gamma_1, \ldots, \gamma_k$ such that the path γ_p starts at the point x_0, approaches the pole a_p, makes a loop around it, and returns to the point x_0. The paths $\gamma_1, \ldots, \gamma_k$ define the monodromy matrices $M_1, \ldots, M_k$. Obviously, the matrices $M_1, \ldots, M_k$ generate the monodromy group. For fixed paths, the monodromy matrices depend on the residue matrices only. Lappo-Danilevsky studied this dependence.

First, he showed that the monodromy matrices M_p are entire functions of the residue matrices A_j. More precisely, there exist special series

$$, M_p = E + 2\pi i A_p + \sum_{1\le i,j\le k} c_{i,j} A_i A_j + \cdots \tag{6.3}$$

in matrices $A_1, \ldots, A_k$ with complex coefficients such that these series represent the monodromy matrices M_p and converge for every choice of matrices $A_1, \ldots, A_k$.

Although a monodromy matrix M_p depends on all residue matrices A_j, its eigenvalues are determined by eigenvalues of the residue matrix A_p only.

Theorem 6.17 ([13, 40]) *Let $\{\mu_m\}$ be the collection of eigenvalues of the matrix A_p. Then $\{e^{2\pi i \mu_m}\}$ is the collection of eigenvalues of the matrix M_p.*

The famous Riemann–Hilbert problem is the question of solvability of the inverse problem, i.e., the question of existence of Fuchsian equations with a given collection of monodromy matrices. The Riemann–Hilbert problem is solvable for almost every collection of monodromy matrices. It was traditionally believed that this classical result carries over to every collection of monodromy matrices. However, as Andrei Bolibrukh [13, 14] discovered, this is not the case. He gave an example of a collection of monodromy matrices for which the Riemann–Hilbert problem is unsolvable.

Lappo-Danilevsky showed further that under the assumption that the residue matrices A_j are small, these matrices are single-valued analytic functions of the monodromy matrices M_p. Namely, he showed that if we confine ourselves to Fuchsian equations with sufficiently small residue matrices $\|A_j\| < \varepsilon$, $\varepsilon = \varepsilon(n, a_1, \ldots, a_k)$, then for monodromy matrices M_p sufficiently close to E, $\|M_p - E\| < \varepsilon$, the Riemann–Hilbert problem has a unique solution. Furthermore, there exist special series

$$A_p = -\frac{1}{2\pi i} E + \frac{1}{2\pi i} M_p + \sum_{1\le i,j\le k} b_{ij} M_i M_j + \cdots \tag{6.4}$$

in matrices $M_1, \dots, M_k$ with complex coefficients such that these series represent the residue matrices A_p and converge for $\|M_p - E\| < \varepsilon$.

Series (6.4) are obtained by inversion of the series (6.3). This result is a peculiar form of the implicit function theorem (for analytic maps with noncommuting variables). We will use Lappo-Danilevsky theory in the form of the following corollary.

Corollary 6.18 *The monodromy matrices lie in the minimal topologically closed algebra with unity containing the residue matrices. Conversely, if the residue matrices are sufficiently small and the monodromy matrices are sufficiently close to E, then the residue matrices lie in the minimal topologically closed algebra with unity containing the monodromy matrices.*

6.2.2 Groups Generated by Matrices Close to the Identity

In this subsection, we prove an analogue of Lie's theorem for linear groups generated by matrices close to the identity. We first recall the statement of Jordan's theorem. A group of linear transformations of a finite-dimensional vector space is said to be *diagonal* if there is a basis in which all elements of the group are represented by diagonal matrices.

Theorem 6.19 (Jordan's theorem) *Every finite subgroup G of the group $\mathrm{GL}(V)$ of all linear transformations of an n-dimensional vector space V has a diagonal normal subgroup G_d of bounded index, $\mathrm{ind}(G, G_d) \leq J(n)$.*

Various explicit upper bounds for the numbers $J(n)$ are known. For example, Schur showed that

$$J(n) \leq \left(\sqrt{8n} + 1\right)^{2n^2} - \left(\sqrt{8n} - 1\right)^{2n^2}$$

(see [92]). Recall that a group H of linear transformations of a finite-dimensional vector space is said to be *triangular* if there exists a basis in which the matrices of all elements of H are triangular. Equivalently, a group H is triangular if there exists a complete flag of vector subspaces invariant under the action of H.

The projective space P_V is by definition the quotient space of the space $V \setminus \{0\}$ by the action of the group $\mathbb{C}^*$, under which $\lambda \in \mathbb{C}^*$ maps $x \in V \setminus \{0\}$ to λx. Let $\pi : \mathrm{GL}(V) \to \mathrm{GP}_V$ denote the natural map of $\mathrm{GL}(V)$ onto the group GP_V of projective transformations of the space P_V. For $A, B \in \mathrm{GL}(V)$, the equality $\pi(A) = \pi(B)$ holds if and only if $A = \mu B$ for some $\mu \in \mathbb{C}$.

We have the following corollary to Jordan's theorem.

Corollary 6.20 *If the projectivization $\pi(G)$ of the group $G \subset \mathrm{GL}(V)$ is finite and $\dim V = n$, then G has a diagonal normal subgroup G_d of index $\leq J(n)$.*

Proof Let $\mathrm{GL}_1(V)$ be the group of linear transformations of the space V with determinant equal to 1, and let $\pi_1 : \mathrm{GL}_1(V) \to \mathrm{GP}_V$ be the restriction of the homomorphism π to $\mathrm{GL}_1(V)$. The map π_1 has finite kernel $T = \{\lambda E\}$, where $\lambda^n = 1$. The group $\pi(G)$ is finite, whence the group $\tilde{G} = \pi_1^{-1} \circ \pi(G)$ is also finite. By Jordan's theorem, the group $\tilde{G}$ has a diagonal normal subgroup $\tilde{G}_d$ of index $\leq J(n)$. We may assume that $T \subset \tilde{G}_d$: if this is not the case, then we can replace the subgroup $\tilde{G}_d$ with the subgroup generated by T and $\tilde{G}_d$. As a normal subgroup $G_d \subset G$, we can take the preimage of $\pi_1(\tilde{G}_d)$ under the projectivization map $\pi : G \to \mathrm{GP}_V$ of G. □

Proposition 6.21 *There exists an integer $T(n)$ such that a subgroup G in $\mathrm{GL}(n)$ has a solvable normal subgroup of finite index if and only if it has a triangular normal subgroup of index $\leq T(n)$.*

Proof Suppose that G' is a solvable normal subgroup of G of finite index. Lie's theorem guarantees the existence of a triangular normal subgroup G_l of finite index in the group G. Indeed, it suffices to set $G_l = G' \cap \overline{G}_0$, where $\overline{G}_0$ is the connected component of the identity in the algebraic closure $\overline{G}$ of the group G. However, the index of G_l can be arbitrarily large. Thus, for example, for the group $\mathbb{Z}_k$ of kth roots of unity, this index equals k for $n = 1$. We will enlarge the normal subgroup G_l while keeping it triangular. Note that it suffices to prove the existence of a triangular subgroup of bounded index, since a subgroup of index k contains a normal subgroup of index at most $k!$.[2] We will argue by induction on the dimension n. If the group G has an invariant subspace V^k of dimension k, $0 < k < n$, then we can perform the induction step. Indeed, in this case, the group G acts on the space V^k of dimension k and on the quotient space V^n / V^k of dimension $(n-k)$. By the induction hypothesis, the group G has a normal subgroup of index $\leq T(k)T(n-k)$ that is triangular both in V^k and in V^n / V^k; hence it is triangular in V^n.

The normal subgroup G_l is triangular and therefore has a nonzero maximal eigenspace V^k. There are two cases: $V^k \subset V^n$ and $V^k = V^n$. Consider the case $V^k \subset V^n$. Let $\tilde{G}_l$ denote the subgroup of G consisting of all transformations such that V^k is invariant (it is here where the expansion of the normal subgroup G_l takes place). Let us prove that $\mathrm{ind}(G, \tilde{G}_l) \leq n$. Indeed, the group G permutes maximal eigenspaces of each of its normal subgroups, in particular G_l. However, there are at most n maximal eigenspaces. Now the desired inequality $\mathrm{ind}(G, \tilde{G}_l) \leq n$ follows. To conclude the proof, it suffices to apply the induction step to the group $\tilde{G}_l$.

Consider the second case: $V^k = V^n$, i.e., the group G_l consists of matrices λE. In this case, the projectivization of the group G is finite. To conclude the proof, it remains to use Corollary 6.20. □

[2] Indeed, a subgroup of index k defines an action of the group on a k-element set such that the subgroup is the stabilizer of some element. The desired normal subgroup of index $\leq k!$ is the kernel of this action.

Proposition 6.22 *There exists an integer $D(n)$ such that a subgroup G of* GL(n) *has a diagonal normal subgroup of finite index if and only if it has a diagonal normal subgroup of index $\leq D(n)$.*

Proposition 6.22 can be proved in the same way as Proposition 6.21, and we will not give the proof here. The numbers $T(n)$ and $D(n)$ also admit an explicit upper bound (cf. [92]).

Lemma 6.23 *The equation $X^N = A$, $\|A - E\| < \varepsilon$, $\|X - E\| < \varepsilon$, where X and A are complex $n \times n$ matrices close to E, the matrix A known and the matrix X unknown, has a unique solution, provided that $\varepsilon = \varepsilon(n, N)$ is sufficiently small. Furthermore, each invariant subspace V of the matrix A is invariant for the matrix X as well.*

Proof Set $B = A - E$ and

$$X = E + \frac{1}{N}B + \frac{1}{2}\frac{1}{N}\left(\frac{1}{N} - 1\right)B^2 + \cdots.$$

For $\|B\| < 1$, the series converges, and $X^N = A$. Now choose $\varepsilon = \varepsilon(n, N)$ to be small enough for the implicit function theorem to guarantee the uniqueness of a solution. The space V is invariant under $B = A - E$ and therefore under X. □

Lemma 6.24 *Suppose that the Nth powers of all matrices from the group G lie in some algebraic group L. Then the group $G \cap L$ has finite index in G.*

Proof Consider the algebraic closure $\overline{G}$ of the group G. It is easy to see that $X^N \in L$ for all $X \in \overline{G}$. Let $\overline{G}_0$ and L_0 denote the connected components of the identity in the groups $\overline{G}$ and L. If A is in the group L_0 and $A = e^M$, then the equation $X^N = A$ has a solution in the same group. Indeed, it suffices to set $X = e^{M/N}$. But the equation $X^N = A$ has a unique solution for matrices A and X close to E. It follows that $\overline{G}_0 \subseteq L_0 \subseteq L$. The lemma now follows from the fact that $\operatorname{ind}(\overline{G}, \overline{G}_0) < \infty$.

□

Remark 6.25 For $L = e$, Lemma 6.24 coincides with Burnside's theorem, which states that a linear algebraic group all of whose elements satisfy the identity $X^N = e$ is finite.

Proposition 6.26 *There exists an integer $N(n)$ such that a subgroup G of* GL(n) *has a solvable normal subgroup of finite index if and only if all matrices $A^{N(n)}$, $A \in G$, reduce simultaneously to triangular form.*

Proof In one direction, the result follows from Proposition 6.21, where one sets $N(n) = T(n)!$. For the proof in the other direction, we need to apply Lemma 6.24 to the group G and the group L of triangular matrices. □

Similarly, we can prove the following result.

Proposition 6.27 *There exists an integer $\overline{N}(n)$ such that a subgroup G of $\mathrm{GL}(n)$ has a diagonal normal subgroup of finite index if and only if all matrices $A^{\overline{N}(n)}$, $A \in G$, reduce simultaneously to diagonal form.*

Theorem 6.28 *There exists a positive number $\varepsilon(n) > 0$ such that the subgroup G of $\mathrm{GL}(n)$ generated by matrices A_α close to the identity, $\|E - A_\alpha\| < \varepsilon(n)$, has a solvable normal subgroup of finite index if and only if all matrices A_α reduce simultaneously to triangular form.*

Proof Choose $\varepsilon(n) > 0$ to be small enough for the equation

$$X^{N(n)} = A,$$

$\|E - X\| < \varepsilon(n)$, to satisfy the conditions of Lemma 6.23. By Proposition 6.26, all matrices $A_\alpha^{N(n)}$ reduce to triangular form. But by Lemma 6.23, the invariant subspaces of the matrices $A_\alpha^{N(n)}$ and A_α coincide. Hence the matrix A_α also reduces to triangular form. □

Similarly, we can prove the following proposition.

Proposition 6.29 *There exists a positive number $\varepsilon(n) > 0$ such that the subgroup G of $\mathrm{GL}(n)$ generated by matrices A_α close to the identity, $\|E - A_\alpha\| < \varepsilon(n)$, has a diagonal normal subgroup of finite index if and only if all matrices A_α reduce simultaneously to diagonal form.*

Remark 6.30 In Theorem 6.28 and Proposition 6.29, we can weaken the assumption that the matrices A_α are close to the identity. It suffices to speak about closeness in the Zariski topology. Let us say that a matrix A is k-resonant if it has distinct eigenvalues λ_1 and λ_2 that are subject to the relation $\lambda_1 = \varepsilon_k \lambda_2$, $\varepsilon_k^k = 1$, $\varepsilon_k \neq 1$. All k-resonant matrices form an algebraic set not containing the identity. It suffices to assume that the matrices A_α are not $N(n)$-resonant.

6.2.3 *Explicit Criteria for Solvability*

Let us proceed with an explicit criterion for solvability. We begin with two simple lemmas.

Lemma 6.31 *A Fuchsian system*

$$\dot{\mathbf{y}} = \sum_{i=1}^{k} \frac{A_i}{x - a_i} \mathbf{y}$$

of order n with sufficiently small coefficients $\|A_i\| < \varepsilon = \varepsilon(n, a_1, \ldots, a_k)$ is solvable by generalized quadratures if and only if its monodromy matrices M_i are triangular.

Proof The monodromy group of the system is generated by the monodromy matrices M_i. If the residue matrices A_i are small, $\|A_i\| < \varepsilon$, then the matrices M_i are close to E. Choose $\varepsilon = \varepsilon(n, a_1, \ldots, a_k)$ to be small enough for the monodromy matrices $M_1, \ldots, M_k$ to satisfy the conditions of Theorem 6.28. By this theorem, the monodromy group has a solvable normal subgroup of finite index if and only if the matrices $M_1, \ldots, M_k$ are triangular in the same basis. It now remains to use Theorem 6.16. □

Lemma 6.32 *For a Fuchsian system, the triangularity and the diagonality of the Galois group are equivalent to the same conditions on the monodromy matrices* $M_1, \ldots, M_k$.

Proof The monodromy group is generated by the monodromy matrices $M_1, \ldots, M_k$, and it is triangular or diagonal whenever they are. The lemma now follows from the fact that for a Fuchsian equation, the Galois group coincides with the algebraic closure of the monodromy group (see Sect. 6.1.3). □

Proposition 6.33 (Criteria for solvability) *Given a set of poles* $a_1, \ldots, a_k$ *and a positive integer* n, *one can find a number* $\varepsilon(n, a_1, \ldots, a_k) > 0$ *such that the solvability conditions*

$$\dot{\mathbf{y}} = \sum_{i=1}^{k} \frac{A_i}{x - a_i} \mathbf{y}$$

for Fuchsian systems of order n *with small coefficients,* $\|A_i\| \leq \varepsilon(n, a_1, \ldots, a_k)$, *have an explicit form. Namely:*

1. *These systems are solvable by quadratures or generalized quadratures*[3] *if and only if the matrices* A_i *are triangular (in some basis).*
2. *These systems are solvable by integrals and algebraic functions or by integrals and radicals if and only if the matrices* A_i *are triangular and their eigenvalues are rational.*
3. *These systems are solvable by integrals if and only if the matrices* A_i *are triangular and their eigenvalues are zero.*
4. *These systems are solvable by exponentials of integrals and by algebraic functions or by exponentials of integrals if and only if the matrices* A_i *are diagonal.*
5. *These systems are solvable by algebraic functions or by radicals if and only if the matrices* A_i *are diagonal and their eigenvalues are rational.*
6. *These systems are solvable by rational functions if and only if all matrices* A_i *are zero.*

[3]These forms of solvability are different unless we restrict the coefficients. The same holds for the forms of solvability appearing in items 2, 4, and 5 below.

Proof Choose $\varepsilon(n, a_1, \dots, a_k)$ to be small enough for the conditions of Lemma 6.23 to hold and for the residue matrices to be expressible through the monodromy matrices (see Sect. 6.2.1).

Each form of solvability implies solvability by generalized quadratures. Solvability by generalized quadratures implies that the monodromy matrices are triangular (Lemma 6.31), and therefore, the Galois group is triangular (Lemma 6.32). Hence the criteria of Theorem 3.36, at the end of Chap. 3, apply. We need to translate the conditions on the Galois group from this criterion to conditions on the residue matrices A_i.

The conditions on the Galois group from the criterion of Theorem 3.36 are equivalent to the same conditions on the monodromy matrices $M_1, \dots, M_k$. This was partially verified in Lemma 6.32. The rest is as simple.

Under the assumptions of our theorem, the condition on the monodromy matrices $M_1, \dots, M_k$ to be in some Zariski closed algebra with unity, for example the algebra of triangular or diagonal matrices, is equivalent to the same condition on the residue matrices $A_1, \dots, A_k$ (Corollary 6.18). The eigenvalues of the matrices M_i are roots of unity or are equal to 1 if and only if the eigenvalues of the matrices A_i are rational numbers or integers (see Sect. 6.2.1). Our criterion now follows from the criterion of Theorem 3.36. □

Remark 6.34 Not long before his death, Andrei Bolibrukh communicated to me that in the criteria for solvability, the requirement that matrices A_i be small can be weakened: it suffices to require only that the eigenvalues of these matrices be small. Bolibrukh's former students completed his arguments; see [102].

6.2.4 Strong Unsolvability of Equations

The topological version of Galois theory allows us to refine the classical results on unsolvability of equations in finite terms. The monodromy group of an algebraic function coincides with the Galois group of the corresponding Galois extension of the field of rational functions. Therefore, by Galois theory:

1. An algebraic function is representable by radicals if and only if its monodromy group is solvable.
2. An algebraic function is expressible through rational functions with the help of radicals and solutions of algebraic equations of degree k if and only if its monodromy group is k-solvable.

From our results (see Sect. 5.7.2), we can deduce the following corollary.

Corollary 6.35 *The following statements hold:*

1. *If the monodromy group of an irreducible algebraic equation over the field of rational functions is not solvable, then its solutions do not belong to the class of functions representable by single-valued $\mathscr{S}$-functions and quadratures.*

2. *If the monodromy group of an irreducible algebraic equation is not k-solvable, then its solutions do not belong to the class of functions representable by single-valued $\mathscr{S}$-functions and k-quadratures.*

In a similar way, one can refine the results on unsolvability in finite terms from Sects. 6.1.1, 6.1.3, and 6.2.3.

Corollary 6.36 *If the monodromy group of a linear differential equation over the field of rational functions is not solvable (k-solvable, almost solvable), then a generic solution of this equation does not belong to the class of functions representable by single-valued $\mathscr{S}$-functions and quadratures (k-quadratures, generalized quadratures).*

Corollary 6.37 *If the monodromy group of a system of linear differential equations over the field of rational functions is not solvable (k-solvable, almost solvable), then at least one component of almost every solution does not lie in the class of functions representable by single-valued $\mathscr{S}$-functions and quadratures (k-quadratures, generalized quadratures).*

Corollary 6.38 *If a Fuchsian system of differential equations with small coefficients is not triangular, then at least one component of almost every solution does not lie in the class of functions representable by single-valued $\mathscr{S}$-functions and quadratures (k-quadratures, generalized quadratures).*

6.3 Maps of the Half-Plane onto Polygons Bounded by Circular Arcs

In this section, we classify all polygons G bounded by circular arcs for which the function f_G establishing the Riemann mapping of the upper half-plane onto the polygon G is representable in explicit form. We will use the Riemann–Schwarz reflection principle and the description of finite subgroups in the group of all fractional linear transformations.

6.3.1 Using the Reflection Principle

Consider a polygon G in the plane of complex numbers bounded by circular arcs. By the Riemann mapping theorem, there exists a holomorphic isomorphism f_G between the upper half-plane and the polygon G. This mapping was studied by Riemann, Schwarz, Christoffel, Klein, and others. We now recall the classical results that we need.

Let $B = \{b_j\}$ denote the preimage of the set of all vertices of the polygon G under the multivalued correspondence generated by f_G; let $H(G)$ denote the group of conformal transformations of the sphere generated by the reflections in the sides of the polygon; and let $L(G)$ denote the subgroup of index 2 in $H(G)$ consisting of fractional linear transformations. The Riemann–Schwarz reflection principle implies the following statements.

Proposition 6.39 *The following properties hold:*

1. *The function f_G (we let the same notation f_G stand for the Riemann map and for the multivalued analytic function generated by it) can be continued meromorphically along all paths avoiding the set B.*
2. *All germs of the multivalued function f_G at a nonsingular point $a \notin B$ are obtained by the action of the group $L(G)$ of fractional linear transformations on a fixed germ f_a.*
3. *The monodromy group of the function f_G is isomorphic to the group $L(G)$.*
4. *At the points b_j, the function f_G has singularities of the following kind. If at the vertex a_j of the polygon G corresponding to the point b_j, the angle α_j is different from 0, then the function f_G reduces to the form $f_G(z) = (z - b_j)^{\beta_j}\varphi(z)$ by a fractional linear transformation, where $\beta_j = \alpha_j/2\pi$, and the function $\varphi(z)$ is holomorphic near the point b_j. If the angle α_j is equal to 0, then the function f_G reduces to the form $f_G(z) = \log(z) + \varphi(z)$ by a fractional linear transformation, where $\varphi(z)$ is holomorphic near b_j.*

It follows from our results that if the function f_G is representable by generalized quadratures, then the group $L(G)$, hence also the group $H(G)$, belongs to the class $\mathcal{M}\langle\mathbb{C},\mathcal{K}\rangle$.

6.3.2 Groups of Fractional Linear and Conformal Transformations of the Class $\mathcal{M}\langle\mathbb{C},\mathcal{K}\rangle$

Let π be the epimorphism of the group SL(2) of 2×2 matrices with determinant 1 onto the group of fractional linear transformations L,

$$\pi : \begin{pmatrix} a & b \\ c & d \end{pmatrix} \to \frac{az+b}{cz+d}.$$

Since $\ker\pi = \mathbb{Z}_2$, a group $\tilde{L} \subseteq L$ and the group $\pi^{-1}(\tilde{L}) = \Gamma \subseteq \mathrm{SL}(2)$ belong simultaneously to the class $\mathcal{M}\langle\mathbb{C},\mathcal{K}\rangle$. The group Γ is a linear algebraic group; hence it belongs to the class $\mathcal{M}\langle\mathbb{C},\mathcal{K}\rangle$ if and only if it has a normal subgroup Γ_0 of finite index that can be reduced to triangular form by a linear change of coordinate. (This version of Lie's theorem holds even in multidimensional spaces, and it plays an important role in differential Galois theory; see Sect. 3.6). A group Γ_0 reduces to triangular form in one of the following cases:

1. The group Γ_0 has a unique one-dimensional eigenspace.
2. The group Γ_0 has exactly two one-dimensional eigenspaces.
3. The group Γ_0 has a two-dimensional eigenspace.

We now turn to the group of fractional linear transformations $\tilde{L} = \pi(\Gamma)$. The group $\tilde{L}$ belongs to the class $\mathscr{M}\langle\mathbb{C},\mathscr{K}\rangle$ if and only if it has a normal subgroup $\tilde{L}_0 = \pi(\Gamma_0)$ of finite index whose set of fixed points is a singleton, a pair of points, or the entire Riemann sphere.

A group $\tilde{H}$ of conformal transformations has a normal subgroup $\tilde{L}$ of index 2 (or of index 1) consisting of fractional linear transformations. Therefore, for a group $\tilde{H}$ of class $\mathscr{M}\langle\mathbb{C},\mathscr{K}\rangle$, a similar statement holds.

Lemma 6.40 (On conformal transformations of the class $\mathscr{M}\langle\mathbb{C},\mathscr{K}\rangle$) *A group of conformal transformations of the sphere belongs to the class $\mathscr{M}\langle\mathbb{C},\mathscr{K}\rangle$ if and only if one of the following three conditions holds:*

1. *The group has a fixed point.*
2. *The group has an invariant two-point set.*
3. *The group is finite.*

This lemma follows from the above discussion, since the set of fixed points of a normal subgroup is invariant under the action of the group. It is well known that a finite group $\tilde{L}$ of fractional linear transformations of the sphere is conjugate to a group of rotations by means of a fractional linear change of coordinate.

It is not hard to prove that if the composition of reflections in two circles corresponds to a rotation of the sphere under stereographic projection from the plane of complex numbers to the sphere, then these two circles correspond to great circles. Therefore, every finite group $\tilde{H}$ of conformal transformations generated by reflections in circles can be reduced by a fractional linear change of coordinate to a group of isometries of the sphere generated by reflections in great circles.

All finite groups of isometries generated by reflections in great circles are well known. Each such group is the isometry group of one of the following bodies:

1. A regular n-gonal pyramid
2. A regular n-gonal dihedron or the body formed by two equal regular n-gonal pyramids sharing the base
3. A regular tetrahedron
4. A regular cube or icosahedron
5. A regular dodecahedron or icosahedron

All these groups of isometries, except for the groups of the dodecahedron and icosahedron, are solvable. The intersections of the sphere whose center coincides with the barycenter of the body with the mirrors in which the body is symmetric is a certain net of great circles. The nets corresponding to the bodies listed above will be called the *finite nets of great circles*. Stereographic projections of finite nets are shown in Fig. 6.1.

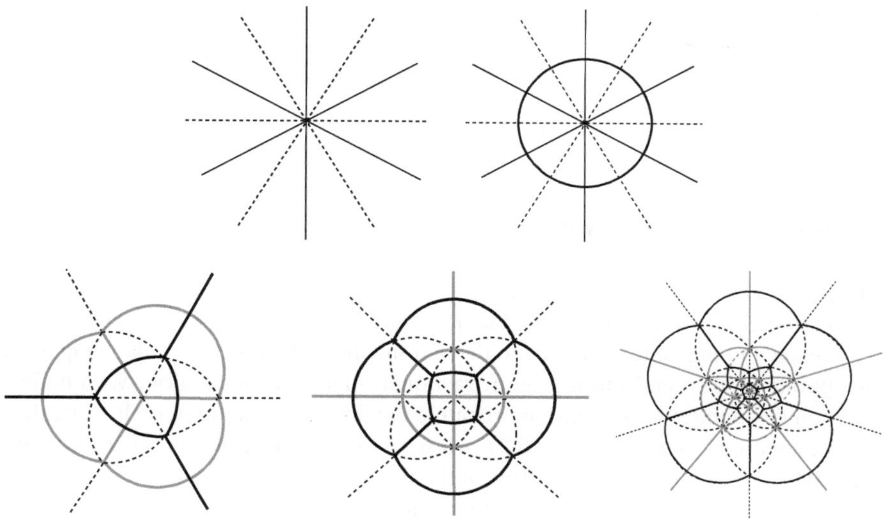

Fig. 6.1 Finite nets of great circles

6.3.3 Integrable Cases

We now return to the question of representability of the function f_G by generalized quadratures. We will consider all possible cases and show that the conditions on the monodromy group we have found above are not only necessary but also sufficient for representability of f_G by generalized quadratures.

The First Case of Integrability

The group $H(G)$ has a fixed point. This means that the continuations of all sides of the polygon G intersect at one point. Mapping this point to infinity by a fractional linear transformation, we obtain a polygon $\overline{G}$ bounded by straight line segments.

All transformations in the group $L(\overline{G})$ have the form $z \mapsto az + b$. All germs of the function $\overline{f} = f_{\overline{G}}$ at a nonsingular point c are obtained from a fixed germ $\overline{f}_c$ by the action of the group $L(\overline{G})$. The germ $R_c = \overline{f}''_c/\overline{f}'_c$ is invariant under the action of the group $L(\overline{G})$. Therefore, the germ R_c is a germ of a single-valued function. The singular points b_j of the function R_c can only be poles (see Proposition 6.39). Hence the function R_c is rational. The equation $\overline{f}''/\overline{f}' = R$, where R is a given rational function and f is an unknown function, is integrable by quadratures. This integrability case is well known. The function $\overline{f}$ in this case is called the Schwarz–Christoffel integral.

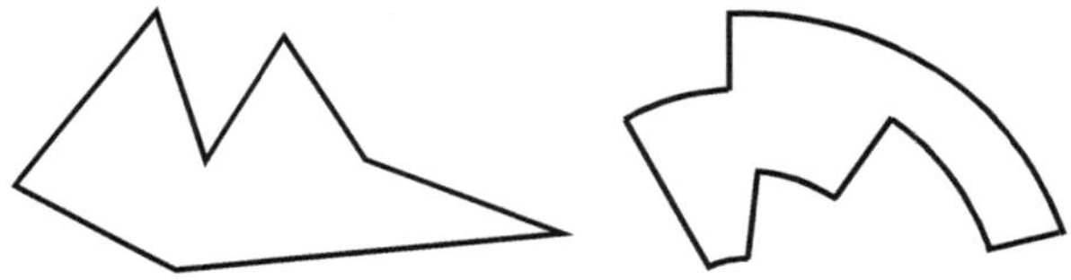

Fig. 6.2 The first and the second cases of integrability

The Second Case of Integrability

The group $H(G)$ has an invariant two-point set. This means the existence of a pair of points such that for every side of the polygon G, these points are either symmetric with respect to that side or belong to the continuation of the side. We can map these two points to zero and infinity by a fractional linear transformation. We obtain a polygon $\overline{G}$ bounded by circular arcs centered at the point 0 and intervals of straight rays emanating from 0 (see Fig. 6.2). All transformations in the group $L(\overline{G})$ have the form $z \mapsto az$, $z \mapsto b/z$. All germs of the function $\overline{f} = f_{\overline{G}}$ at a nonsingular point c are obtained from a fixed germ $\overline{f}_c$ by the action of the group $L(\overline{G})$:

$$\overline{f}_c \to a\overline{f}_c, \quad \overline{f}_c \to \frac{b}{\overline{f}_c}.$$

The germ $R_c = (\overline{f}'_c/\overline{f}_c)^2$ is invariant under the action of the group $L(\overline{G})$; hence it is a germ of a single-valued function R. The singularities of the function R can only be poles (see Proposition 6.39). Therefore, the function R is rational. The equation $R = (\overline{f}\,'/\overline{f})^2$ is integrable by quadratures.

The Third Case of Integrability

The group $H(G)$ is finite (see Fig. 6.1). This means that the polygon G can be mapped by some fractional linear transformation to a polygon $\overline{G}$ whose sides belong to some net of great circles. The group $L(G)$ is finite, and hence the function f_G has finitely many branches. Since all singularities of the function f_G are of polynomial type (see Proposition 6.39), the function f_G is an algebraic function.

Let us consider the case of a finite solvable group $H(G)$. This case is possible if and only if the polygon G can be mapped by a fractional linear transformation to a polygon $\overline{G}$ whose sides belong to some finite net different from that of the dodecahedron or icosahedron. In this case, the group $L(G)$ is solvable. Using Galois theory, it is easy to prove that in this case, the function f_G can be expressed through rational functions by arithmetic operations and radicals.

Our results (see Sect. 5.7.2) imply the following result.

Theorem 6.41 (Theorem on polygons bounded by circular arcs) *For every polygon G not described by the three integrability cases considered above, the function f_G is not only unrepresentable by generalized quadratures, but also cannot be expressed through single-valued $\mathcal{S}$-functions by generalized quadratures, compositions, and meromorphic operations.*

Chapter 7
Multidimensional Topological Galois Theory

7.1 Introduction

In topological Galois theory for functions of one variable (see Chap. 5), it is proved that the way the Riemann surface of a function is positioned over the complex line can obstruct the representability of that function by quadratures. This not only explains why many differential equations are not solvable by quadratures, but also gives the strongest known results on their unsolvability.

I was always under impression that a full-fledged multidimensional version of topological Galois theory was impossible. The reason was that to construct such a version for the case of many variables, one would need to have information on the extendability of function germs not only outside their ramification sets but also along those sets. It seemed that there was nothing from which one could extract such information. It was spring 1999 when I suddenly realized that function germs can sometimes be automatically extended along their ramification sets. This is exactly the reason for the existence of multidimensional topological Galois theory. We discuss this theory in this chapter (see also [51–53]). In Sect. 7.3, we describe the property of extendability of functions along their ramification sets, which, in my view, is interesting in its own right.

Let f be a multivalued analytic function on $\mathbb{C}^n$ for which the monodromy group is defined. Let $\pi : (Y, y_0) \to (\mathbb{C}^n, a)$ be an analytic map of a complex analytic manifold Y into $\mathbb{C}^n$. The germ $\pi^*(f_a)_{y_0}$ can be a germ of a multivalued function on the manifold Y for which the monodromy group is defined. This situation is possible even if the point a belongs to the set of singular points of the function f (some of the germs of the multivalued function f may appear to be nonsingular at singular points of this function), and the manifold Y maps to the set of singular points. Can one estimate the monodromy group of the considered pullback of f knowing the monodromy group of the initial function f (is it true, for example, that if the monodromy group of f is solvable, then the monodromy group of every

A. Khovanskii, *Topological Galois Theory*, Springer Monographs in Mathematics,
DOI 10.1007/978-3-642-38871-2_7

pullback of it is also solvable)? In Sect. 7.3, we pose this question more precisely and give an affirmative answer to it (see Sects. 7.3.4 and 7.3.5).

To describe the connection between the monodromy group of the function f and the monodromy groups of its pullbacks, we introduce the notion of pullback closure for groups (see Sect. 7.3.2). The use of this operation, in turn, forces us to reconsider the definition of various group pairs (see Sect. 7.3.5) that appeared in the one-dimensional version of topological Galois theory (see Chap. 5). In Sect. 7.3, we introduce definitions that allow us to work with multivariate functions having an everywhere dense set of singular points and monodromy groups of cardinality of the continuum.

In Chap. 5, we described a wide class of one-variable functions with infinitely many branches for which the monodromy group is defined. Is there a sufficiently wide class of multivariate function germs (containing the germs of functions representable by generalized quadratures and the germs of entire functions of several variables and stable under some natural operations such as composition) with a similar property? For a long time, I thought that the answer to this question was negative. In Sect. 7.3, we define the class of $\mathcal{SC}$-germs that provides an affirmative answer to this question. The proof uses results on extendability of multivalued analytic functions along their sets of singular points (see Sect. 7.2).

The main theorem (see Sect. 7.4.5) describes how the monodromy groups of $\mathcal{SC}$-germs change as natural operations apply to these germs. This theorem is very close to the corresponding one-dimensional theorem (see Sect. 5.6) but uses also new results of an analytic (see Sect. 7.2) and group-theoretic (see Sect. 7.3) nature. As a consequence, we obtain topological results on the unsolvability of equations in explicit form that are stronger than the analogous classical theorems.

In Sect. 7.1.1, we define operations on multivalued functions of several variables (which are understood in a slightly more restrictive sense than operations on multivalued functions of a single variable). In Sects. 7.1.2–7.1.4, we define Liouvillian function classes and Liouville extensions of functional differential fields for the case of multivariate functions.

7.1.1 Operations on Multivariate Functions

In this chapter, operations on multivariate functions are understood as operations on their single-valued germs (cf. Sect. 1.2). Fix a class of basic functions and a supply of admissible operations. Can a given function (obtained, say, by solving a certain algebraic or differential equation) be expressed through the basic functions by means of admissible operations? We are interested in various *single-valued branches* of multivalued functions over various domains. Every function, even if it is multivalued, will be considered a collection of all its single-valued branches. We will apply admissible operations (such as arithmetic operations and composition) only to single-valued branches of the function over various domains. Since we shall be dealing with analytic functions, it suffices to consider only small neighborhoods of points as domains.

We can now rephrase the question in the following way:

Can a given function germ at a given point be expressed through the germs of basic functions with the help of admissible operations?

Of course, the answer depends on the choice of a point and on the choice of a single-valued germ at that point belonging to the given multivalued function. It turns out, however, that (for the classes of functions of interest to us) either the desired expression is impossible for every germ of a given multivalued function at every point, or the "same" expression serves all germs of a given multivalued function at almost every point of the space. In the first case, we say that no branch of a given multivalued function is expressible through the germs of basic functions by means of admissible operations. In the second case, we say that such an expression exists. Throughout this chapter, operations on multivalued functions will be understood in the sense described above.

For many functions of one variable, we have used a different, extended, definition of operations on multivalued functions in which the multivalued function was viewed as a single object (see Sect. 1.1). This definition is essentially equivalent to including the operation of analytic continuation in the list of admissible operations on analytic germs. For functions of many variables, we need to adopt the more restrictive understanding of operations on multivalued functions, which is, however, no less (and perhaps even more) natural.

7.1.2 Liouvillian Classes of Multivariate Functions

In Sects. 7.1.2–7.1.4, we define Liouvillian classes of functions and Liouville extensions of functional differential fields for the case of several variables. These classes and these extensions are defined in the same way as the corresponding classes and field extensions for functions of one variable (see Sects. 1.2 and 1.4). The only difference is in some of the details.

We fix an ascending chain of standard coordinate subspaces of strictly increasing dimension: $0 \subset \mathbb{C}^1 \subset \cdots \subset \mathbb{C}^n \subset \cdots$ with coordinate functions $x_1, \ldots, x_n, \ldots$ (for every $k > 0$, the functions $x_1, \ldots, x_k$ are coordinate functions on $\mathbb{C}^k$). Below, we define Liouvillian classes of functions for each of the standard coordinate subspaces $\mathbb{C}^k$.

Functions of *n* Variables Representable by Radicals

List of basic functions:

- All complex constants
- All coordinate functions on every standard coordinate subspace

List of admissible operations:

- Arithmetic operations
- The operation of taking the mth root $\sqrt[m]{f}$, $m = 2, 3, \ldots$, of a given function f

A *function of m variables representable by radicals* is any function of the variables $x_1, \ldots, x_n$ that can be obtained from the basic functions listed above with the help of the admissible operations listed above.

The function

$$g(x_1, x_2, x_3) = \sqrt[3]{5x_1 + 2\sqrt[2]{x_2} + \sqrt[7]{x_3^3 + 3}}$$

gives an example of a function representable by radicals.

To define other classes, we will need a list of basic elementary functions.
List of basic elementary functions:

1. All complex constants and all coordinate functions $x_1, \ldots, x_n$ for every standard coordinate subspace $\mathbb{C}^n$.
2. The exponential, the logarithm, and the power x^α, where α is any complex constant.
3. Trigonometric functions: sine, cosine, tangent, cotangent.
4. Inverse trigonometric functions: arcsine, arccosine, arctangent, arccotangent.

Let us now turn to the list of classical operations on functions. We give here the beginning of the list. It will be continued in Sect. 7.1.3.
List of classical operations:

1. The operation *composition* takes a function f of k variables and functions $g_1, \ldots, g_k$ of n variables to the function $f(g_1, \ldots, g_n)$ of n variables.
2. The *arithmetic operations* take functions f and g to the functions $f + g$, $f - g$, fg, and f/g.
3. The *operations of partial differentiation with respect to independent variables.* For functions of n variables, there are n such operations: the ith operation assigns the function $\partial f/\partial x_i$ to a function f of the variables $x_1, \ldots, x_n$.
4. The *operation of integration* takes k functions $f_1, \ldots, f_k$ of the variables $x_1, \ldots, x_n$, for which the differential one-form $\alpha = f_1\,dx_1 + \cdots + f_k\,dx_k$ is closed, to the indefinite integral y of the form α (i.e., to any function y such that $dy = \alpha$). The function y is determined by the functions $f_1, \ldots, f_k$ up to an additive constant.
5. The *operation of solving an algebraic equation* takes functions $f_1, \ldots, f_n$ to the function y such that $y^n + f_1 y^{n-1} + \cdots + f_n = 0$. The function y may not be quite uniquely determined by the functions $f_1, \ldots, f_n$, since an algebraic equation of degree n can have n solutions.
6. The *operation of exponentiating the integral* takes k functions $f_1, \ldots, f_k$ of the variables $x_1, \ldots, x_k$ for which the differential one-form $\alpha = f_1\,dx_1 + \cdots + f_k\,dx_k$ is closed to the exponential of an antiderivative of the form α (i.e., any function

z such that $dz = \alpha z$). The functions $f_1, \ldots, f_k$ determine the function z up to a multiplicative constant.

We now resume defining Liouvillian classes of functions.

Elementary Functions of *n* Variables

List of basic functions:

- Basic elementary functions

List of admissible operations:

- Composition
- Arithmetic operations
- Differentiation

An *elementary function of n variables* is any function of the variables $x_1, \ldots, x_n$ that can be obtained from the basic functions listed above by means of the admissible operations listed above. All elementary functions are given by formulas such as, for example, the following:

$$f(x_1, x_2) = \arctan(\exp(\sin x_1) + \cos x_2).$$

The other Liouvillian classes of functions are defined similarly. As definitions of these classes, we just give the lists of basic functions and admissible operations.

Functions of *n* Variables Representable by Quadratures

List of basic functions

- Basic elementary functions

List of admissible operations:

- Composition
- Arithmetic operations
- Differentiation
- Integration

Generalized Elementary Functions of *n* Variables

This class of functions is defined in the same way as the class of elementary functions. We only need to add the operation of solving algebraic equations to the list of admissible operations.

Functions of n Variables Representable by Generalized Quadratures

This class of functions is defined in the same way as the class of functions representable by quadratures. We only need to add the operation of solving algebraic equations to the list of admissible operations.

Functions of n Variables Representable by k-Radicals

This class of functions is defined in the same way as the class of functions representable by radicals. We only need to add the operation of solving algebraic equations of degree at most k to the list of admissible operations.

Functions of n Variables Representable by k-Quadratures

This class of functions is defined in the same way as the class of functions representable by quadratures. We only need to add the operation of solving algebraic equations of degree at most k to the list of admissible operations.

7.1.3 New Definitions of Liouvillian Classes of Multivariate Functions

All basic elementary functions can be reduced to the logarithm and the exponential (see Lemma 1.2). The compositions $y = \exp f$ and $z = \log f$ can be regarded as solutions of the differential equations $dy = y\,df$ and $dz = df/f$. Thus, within Liouvillian classes of functions, it suffices to consider operations of solving some simple differential equations. After that, the solvability problem for Liouvillian classes of functions becomes differential-algebraic, and carries over to abstract differential fields.

We will now continue the list of classical operations (the beginning of the list is given in Sect. 7.1.2).

7. The *operation of exponentiation* takes a function f to the function $\exp f$.
8. The *operation of logarithmation* takes a function f to the function $\log f$.

We will now give new definitions of transcendental Liouvillian classes of functions.

Elementary Functions of n Variables

List of basic functions

- All complex constants
- An independent variable x

List of admissible operations:

- Exponentiation
- Logarithmation
- Arithmetic operations
- Differentiation

Functions of *n* Variables Representable by Quadratures

List of basic functions:

- All complex constants

List of admissible operations:

- Exponentiation
- Arithmetic operations
- Differentiation
- Integration

Generalized Elementary Functions of *n* Variables and Functions of *n* Variables Representable by Generalized Quadratures and *k*-Quadratures

These functions are defined in the same way as the corresponding nongeneralized classes of functions; we have only to add the operation of solving algebraic equations or the operation of solving algebraic equations of degree at most k to the list of admissible operations.

The following statement holds.

Proposition 7.1 *For each of the transcendental Liouvillian function classes, the new definition is equivalent to the old definition.*

For more on this, see this section and Sect. 7.1.2. We will not prove this statement here: the proof is very similar to that of Theorem 1.3.

A field K is said to be a field *with n commuting differentiations* if there are n additive maps $\delta_i : K \to K$, $i = 1, \dots, n$, that satisfy the Leibniz rule $\delta_i(ab) = (\delta_i a)b + a(\delta_i b)$ and commute, i.e., $\delta_i \delta_j = \delta_j \delta_i$. In the sequel, we will refer to a field equipped with commuting differentiations as a *differential field* (provided that this abridged terminology leads to no ambiguity).

An element y of a differential field K is called a *constant* if $\delta_i y = 0$ for all $i = 1, \dots, n$. All constants form a subfield that is called the *field of constants*. In all cases of interest to us, the field of constants will coincide with the field of complex numbers. Thus we will always assume in the sequel that the field of constants for our differential field coincides with the field of complex numbers. An element y of a differential field is said to be:

1. An *exponential* of an element a if $\delta_i y = y\delta_i a$ for all $i = 1, \dots, n$;
2. An *exponential of integral* of a collection of elements $a_1, \dots, a_n$ if $\delta_i y = a_i y$ for all $i = 1, \dots, n$;
3. A *logarithm* of an element a if $\delta_i y = \delta_i a/a$ for all $i = 1, \dots, n$;
4. An *integral* of a collection of elements $a_1, \dots, a_n$ if $\delta_i y = a_i$ for all $i = 1, \dots, n$.

Suppose that a differential field K and a set M lie in some differential field F. The *adjunction* of the set M to the differential field K is the minimal differential field $K\langle M\rangle$ containing both the field K and the set M.

A differential field F containing a differential field K and having the same field of constants is said to be an *elementary extension* of the field K if there exists a chain of differential fields $K = F_1 \subseteq \cdots \subseteq F_n = F$ such that for every $i = 1, \dots, n-1$, the field $F_{i+1} = F_i\langle x_i\rangle$ is obtained by adjoining an element x_i to the field F_i, and x_i is an exponential or a logarithm of some element a_i from the field F_i. An element $a \in F$ is said to be *elementary* over K, $K \subset F$, if it is contained in a certain elementary extension of the field K.

A *generalized elementary extension*, a *Liouville extension*, a *generalized Liouville extension*, and a *k-Liouville extension* of a field K are defined similarly. In the construction of generalized elementary extensions, it is permissible to adjoin exponentials and logarithms and to take algebraic extensions. In the construction of Liouville extensions, one is allowed to adjoin integrals and exponentials of integrals. In generalized Liouville extensions and k-Liouville extensions, one is also allowed to take algebraic extensions and to adjoin solutions of algebraic equations of degree at most k, respectively. An element $a \in F$ is said to be *generalized elementary* over K, $K \subset F$ (*representable by quadratures, by generalized quadratures, by k-quadratures* over K) if a is contained in some generalized elementary extension (Liouville extension, generalized Liouville extension, k-Liouville extension) of the field K.

7.1.4 Liouville Extensions of Differential Fields Consisting of Multivariate Functions

We now turn to functional differential fields whose elements are functions of n variables. These are the fields that we will deal with in this chapter.

Every subfield K in the field of all meromorphic functions on a connected domain U of the space $\mathbb{C}^n$ containing all complex constants and stable under differentiation by each of the variables (i.e., if $f \in K$, then $\partial f/\partial x_i \in K$ for all $i = 1, \dots, n$) gives an example of a functional differential field with n commuting differentiations.

We now give a general definition. Let V, v be a pair consisting of a connected n-dimensional complex analytic manifold V with n commuting meromorphic vector fields $v = v_1, \dots, v_n$ on it. The Lie derivative L_{v_i} along the vector field v_i acts on the field F of meromorphic functions on the manifold V and defines the differentiation $\delta_i f = L_{v_i} f$ of this field. A *functional differential field* is any differential subfield of the field F containing all complex constants.

The following construction helps to extend functional fields. Let K be a subfield in the field of meromorphic functions on a connected analytic manifold V equipped with n commuting meromorphic vector fields $v = v_1, \dots, v_n$ such that differentiation along these fields takes the field K to itself (i.e., if $f \in K$, then $L_{v_i} f \in K$). Consider any connected complex analytic manifold W together with an analytic map $\pi : W \to V$ that is a local homeomorphism. On W, fix meromorphic vector fields $w = w_1, \dots, w_n$ such that $v_i = d(\pi)w_i$ (here $d(\pi)$ denotes the first differential of the map π). The differential field F of all meromorphic functions on W with the differentiations $\delta_i = L_{w_i}$ contains the differential subfield $\pi^* K$ consisting of functions of the form $\pi^* f$, where $f \in K$. The differential field $\pi^* K$ is isomorphic to the differential field K, and it lies in the differential field F. For a suitable choice of the manifold W, an extension of the field $\pi^* K$, which is isomorphic to K, can be done within the field F.

Suppose that we need to extend the field K, say by an integral y of some collection of functions $f_1, \dots, f_n \in K$. This can be done in the following way. Since the vector fields $w_1, \dots, w_n$ are meromorphic and commuting, there exist meromorphic one-forms $\alpha_1, \dots, \alpha_n$ on W that are defined by the relations $\alpha_i(w_j) = 0$ for $i \neq j$ and $\alpha_i(w_i) = 1$. The one-form $\alpha = f_1\alpha_1 + \dots + f_n\alpha_n$ must be closed (otherwise, the integral y does not exist), and y is an antiderivative of the form α.

Consider the Riemann surface W of the multivalued function y.[1] By definition of the Riemann surface, there exists a natural projection $\pi : W \to V$, and the function y is a single-valued meromorphic function on the variety W. The differential field F of meromorphic functions on W with the differentiations along the vector fields[2] $w_i = d^{-1}(\pi)v_i$ contains the element y as well as the field $\pi^* K$ isomorphic to K. That is why the extension $\pi^* K\langle y\rangle$ is well defined as a subfield of the differential field F. We mean this particular construction of the extension whenever we talk about extensions of functional differential fields.

The same construction allows us to adjoin (to the functional differential field K) a logarithm or an exponential of any function f from the field K, or an integral or an exponential of integral of any collection of functions $f_1, \dots, f_n$ for which the one-form $\alpha = f_1\alpha_1 + \dots + f_n\alpha_n$ is closed. Similarly, for any functions $f_1, \dots, f_n \in K$, one can adjoin a solution y of the algebraic equation $y^n + f_1 y^{n-1} + \dots + f_n = 0$ or all solutions $y_1, \dots, y_n$ of this equation to K (the

[1]The Riemann surface of a multivariate analytic function is defined similarly to the single-variable case; note, however, that a Riemann surface of a function is in general not a surface.

[2]Here $d^{-1}(\pi)v_i$ denotes the pullback of the vector field v_i under the map π; we use that π is a local homeomorphism.

adjunction of all solutions $y_1, \dots, y_n$ can be implemented on the Riemann surface of the multivalued vector function $\mathbf{y} = y_1, \dots, y_n)$. In the same way, for any functions $f_1, \dots, f_{n+1} \in K$, one can adjoin (to K) the finite-dimensional $\mathbb{C}$-affine space of all solutions of any holonomic system of linear differential equations with coefficients in the field K. (Recall that a germ of any solution of a holonomic system admits an analytic continuation along any path on the surface V avoiding a certain analytic subvariety of positive codimension.)

Thus, all above-mentioned extensions of functional differential fields can be implemented without leaving the class of functional differential fields. In talking about extensions of functional differential fields, we always mean this particular procedure.

The differential field of all complex constants and the differential field of all rational functions of n variables can be regarded as differential fields of functions defined on the space $\mathbb{C}^n$. The following claims can be verified in the same way as in the one-dimensional case.

Proposition 7.2 *A function of n complex variables (possibly multivalued) belongs to:*

1. *The class of elementary functions if and only if it belongs to some elementary extension of the field of all rational functions of n variables.*
2. *The class of generalized elementary functions if and only if it belongs to some generalized elementary extension of the field of rational functions of n variables.*
3. *The class of functions representable by quadratures if and only if it belongs to some Liouville extension of the field of all complex constants.*
4. *The class of functions representable by k-quadratures if and only if it belongs to some k-Liouville extension of the field of all complex constants.*
5. *The class of functions representable by generalized quadratures if and only if it belongs to a generalized Liouville extension of the field of all complex constants.*

7.2 Continuation of Multivalued Analytic Functions to an Analytic Subset

Let M be an analytic variety, and Σ an analytic subset of it. Suppose that at a point $b \in M$, we are given a germ f_b of an analytic function that has analytic continuation along any path $\gamma : [0, 1] \to M$, $\gamma(0) = b$, avoiding the set Σ except possibly at the initial point. What can we say about the possibility of continuing the germ f_b along the paths that beginning at some moment, belong to Σ? In this section, we will deal with this question. In Sect. 7.2.1, we consider the classical case, for which it is known additionally that the continuations of the germ f_b define a single-valued analytic function on the set $M \setminus \Sigma$. In this case, the only obstructions to the continuation of the germ f_b are irreducible components of the set Σ that have codimension 1 in the variety M and whose closures do not contain the given point b (see Proposition 7.6, which is a version of Riemann's theorem and Hartogs's theorem on continuation of analytic functions). The germ f_b extends to the complement of the union of such components and no further in general. However, as the following simple example shows, this claim does not carry over to the case of multivalued analytic functions, at least not in a straightforward way.

Example 7.3 Consider a cubic equation

$$y^3 + py + q = 0,$$

whose coefficient with y^2 is zero. In the complement of the discriminant curve Σ, this equation defines a three-valued analytic function $y(p, q)$. The discriminant curve of the equation is a semicubic parabola, an irreducible curve with a unique singular point at the origin. At the origin, all three roots of the equation merge, and this is the only point of the (p, q)-plane with this property. Over each point of the set $\Sigma \setminus \{0\}$, exactly two roots of the equation merge. Let b be any point in the complement of the discriminant curve, and a any point in the discriminant curve different from the origin; let $\gamma : [0, 1] \to \mathbb{C}^2$ be any path originating at the point b, terminating at the point a, and intersecting the set Σ only at the final moment, that is, $\gamma(0) = b$, $\gamma(1) = a$, $\gamma(t) \notin \Sigma$ for $t \neq 1$. Choose the germ of the function $y(p, q)$ over the point b that does not collide at the point a with another germ after continuation along the path γ. Such a germ is unique; we let f_b denote this germ. Firstly, the germ f_b admits an analytic continuation along any path avoiding the set Σ. Secondly, it admits a continuation to the point a along the path γ. Thirdly, the germ f_a obtained by this continuation admits an analytic continuation along any path in the set Σ avoiding the origin. At the origin, there is no analytic germ of the function $y(p, q)$ (otherwise, we would have a globally defined branch of $y(p, q)$, which is impossible). In this example, the obstruction to continuation of the germ along the set Σ is the point 0. At this point, no other branch of the discriminant intersects Σ, but the local topology of the curve Σ changes (at the origin, the semicubic parabola has a singularity; at all other points, it is smooth).

This example leads to the following natural guess. Let B be any stratum (an analytic submanifold) lying in the set Σ and containing the point a. Suppose that a germ f_a of an analytic function admits an analytic continuation along every path never intersecting the set Σ except possibly at the initial moment. If the topology of the pair (Σ, B) does not change as we move along the path $\gamma(t) \in B$, $\gamma(0) = a$, then the germ f_a admits an analytic continuation along such paths.

This guess turns out to be correct. First, in Sect. 7.2.3, we prove it for functions f that are single-valued on the complement $M \setminus \Sigma$ of the set Σ in the manifold M. By the results of Sect. 7.2.1, it suffices to prove that as the stratum B intersects the closure of an irreducible component of Σ whose codimension is 1, the topology of the pair (Σ, B) changes. In the proof, we make essential use of Whitney's results on the existence of analytic stratifications of analytic sets that are nicely correlated with the topology. We recall these results of Whitney's in Sect. 7.2.2. The case of a multivalued function f on $M \setminus \Sigma$ can be reduced by a simple topological construction to the case of a single-valued function (see Sect. 7.2.6). This construction generalizes the classical construction of a locally trivial covering (see Sect. 7.2.4) and also makes essential use of the Whitney stratifications (see Sect. 7.2.5).

7.2.1 Continuation of a Single-Valued Analytic Function to an Analytic Subset

Let us represent the space $\mathbb{C}^n$ as the direct product of the $(n-1)$-dimensional space $\mathbb{C}^{n-1}$ and the complex line $\mathbb{C}^1$. We will identify the space $\mathbb{C}^{n-1}$ with the hyperplane $z = 0$, where z is one of the coordinate functions on the space $\mathbb{C}^n$.

Lemma 7.4 *Suppose that a neighborhood U of the origin in the space $\mathbb{C}^n$ is the direct product of a connected neighborhood U_1 in the space $\mathbb{C}^{n-1}$ and a connected neighborhood U_2 on the complex line $\mathbb{C}^1$, $U = U_1 \times U_2$. Then every function f analytic in the complement of the hyperplane $z = 0$ in the neighborhood U and bounded in some neighborhood of the origin has an analytic extension to the entire neighborhood U.*

Proof The lemma follows from the Cauchy integral formula. Indeed, define the function $\tilde{f}$ on the domain U as the Cauchy integral

$$\tilde{f}(x,z) = \frac{1}{2\pi i}\int_{\gamma(x,z)} \frac{f(x,u)\,du}{u-z},$$

where x and z are points in the domains U_1 and U_2, $f(x,u)$ is the given function, and $\gamma(x,z)$ is a contour in the domain U lying on the complex line $\{x\} \times \mathbb{C}^1$, going around the points (x,z) and $(x,0)$ and depending continuously on the point (x,z). The function $\tilde{f}(x,z)$ provides the desired analytic continuation. Indeed, the function $\tilde{f}$ is analytic on the entire domain U. In a neighborhood of the origin, it coincides with the given function f by the removable singularities theorem. □

Proposition 7.5 *Let M be an n-dimensional complex analytic manifold, Σ an analytic subset of M, and $a \in \Sigma$ a point in this subset such that every irreducible component of the set Σ of dimension $n-1$ contains the point a. Then every function f that is analytic on the complement $M \setminus \Sigma$ of the set Σ in the manifold M and that is bounded in some neighborhood of the point a admits an analytic extension to the entire manifold M.*

Proof The assertion of the proposition can be reduced to Lemma 7.4. Indeed, let Σ_H denote the subset of the set Σ defined by the following condition: in a neighborhood of every point of the set Σ_H, the analytic set Σ is a nonsingular $(n-1)$-dimensional analytic hypersurface in the manifold M. The intersection of every irreducible $(n-1)$-dimensional component D_i of the set Σ with the set Σ_H is a connected

$(n-1)$-dimensional manifold. Let us prove that the function f admits an analytic extension to the set $D_i \cap \Sigma_H$.

Let A_i denote the maximal subset in $D_i \cap \Sigma_H$ into which the function f can be analytically continued. It is obvious that A_i is open in the topology of the space $D_i \cap \Sigma_H$. The set A_i is nonempty, since by Riemann's theorem on continuation of a holomorphic function (see [33]), it contains all nonsingular points of the component D_i that are sufficiently close to the point a. Let us prove that the set A_i is closed in the topology of the set $D_i \cap \Sigma_H$. Indeed, let b be a limit point of this set. By definition of the set $D_i \cap \Sigma_H$, there exists near the point b a local coordinate system on the manifold M such that $D_i \cap \Sigma_H$ coincides with a coordinate hyperplane in this coordinate system. The desired claim now follows from Lemma 7.4. Moreover, since the set $D_i \cap \Sigma_H$ is connected, we obtain that the set A_i coincides with the set $D_i \cap \Sigma_H$, i.e., that the function f admits an analytic extension to the entire set. Thus the function f admits an analytic extension to the set $\Sigma_H = \bigcup(D_i \cap \Sigma_H)$. But the set $\Sigma \setminus \Sigma_H$ has codimension at least 2 in the manifold M. By Hartogs's extension theorem (see [33]), the result is proved. □

Proposition 7.6 *Let f be an analytic function on the complement of an analytic set Σ in an n-dimensional analytic manifold M. If the function f is bounded in some neighborhood of a point $a \in \Sigma$, then it admits an analytic extension to the set $M \setminus D_a$, where D_a is the union of all $(n-1)$-dimensional irreducible components of the set Σ not containing the point a.*

Proof The result follows from Proposition 7.5 applied to the manifold $M \setminus D_a$, its analytic subset $\Sigma \setminus D_a$, and the function f. □

7.2.2 Admissible Stratifications

Let Σ be a proper analytic subset of a complex analytic manifold M. A stratification of the set Σ is defined as a partition of the set into disjoint submanifolds called *strata* (and having, in general, different dimensions) such that the following properties hold:

1. Each stratum Σ_i is a connected analytic manifold.
2. The closure $\overline{\Sigma_i}$ of each stratum is an analytic subset of M; moreover, the boundary $\overline{\Sigma_i} \setminus \Sigma_i$ is a union of strata of smaller dimension.

We say that a pair consisting of an analytic manifold M and its analytic subset Σ has *constant topology along a stratum* $B \subset \Sigma$ if the following two requirements are satisfied.

Requirement 1 For every point $a \in B$ and every analytic submanifold L of the manifold M that is transverse to the stratum B at the point a, there exists a small neighborhood V_a of the point a in the manifold L such that the topology of the pair

(V_a, F_a), where $F_a = V_a \cap \Sigma$, depends neither on the choice of the point a nor on the choice of the section L but is determined by the stratum B and the subset Σ.

Requirement 2 The stratum B has a neighborhood U in the manifold M together with a projection $\pi : U \to B$, whose restriction to the set B is the identity, such that for every point $a \in B$, the pair $(\pi^{-1}(a), \pi^{-1}(a) \cap \Sigma)$ is homeomorphic to the pair (V_a, F_a). Moreover, for every point $a \in B$, there exists a neighborhood K_a of a in the manifold B such that the pair $(\pi^{-1}(K_a), \pi^{-1}(K_a) \cap \Sigma)$ is homeomorphic to the pair $(V_a \times K_a, F_a \times K_a)$, and some homeomorphism mapping one pair to the other transforms the projection π into the direct product projection of $V_a \times K_a$ onto the subset $\{a\} \times K_a$, while the restriction of this homeomorphism to the subset $K_a \subset \pi^{-1}(K_a)$ is the identity (more precisely, it maps a point $b \in K_a$ to the point $a \times b$ in the direct product $V_a \times K_a$).

We say that a stratification of an analytic subset $\Sigma \subset M$ is *admissible* if the pair (M, Σ) has constant topology along every stratum Σ_i of this stratification.

As Whitney discovered, admissible stratifications exist for every complex analytic subset in every complex analytic manifold (see [35]). We will use this result.

7.2.3 How the Topology of an Analytic Subset Changes at an Irreducible Component

According to the following lemma, a real topological submanifold in M lying in an analytic hypersurface Σ whose complement in this hypersurface has small dimension has the same number of connected components as the number of irreducible $(n-1)$-dimensional components in Σ.

Lemma 7.7 *Suppose that a subset T of an analytic $(n-1)$-dimensional set Σ lying in an n-dimensional analytic manifold M has the following properties:*

1. *The set T is a real topological submanifold of the manifold M of codimension 2, i.e., every point $a \in T$ has a neighborhood U_a in the manifold M such that the set $U_a \cap T$ is a real topological submanifold in the manifold U_a of real dimension $2n-2$.*
2. *The set $\Sigma \setminus T$ is a closed subset of Σ of real codimension at least 2 (i.e., it is a union of finitely many real topological submanifolds of M of real dimension at most $2n-4$).*

Then every $(n-1)$-dimensional irreducible component of the set Σ intersects exactly one connected component of the topological manifold T. Moreover, every connected component of the manifold T is dense in the corresponding irreducible $(n-1)$-dimensional component of the analytic set Σ.

Proof The result is a consequence of the following facts: (1) a set of codimension 2 cannot separate a topological manifold, and (2) if we remove all singular points from an irreducible component of an analytic set, then what is left is a connected manifold.

Let us first prove that every connected component T^0 of the set T intersects exactly one irreducible component of the set Σ. Indeed, the real dimension of the set $\Sigma \setminus \Sigma_H$ is at most $2n-4$; hence it cannot divide the connected $(2n-2)$-dimensional real manifold T^0 into nontrivial parts. Therefore, the complement in the set T^0 of its intersection with the set $\Sigma \setminus \Sigma_H$ is covered by exactly one set $D_i \cap \Sigma_H$. Since the set $D_i \cap \Sigma_H$ is dense in the component D_i, and the set D_i is closed, the set T^0 lies entirely in the irreducible component D_i of the set Σ.

Suppose that some point a of the set T^0 lies also in some other $(n-1)$-dimensional irreducible component D_j, $D_j \neq D_i$, of the set Σ. However, by the assumption, the set T, hence also its component T^0, is open in the topology of the set Σ. Since the set $D_j \cap \Sigma_H$ is dense in D_j, the set T^0 must contain some points of the set $D_j \cap \Sigma_H$, which is impossible. This contradiction proves the desired result.

Let us now prove that different connected components of the manifold T cannot lie in the same $(n-1)$-dimensional irreducible component of the set Σ. Indeed, if we remove all singular points and also points not lying in the manifold T from this $(n-1)$-dimensional irreducible component, then what is left is a connected manifold. Therefore, it is covered by exactly one connected component of the manifold T. This completes the proof of the lemma. □

Proposition 7.8 *Suppose that a pair consisting of an n-dimensional analytic manifold and its analytic subset Σ has constant topology along a connected stratum $B \subset \Sigma$ (see Sect. 7.2.2). Then every $(n-1)$-dimensional irreducible component of the set Σ is either disjoint from the stratum B or contains the entire stratum.*

Proof We first consider the local case. Suppose that the manifold B coincides with the set K_a, and the manifold M coincides with its neighborhood $\pi^{-1}(K_a)$ (in the notation of Requirement 2 from Sect. 7.2.2). We will show that in this case, the closure in M of every $(n-1)$-dimensional irreducible component of the set Σ coincides with the set K_a.

We will use notation introduced in Sect. 7.2.2. Let $F_a^0 \subset F_a$ be the set consisting of the points of the analytic set F_a in a neighborhood of which the set F_a is an analytic hypersurface in the manifold V_a. The set F_a^0 splits into connected components $F_a^{0,i}$. The complement $F_a \setminus F_a^0$ has smaller complex dimension than the set F_a.

The homeomorphism that appears in Requirement 2 takes the set F_a^0 to the set Σ. It follows from Lemma 7.7 that this homeomorphism takes the sets $F_a^{0,i} \times K_a$ to different irreducible $(n-1)$-dimensional components of the set Σ; moreover, the image of each of the sets $F^{0,i} \times K_a$ is dense in the corresponding irreducible component of the set Σ, and each of the $(n-1)$-dimensional components of Σ contains the image of some set $F^{0,i} \times K_a$.

Furthermore, for every connected component $F_a^{0,i}$, the point a is a limit point of that component (components for which this is not the case do not intersect a small neighborhood of a and thus are not in the set F_a^0). Thus the closure of each of the sets[3] $F^{0,i} \times K_a$ contains the set K_a. Therefore, each of the $(n-1)$-dimensional irreducible components of the set Σ contains the set K_a (the homeomorphism that appears in Requirement 2 from Sect. 7.2.2 is the identity map on the base K_a).

We have worked out the local case. Suppose now that the manifold M is in a small neighborhood of a stratum B. Namely, suppose that M coincides with the neighborhood U of the stratum B as in Requirement 2. In this case, the stratum B is covered by domains K_{a_j}. In each of these domains, the argument given above works. Hence if an irreducible $(n-1)$-dimensional component D_i of the set Σ intersects the set $\pi^{-1}(K_{a_j})$, then its closure contains the entire neighborhood K_{a_j}. Thus, the set of limit points of the component D_i lying in the stratum B is open in the topology of the stratum B. Hence, since the stratum B is connected, it must be contained in the closure of the component D_i.

Let us now proceed with the general case. If an irreducible $(n-1)$-dimensional component of the set Σ does not intersect the neighborhood U of the stratum B, then the stratum B contains no limit points of this component. If it intersects the domain U, then the preceding argument applies, which shows that the closure of the component contains the entire stratum B.

This completes the proof of the proposition. □

Theorem 7.9 *Suppose that a pair consisting of an n-dimensional analytic manifold M and its analytic subset Σ has constant topology along a connected stratum $B \subset \Sigma$. Then every function f that is analytic in the complement $M \setminus \Sigma$ of the set Σ in the manifold M and bounded in some neighborhood of a point $a \in B$ admits an analytic extension to a neighborhood of the stratum B.*

Proof Every $(n-1)$-dimensional irreducible component D_i of the set Σ not containing the point a does not intersect the stratum B (see Proposition 7.8). Therefore, the union D_a of irreducible $(n-1)$-dimensional components of the set not containing the point a is a closed set not intersecting the stratum B. The theorem now follows from Proposition 7.6. □

7.2.4 *Covers Over the Complement of a Subset of Hausdorff Codimension Greater Than 1 in a Manifold*

In topological Galois theory, Riemann surfaces play the role of fields, and their monodromy groups play the role of Galois groups. For this, we need to require that the Riemann surfaces we consider have some reasonable topological properties. For example, Riemann surfaces that are locally trivial covers have these properties. However, the class of locally trivial covers is too narrow and is not sufficient for our purposes. In this subsection, we describe a class of covering manifolds over $M \setminus \Sigma$,

---

[3] More precisely, of their images under the homeomorphism of Requirement 2.

where M is a manifold with a distinguished subset Σ, that is in a sense small. In the one-dimensional version of topological Galois theory (see Chap. 5), we talk about functions whose Riemann surfaces cover the complex line, on which a countable (possibly dense) set of points Σ is marked. In this section, covering manifolds over a manifold M with a marked analytic subset Σ play a key role (see Sect. 7.2.5).

Let (M, Σ) be a pair consisting of a connected real manifold M and a subset $\Sigma \subset M$ such that the complement $M \setminus \Sigma$ is locally path connected and dense in the manifold M. As an example of such a set Σ, we can take any subset in the manifold M whose Hausdorff codimension is greater than 1. Mark a point b lying in the complement of the set Σ.

Definition 7.10 A connected manifold R together with a marked point c and a projection $\pi : R \to M$ is said to be a *covering manifold over* $M \setminus \Sigma$ with a marked point b if firstly, the map π is a local homeomorphism; secondly, it takes the marked point c to the marked point b, $\pi(c) = b$; and thirdly, for every continuous path γ in the set $M \setminus \Sigma$ originating at the point b, $\gamma : [0,1] \to M \setminus \Sigma$, $\gamma(0) = b$, there exists a lift $\tilde{\gamma}$ of the path γ, $\tilde{\gamma} : [0,1] \to R$, $\pi \circ \tilde{\gamma} = \gamma$, originating at the point c, $\tilde{\gamma}(0) = c$. The uniqueness of the lift follows from the assumption that π is a local homeomorphism.

To make things easier, we will fix some Riemannian metric on the manifold M.

Definition 7.11 We say that a subgroup Γ of the fundamental group $\pi_1(M \setminus \Sigma, b)$ is *open* if for every continuous path $\gamma : [0,1] \to M \setminus \Sigma$, $\gamma(0) = \gamma(1) = b$, whose class belongs to the subgroup Γ, there exists a real number $\varepsilon > 0$ such that every continuous path $\tilde{\gamma} : [0,1] \to M \setminus \Sigma$, $\tilde{\gamma}(0) = \tilde{\gamma}(1) = b$, with the property that for every t, $0 \le t \le 1$, the distance between the points $\gamma(t)$ and $\tilde{\gamma}(t)$ does not exceed ε, also defines an element of Γ.

We associate to every covering manifold $\pi : (R, c) \to (M, b)$ over the set $M \setminus \Sigma$, a subgroup of the fundamental group of the space $(M \setminus \Sigma, b)$. We say that a continuous path $\gamma : [0,1] \to M \setminus \Sigma$, $\gamma(0) = \gamma(1) = b$, is *admissible* for a covering manifold (R, c) if the lift $\tilde{\gamma} : [0,1] \to R$ of this path originating at the point c is a loop, i.e., if $\tilde{\gamma}(1) = c$. It is clear that all admissible paths for a given covering manifold form a subgroup of the fundamental group $\pi_1(M \setminus \Sigma, b)$. We say that this subgroup *corresponds* to the covering manifold (R, c).

Definition 7.12 A covering manifold $\pi : (R, c) \to (M, b)$ over the set $M \setminus \Sigma$ is called *maximal* if it cannot be embedded in a bigger covering manifold, in other words, if the existence of another covering manifold $\pi^1 : (R_1, c_1) \to (M, b)$ over the set $M \setminus \Sigma$ and the existence of an embedding $i : (R, c) \to (R_1, c_1)$ compatible with the projections $\pi = \pi^1 \circ i$ imply that the embedding i is a homeomorphism.

Theorem 7.13 *The following properties hold:*

1. *If a subgroup Γ of the fundamental group of the set $M \setminus \Sigma$ with a marked point b corresponds to a covering manifold $\pi : (R, c) \to (M, b)$ over $M \setminus \Sigma$ with a marked point c, $\pi(c) = b$, then the subgroup Γ is open in $\pi_1(M \setminus \Sigma, b)$.*

2. *For every open subgroup* Γ *of the group* $\pi_1(M \setminus \Sigma, b)$*, there exists a unique maximal covering manifold* $\tilde{\pi}(\Gamma) : (\tilde{R}(\Gamma), c) \to (M, b)$ *over the set* $M \setminus \Sigma$ *corresponding to the subgroup* Γ.
3. *An arbitrary open subset* U *of the manifold* $\tilde{R}(\Gamma)$ *containing the full preimage of the set* $M \setminus \Sigma$ *under the map* $\tilde{\pi}(\Gamma)$ *together with the restriction of the covering projection* $\tilde{\pi}(\Gamma) : U \to M$ *is a covering manifold over* $M \setminus \Sigma$ *corresponding to the group* $\Gamma \subset \pi_1(M \setminus \Sigma, b)$*. Conversely, every covering manifold over* $M \setminus \Sigma$ *corresponding to the subgroup* Γ *can be obtained in this way.*

We will now sketch a proof of this theorem. Let us first prove assertion 1. Suppose that a loop γ in the set $M \setminus \Sigma$ lifts to R, starting at the point c, as a loop. The map $\pi : R \to M$ is a local homeomorphism. Therefore, all loops $\tilde{\gamma}$ lying in the manifold M sufficiently close to the loop γ also lift as loops originating at c. (This is true even if a nearby path $\tilde{\gamma}$ intersects the set Σ.) Therefore, the subgroup Γ corresponding to the covering manifold over the set $M \setminus \Sigma$ is an open subgroup in $\pi_1(M \setminus \Sigma, b)$.

To prove the second assertion, we first of all need to construct the maximal covering manifold $\tilde{\pi}(\Gamma) : (\tilde{R}(\Gamma), c) \to (M, b)$ over $M \setminus \Sigma$ corresponding to an open subgroup Γ of the group $\pi_1(M \setminus \Sigma, b)$.

Definition 7.14 Let Γ be an open subgroup in $\pi_1(M \setminus \Sigma, b)$. A loop γ in the manifold M that starts and ends at the point b is said to be *Γ-admissible* if it has the following property. There exists a number $\varepsilon > 0$ such that every loop in the set $M \setminus \Sigma$ beginning and ending at the point b, $\tilde{\gamma} : [0, 1] \to M \setminus \Sigma$, $\tilde{\gamma}(0) = \tilde{\gamma}(1) = b$, with the property that for every t, $0 \le t \le 1$, the distance between the points $\gamma(t)$ and $\tilde{\gamma}(t)$ does not exceed ε belongs to the group Γ.

To every (not necessarily closed) path $\gamma : [0, 1] \to M$, $\gamma(0) = b$, we can associate the doubly traversed path γ, i.e., a loop obtained as the composition of γ and γ^{-1}.

Definition 7.15 We say that a (not necessarily closed) path $\gamma : [0, 1] \to M$ is *Γ-good* if the following conditions are satisfied:

1. The path γ originates at the marked point $\gamma(0) = b$.
2. The doubly traversed path γ is Γ-admissible.

Let $\Pi(\Gamma, b)$ denote the set of all Γ-good paths. On the set $\Pi(\Gamma, b)$, one can introduce the following equivalence relation. Two Γ-good paths γ_1 and γ_2 are called *Γ-equivalent* if the following conditions are satisfied:

1. The paths γ_1 and γ_2 terminate at the same point, $\gamma_1(1) = \gamma_2(1)$.
2. The composition of the paths γ_1 and γ_2^{-1} is Γ-admissible.

We can now describe the set $\tilde{R}(\Gamma)$ and the map $\tilde{\pi}(\Gamma) : \tilde{R}(\Gamma) \to M$ at the set-theoretic level. The set $\tilde{R}(\Gamma)$ is the quotient set of the set $\Pi(\Gamma, b)$ of all Γ-good paths by the equivalence relation defined above. The map $\tilde{\pi}(\Gamma)$ assigns the terminal point $\gamma(1)$ to each path $\gamma \in \Pi(\Gamma, b)$. The marked point $\tilde{c}$ in the set $\tilde{R}(\Gamma)$ is the equivalence class of the constant path $\gamma(t) = b$.

We now define a topology in the set $\tilde{R}(\Gamma)$: the topology in $\tilde{R}(\Gamma)$ is the smallest (coarsest) topology for which the map $\tilde{\pi}(\Gamma) : \tilde{R}(\Gamma) \to M$ is continuous.

It is easy to see that the manifold $\tilde{R}(\Gamma)$ thus constructed together with the marked point $\tilde{c}$ and the projection $\tilde{\pi}(\Gamma)$ indeed forms a covering manifold over the set $M \setminus \Sigma$ corresponding to the subgroup Γ.

Let us prove that $\tilde{R}(\Gamma)$ is an extension of every other covering manifold $\pi : (R, c) \to (M, b)$ over the set $M \setminus \Sigma$ corresponding to the subgroup Γ. Let $\gamma : [0, 1] \to R$ be any path in the manifold R originating at the point c. Clearly, the projection $\pi \circ \gamma$ of this path is Γ-good in the manifold M.

To every point d in the manifold R, assign the collection $\Pi(c, d, R)$ of all paths $\gamma : [0, 1] \to R$ in the manifold R with initial point c and final point d, $\gamma(0) = c$, $\gamma(1) = d$. It is clear that the projections $\pi \circ \gamma$ of all paths γ in the set $\Pi(c, d, R)$ are Γ-equivalent paths. Therefore, the map assigning to every point d in the manifold R the collection of projections $\pi \circ \gamma$ of all paths γ in the set $\Pi(c, d, R)$ is an embedding of the manifold R into the manifold $\tilde{R}$ defined above.

It is not hard to check the remaining claims of the theorem; this verification is left to the reader.

7.2.5 Covers Over the Complement of an Analytic Set

Proposition 7.16 *Let M be a complex analytic manifold, Σ an analytic subset of M, and $b \in M \setminus \Sigma$ a marked point. Fix some subgroup Γ of the fundamental group $\pi_1(M \setminus \Sigma, b)$. Assume that some Γ-good path (see Definition 7.15) $\gamma_1 : [0, 1] \to M$, $\gamma_1(0) = b$, belongs to the set $M \setminus \Sigma$ for all $0 \le t < 1$, and that its terminal point $a = \gamma(1)$ belongs to the set Σ. Consider any admissible stratification of the set Σ (see Sect. 7.2.2). Let B be a stratum of this stratification that contains the point a, and let $\gamma_2 : [0, 1] \to B$ be any path in this stratum originating at the point a, $\gamma_2(0) = a$. Then the composition of the paths γ_1 and γ_2 is a Γ-good path.*

Proof Let U be a sufficiently small neighborhood of the stratum B that appears in Requirement 2 of Sect. 7.2.2, and $\pi : U \to B$ the corresponding projection. Let $\pi(\Gamma) : (R(\Gamma), c) \to (M \setminus \Sigma, b)$ denote the locally trivial covering corresponding to the group $\Gamma \subset \pi_1(M \setminus \Sigma, b)$. By definition, every path $\gamma_1 : [0, t_1] \to M \setminus \Sigma$, where t_1 is any number strictly less than 1, lifts to the manifold $R(\Gamma)$, so that the lifted path originates at the marked point $c \in R(\Gamma)$. Fix a parameter value t_1 close enough to the value 1 so that the point $b_1 = \gamma(t_1)$ belongs to the set U. Let c_1 denote the point on the lifted path corresponding to the parameter value t_1, $\pi(c_1) = b_1$. Let R_1 be the connected component of the full preimage of the set U under the map $\pi(\Gamma) : R(\Gamma) \to M \setminus \Sigma$ containing the point c_1. The restriction ρ of the map $\pi(\Gamma)$ to the manifold R_1 gives rise to a locally trivial covering $\rho : (R_1, c_1) \to (U \setminus \Sigma, b_1)$. Let Γ_1 denote the subgroup of the fundamental group $\pi_1(U \setminus \Sigma, b_1)$ corresponding to this covering.

Lemma 7.17 *The group Γ_1 contains the kernel of the homomorphism from the fundamental group of the space $U \setminus \Sigma$ to the fundamental group of the stratum B induced by the projection $\pi : U \to B$.*

Proof The restriction of the map $\pi : U \to B$ to the domain $U \setminus \Sigma$ is a locally trivial fibration (see Requirement 2 from Sect. 7.2.2). Let a_1 denote the image of the point b_1 under the projection π, and let $V \setminus F$ denote the fiber of this fibration over the point a_1. From the segment

$$\cdots \to \pi_1(V \setminus F, b_1) \to \pi_1(U \setminus \Sigma, b_1) \to \pi_1(B, a_1) \to \cdots$$

of the long exact sequence of this fibration, it follows that the kernel of the homomorphism we are interested in coincides with the image of the fundamental group of the fiber $\pi_1(V \setminus F, b_1)$. Thus we need to show that the group Γ_1 contains the image of the fundamental group of the fiber. Let $\overline{\gamma} : [0, 1] \to V \setminus F$, $\overline{\gamma}(0) = \overline{\gamma}(1) = b_1$, be any loop contained in the fiber. We will now show that $\overline{\gamma} \in \Gamma_1$. To this end, we need to verify that the composition of the paths $\tilde{\gamma}_1$, $\overline{\gamma}$, and $\tilde{\gamma}_1^{-1}$, where $\tilde{\gamma}_1$ is the restriction of the path γ_1 to the interval $[0, t_1]$, belongs to the group $\Gamma \subset \pi_1(M \setminus \Sigma, b)$. But the composition of the three paths can be regarded as a small perturbation of the path $\gamma_1\gamma_1^{-1}$. By our assumption, the path γ_1 is Γ-good, which means that a small perturbation of the path $\gamma_1\gamma_1^{-1}$ that is disjoint from the set Σ lies in the group Γ. The lemma is thus proved. □

We now continue the proof of Proposition 7.16. Let γ be the composition of the paths γ_1 and γ_2 from the statement of the proposition. We need to show that the path γ is Γ-good, i.e., that every small deformation of the doubly traversed path γ that avoids the set Σ lies in the group Γ. We will first prove this statement for special small deformations not intersecting the set Σ and having the following form. The path $\overline{\gamma}$ must be the composition of paths $\overline{\gamma}_1$, $\overline{\gamma}_2$, and $\overline{\gamma}_3$ that are small deformations of the paths γ_1, $\gamma_2\gamma_2^{-1}$, and γ_1^{-1}, respectively; moreover, the path $\overline{\gamma}_2$ must be a loop. Of course, we assume that $\overline{\gamma}_1(1) = \overline{\gamma}_2(0) = \overline{\gamma}_2(1) = \overline{\gamma}_3(0)$, for otherwise, the composition is not defined. Since the path $\overline{\gamma}_2$ is a loop close to the doubly traversed path γ_2, its projection to the stratum B defines the identity element in the fundamental group of the base. Consider a lift of the path $\overline{\gamma_1}$ to the fibered space $R(\Gamma)$ originating at the point c. Let c_1 denote the terminal point of the lifted path. By the lemma, a lift of the path $\overline{\gamma}_2$ to to the space $R(\Gamma)$ originating at the point c_1 will terminate at the same point c_1. Furthermore, a lift of the path $\overline{\gamma_3}$ to $R(\Gamma)$ originating at the point c_1 must terminate at the point c. Indeed, the composition of the paths $\overline{\gamma}_1$ and $\overline{\gamma}_3$ is a small deformation of the path $\gamma_1\gamma_1^{-1}$. The path γ_1 is Γ-good. Therefore, a lift of the composition of the paths $\overline{\gamma}_1$ and $\overline{\gamma}_3$ to $R(\Gamma)$ originating at the point c must terminate at the same point.

Thus a lift of the composition of the paths $\overline{\gamma}_1$, $\overline{\gamma}_2$, and $\overline{\gamma}_3$ to $R(\Gamma)$ originating at the point c also terminates at the point c. In other words, the composition of these paths belongs to the group Γ. We have proved the desired statement for a special perturbation of the doubly traversed composition of the paths γ_1 and γ_2. It is clear that every small perturbation of this path lying in the set $M \setminus \Sigma$ is homotopic in this set to some special perturbation of the path.

(The doubly traversed composition of the paths γ_1 and γ_2 is the composition of the paths γ_1, $\gamma_2\gamma_2^{-1}$, and γ_1^{-1}. A perturbation of this composition is the composition of three paths l_1, l_2, and l_3, where the path l_2 is close to the loop $\gamma_2\gamma_2^{-1}$ but is not necessarily a loop. Such a composition is clearly homotopic to the composition of close paths $\tilde{l}_1$, $\tilde{l}_2$, and $\tilde{l}_3$, where $\tilde{l}_2$ is a loop.)

This completes the proof of Proposition 7.16. □

7.2.6 The Main Theorem

We now have everything ready for the proof of the main theorem of this section.

Theorem 7.18 (On analytic continuation of a function along an analytic set) *Let M be a complex analytic manifold, Σ an analytic subset of M, and f_b a germ of an analytic function at some point $b \in M$. Suppose that the germ f_b admits an analytic continuation along every path $\gamma : [0,1] \to M$, $\gamma(0) = b$, not intersecting the set Σ for $t > 0$. Suppose also that the germ f_b admits an analytic continuation along some path $\gamma_1 : [0,1] \to M$, $\gamma_1(0) = b$, with the terminal point at a point $a = \gamma_1(1)$ belonging to the set Σ, $a \in \Sigma$. Consider any admissible stratification of the set Σ (see Sect. 7.2.2). Let B be a stratum of this stratification whose closure contains the point a, and let $\gamma_2 : [0,1] \to M$ be any path originating at the point a, $\gamma_2(0) = a$, such that $\gamma_2(t) \in B$ for $t > 0$. Then the germ f_b admits an analytic continuation along the composition of the paths γ_1 and γ_2.*

Proof Suppose that a germ f_b of an analytic function admits an analytic continuation along the path γ. Consider any point $\tilde{b}$ lying within the domain of convergence of the Taylor series for the germ f_b. The germ at the point $\tilde{b}$ of the sum of this series admits a continuation along every path that is sufficiently close to the path γ outside of the domain of convergence of the Taylor series. Therefore, without loss of generality, we can assume in the statement of the theorem that the point b lies in the set $M \setminus \Sigma$, the point a lies in the stratum B, and the path $\gamma_1 : [0,1] \to M$, $\gamma_1(0) = b$, $\gamma_1(1) = a$, does not intersect the set Σ for $0 \leq t < 1$. We will prove the theorem under these assumptions.

Let Γ denote the subgroup of the fundamental group of the space $M \setminus \Sigma$ with the marked point b consisting of all loops in $M \setminus \Sigma$ based at the marked point such that the continuation of f_b along these loops results in the same germ. Consider the maximal covering manifold $\tilde{\pi}(\Gamma) : (\tilde{R}(\Gamma), c) \to (M, b)$ over the set $M \setminus \Sigma$ corresponding to this subgroup Γ (see Definition 7.12). The manifold $\tilde{R}(\Gamma)$ has a natural structure of a complex analytic manifold; this structure is inherited from the analytic structure on M under the map $\tilde{\pi}(\Gamma)$, which is a local homeomorphism. The set $\tilde{\Sigma} = \tilde{\pi}^{-1}(\Gamma)(\Sigma)$ is an analytic subset of this manifold $\tilde{R}(\Gamma)$. By definition, the germ $\tilde{f}_c = \pi^* f_b$ regarded as a germ at the point c of an analytic function on the analytic manifold $\tilde{R}(\Gamma)$ admits an analytic continuation to the entire manifold

$\tilde{R}(\Gamma) \setminus \tilde{\Sigma}$ and defines a single-valued analytic function $\tilde{f}$ there. Every path $\gamma : [0, 1] \to M$, $\gamma(0) = b$, such that the germ f_b has an analytic continuation along it is a Γ-good path. Indeed, the analytic continuation of the germ f_b along the doubly traversed path γ as well as along some loop $\tilde{\gamma} : [0, 1] \to M$, $\tilde{\gamma}(0) = \tilde{\gamma}(1) = b$, close enough to $\gamma\gamma^{-1}$ results in the same germ f_b with which we began.

In particular, the path $\gamma_1 : [0, 1] \to M$, $\gamma_1(0) = b$, $\gamma_1(1) = a \in \Sigma$, along which we continue the germ f_b and which appears in the statement of the theorem is Γ-good. Therefore, there exists a lift of the path γ_1 to $\tilde{R}(\Gamma)$ that originates at the point c. Let $\tilde{a}$ denote the terminal point of this lifted path. By Proposition 7.16, for every path γ_2 originating at the point a and lying in the stratum B, the composition of the paths γ_1 and γ_2 is a Γ-good path. Therefore, there exists a lift of this composition to $\tilde{R}(\Gamma)$ originating at the point c. In other words, this means that every path lying in the stratum B and originating at the point a lifts to $\tilde{R}(\Gamma)$ starting at the point $\tilde{a}$. Let $\tilde{B}$ be the connected component of the full preimage of B under the projection $\tilde{\pi}(\Gamma)$ that contains the point $\tilde{a}$. We have proved that the restriction of the map $\tilde{\pi}(\Gamma)$ to $\tilde{B}$ defines a locally trivial covering over the stratum B. It is clear that the topology of the pair consisting of the manifold $\tilde{R}(\Gamma)$ and the set $\tilde{\Sigma}$, which is the full preimage of the set Σ under the projection $\tilde{\pi}(\Gamma)$, is constant along the stratum $\tilde{B}$. Indeed, locally, the triple $(\tilde{R}(\Gamma), \tilde{\Sigma}, \tilde{B})$ is homeomorphic to the triple (M, Σ, B), and the topology of the pair (M, Σ) is constant along the stratum B by our assumptions.

We can now apply Theorem 7.9 to the germ $\tilde{f}_c = \pi^* f_b$ of a single-valued analytic function on the manifold $\tilde{R}(\Gamma) \setminus \tilde{\Sigma}$, which can be extended to a neighborhood of the point $\tilde{a} \in \tilde{B}$. This concludes the proof of Theorem 7.18. □

7.3 On the Monodromy of a Multivalued Function on Its Ramification Set

A multivalued analytic function on $\mathbb{C}^n$ is called a $\mathscr{S}$-function if the set of its singular points can be covered by a countable union of analytic subsets (and hence occupies only a very small part of the space $\mathbb{C}^n$). Under a map $\pi : (Y, y_0) \to (\mathbb{C}^n, a)$ of a topological space Y to $\mathbb{C}^n$, the germ f_a of an $\mathscr{S}$-function f at a point a can induce a multivalued function on the space Y. For this, we need to require that the germ f_b have analytic continuation along the image of every path in the space Y originating at the point y_0. This is possible even when the point a belongs to the set of singular points of the function f (some of the germs of the multivalued function f can be nonsingular even at singular points of this function) and Y is mapped to this set.

Can we estimate the monodromy groups of the multivalued functions thus obtained through the monodromy group of the initial $\mathscr{S}$-function f (is it true, e.g., that if the monodromy group of the $\mathscr{S}$-function f is solvable, then the monodromy group of every induced function is also solvable)? In this section, we obtain an affirmative answer to this question (see Sects. 7.3.4 and 7.3.5). This is far from being obvious if the set of singular points of f is not closed. Note, by the way, that

$\mathscr{S}$-functions whose sets of singular points are not closed are not anything unusual. Most multivalued elementary functions are like that (see Sect. 5.3).

The description of a connection between the monodromy group of an initial $\mathscr{S}$-function and the monodromy groups of the induced functions has led to the notion of induced closure of groups (see Sect. 7.3.2). The use of this operation, in turn, forces us to revisit the definitions of various classes of group pairs (see Sect. 7.3.5) appearing in the one-dimensional version of topological Galois theory (see Sects. 5.5.5 and 5.7.1). In this section, we give definitions that will allow us to work with multivariate functions having dense sets of singular points and monodromy groups of cardinality of the continuum.

7.3.1 $\mathscr{S}$-Functions

In the one-dimensional version of topological Galois theory, $\mathscr{S}$-functions, i.e., multivalued analytic functions of one variable, whose sets of singular points are at most countable play a key role. We now generalize the notion of an $\mathscr{S}$-function to the multidimensional case. All statements in this subsection can be proved in the same way as the similar one-dimensional statements (see Sect. 5.4); hence we will not give proofs here.

A subset A in a connected k-dimensional analytic manifold M is called *meager* if there exist a countable set of open domains $U_i \subset M$ and a countable collection of proper analytic subsets $A_i \subset U_i$ in those domains such that $A \subset \bigcup A_i$. A multivalued analytic function on the manifold M is called an *$\mathscr{S}$-function* if its set of singular points is meager. Let us make this definition more precise.

Two regular germs f_a and g_b defined at points a and b of the manifold M are called *equivalent* if the germ g_b can be obtained from the germ f_a by regular continuation along some path. Every germ g_b equivalent to a germ f_a is also called the germ of the multivalued analytic function f generated by the germ f_a. A point $b \in M$ is said to be *singular* for the germ f_a if there exists a path $\gamma : [0,1] \to M$, $\gamma(0) = a$, $\gamma(1) = b$, along which the germ f_a has no regular continuation, but for every t, $0 \leq t < 1$, there is a regular continuation of the germ f_a along the truncated path $\gamma : [0,t] \to M$. It is easy to see that equivalent germs have the same sets of singular points. A germ is said to be an $\mathscr{S}$-germ if the set of its singular points is meager. A multivalued analytic function is called an $\mathscr{S}$-function if each of its germs is an $\mathscr{S}$-germ.

Let us fix an arbitrary Riemannian metric ρ on M.

Lemma 7.19 (On releasing a path from a meager set) *Let A be a meager subset of the manifold M, $\gamma : [0,1] \to M$ a continuous path, and φ a continuous positive function on the interval $0 < t < 1$. Then there exists a path $\hat{\gamma} : [0,1] \to M$ such that for $0 < t < 1$, we have $\hat{\gamma}(t) \notin A$ and $\rho(\gamma(t), \hat{\gamma}(t)) < \varphi(t)$.*

Besides the set of singular points, it is also convenient to consider other sets such that the function admits an analytic continuation everywhere in the complement.

A meager set A is called a *forbidden set* for a regular germ f_a if the germ f_a admits a regular continuation along every path $\gamma(t)$, $\gamma(0) = a$, never intersecting the set A except possibly at the initial moment.

Theorem 7.20 (On forbidden sets) *A meager set is a forbidden set of a germ if and only if it contains the set of its singular points. In particular, a germ has a forbidden set if and only if it is a germ of an $\mathcal{S}$-function.*

The monodromy group of an $\mathcal{S}$-function f with a forbidden set A (or the A-monodromy group for short) is the group of all permutations of branches of f that correspond to loops around A. Let us discuss this in more detail.

Let X be the set of all germs of the $\mathcal{S}$-function f at the point a not lying in the set A. Consider a loop γ in $M \setminus A$ based at the point a. The continuation of every germ from the set X along the loop γ gives another germ from X. Thus every loop γ gives rise to a self-map of the set X; moreover, homotopic loops give rise to the same map. The composition of loops corresponds to the composition of maps. In this way, we obtain a homomorphism τ from the fundamental group of the set $M \setminus A$ to the group $S(A)$ of all one-to-one transformations of the set X. (Here and in the rest of the section, we mean the following group structure in the group $S(A)$: if f and g are bijective transformations of the set X, then their product fg in the group $S(A)$ is defined as the composition $g \circ f$ of the maps f and g). The A-monodromy group of an $\mathcal{S}$-function is defined as the image of the fundamental group $\pi_1(M \setminus A, a)$ in the group $S(X)$ under the homomorphism τ.

The A-monodromy group is not only an abstract group but also a transitive group of permutations of the function branches (the transitivity can be easily deduced from the lemma on releasing a path from a meager set). Algebraically, such an object is determined by a pair of groups: a group of permutations and its subgroup that is the stabilizer of some element.

The monodromy pair of an $\mathcal{S}$-function with a forbidden set A is defined as a pair of groups consisting of the A-monodromy group and the stabilizer of some branch of this function. This pair of groups is well defined, i.e., it depends neither on the choice of the point a nor on the choice of a branch of the function, up to isomorphism. When the forbidden set coincides with the set of singular points of the function, we do not mention the forbidden set and speak simply of the *monodromy group* and *monodromy pair* of this function. When the set of singular points of the function is not closed, the A-monodromy group may happen to be a proper subset of the monodromy group of this function.

The group $S(X)$ is equipped with a natural topology (see Sects. 5.5.2 and 7.3.3).

Proposition 7.21 *The closure in the group $S(X)$ of the A-monodromy group of an $\mathcal{S}$-function f does not depend on the choice of a forbidden set A, and it contains the monodromy group of the function f. Moreover, the closure of the stabilizer of some fixed branch f_a under the action of the A-monodromy group contains the stabilizer of this branch under the action of the monodromy group.*

7.3.2 Almost Homomorphisms and Induced Closures

We will need a construction that to every group of transformations of a set X, associates some group of transformations of a subset L of X (see Sect. 7.3.3). To study its properties, it will be convenient to use the notions of an almost homomorphism and an induced closure.

Let T be a topological space, and S a group lying in T.

Definition 7.22 A map $J : G \to T$ of the group G into the space T is said to be an *almost homomorphism near the group S* if the following hold:

1. The map J takes the identity element of G to the identity element of S.
2. For every point a of the group S and every neighborhood V of the point a^{-1} in the space T, there exists a neighborhood U of the point a in the space T such that for every point $\tilde{a} \in G$ for which $J(\tilde{a}) \in U$, we also have $J(\tilde{a}^{-1}) \in V$.
3. For every pair of points a, b of the group S and every neighborhood V of the point ab in the space T, there exist neighborhoods U_1 and U_2 of the points a and b in the space T such that for every pair of points $\tilde{a}, \tilde{b}$ of the group G for which $J(\tilde{a}) \in U_1$, $J(\tilde{b}) \in U_2$, we also have $J(\tilde{a}\tilde{b}) \in V$.

The main example of an almost homomorphism near a group is described in Sect. 7.3.3.

Definition 7.23 The *induced closure* $\overline{G}(S)$ of a group G in the group S with respect to an almost homomorphism $J : G \to T$ near the group S is defined as the intersection of the group S with the closure $\overline{J}(G)$ in the space T of the image of the group G under the map J.

We now describe some properties of induced closure. First of all, the induced closure $\overline{G}(S)$ is a subgroup of the group S. Furthermore, the restriction $J : G_1 \to T$ of an almost homomorphism $J : G \to T$ of the group G near the group S to a subgroup G_1 of the group G is clearly an almost homomorphism of G_1 near the group S. Therefore, the induced closure is defined for all subgroups in the group G simultaneously.

Let $\mathscr{A} = \{A_1, \dots, A_n\}$ be an alphabet with n symbols. A *word* W in this alphabet is an expression of the form $W = A_{i_1}^{k_1} \dots A_{i_N}^{k_N}$, where each A_{i_j}, $j = 1, \dots, N$, belongs to $\mathscr{A}$, and each of the exponents $k_1, \dots, k_N$ is equal to ± 1. The number $k = |k_1| + \dots + |k_N|$ is called the length of the word W. For every group Π and every finite sequence $\Sigma = \{a_1, \dots, a_n\}$ of elements of Π, the *value* $W(a_1, \dots, a_n)$ of the word W evaluated at Σ is defined as the element $a_{i_1}^{k_1} \cdots a_{i_n}^{k_n}$ of the group Π.

Proposition 7.24 *Let $J : G \to T$ be an almost homomorphism of the group G near the group S. Then for every word W, every collection of elements $a_1, \dots, a_n$ in the group S, and every neighborhood V in the space T of the element $W(a_1, \dots, a_n) \in S$, there exist neighborhoods $U_1, \dots, U_n$ in the space T of the elements $a_1, \dots, a_n$*

such that for all points $\tilde{a}_1, \dots, \tilde{a}_n$ for which $J(\tilde{a}_1) \in U_1, \dots, J(\tilde{a}_n) \in U_n$, we also have $J(W(\tilde{a}_1, \dots, \tilde{a}_n)) \in V$.

Proof We will argue by induction on the length k of the word W. For the only nontrivial word $W = A_{i_1}^{-1}$ of length 1, the statement holds by definition of an almost homomorphism. Every word W of length $k > 1$ has either the form $A_{i_1} W_1$ or the form $A_{i_1}^{-1} W_1$, where W_1 is a word of length $k - 1$. In each of these two cases, by definition of an almost homomorphism, for every neighborhood V of the point $W(a_1, \dots, a_n)$ there exist neighborhoods V_1 and V_2 of the points a_{i_1} and $W_1(a_1, \dots, a_n)$ such that if the elements $\tilde{a}_1, \dots, \tilde{a}_n$ of the group G satisfy the inclusions $J(\tilde{a}_{i_1}) \in V_1$ and $J(W_1(\tilde{a}_1, \dots, \tilde{a}_n)) \in V_2$, then $J(W(\tilde{a}_1, \dots, \tilde{a}_n)) \in V$. By the induction hypothesis, there exist neighborhoods $U_1, \dots, U_n$ of the points $a_1, \dots, a_n$ such that if elements $\tilde{a}_1, \dots, \tilde{a}_n$ satisfy the inclusions $J(\tilde{a}_1) \in U_1, \dots, J(\tilde{a}_n) \in U_n$, then $J(W(\tilde{a}_1, \dots, \tilde{a}_n)) \in V_2$. Hence, if a point $\tilde{a}_{i_1}$ satisfies the inclusion $J(\tilde{a}_{i_1}) \in U_{i_1} \cap V_1$, and the elements $\tilde{a}_j$ for $j \neq i_1$ satisfy the inclusions $J(\tilde{a}_j) \in U_j$, then $J(W(\tilde{a}_1, \dots, \tilde{a}_n)) \in V$. The proposition is proved. □

Theorem 7.25 *For every almost homomorphism $J : G \to T$ of the group G near the group S and every normal subgroup G_1 of the group G for which the quotient group G/G_1 is abelian, the induced closures $\overline{G}_1(S), \overline{G}(S)$ of the groups G_1, G in the group S with respect to the homomorphism J satisfy the following conditions: the group $\overline{G}_1(S)$ is a normal subgroup of the group $\overline{G}(S)$, and the quotient group $\overline{G}(S)/\overline{G}_1(S)$ is abelian.*

Proof We need to prove that for every pair of elements a, b of the group $\overline{G}(S)$, the element $aba^{-1}b^{-1}$ belongs to the group $\overline{G}_1(S)$, i.e., that in every neighborhood V of the element $aba^{-1}b^{-1}$, there are elements lying in the image $J(G_1)$ of the group G_1. By Proposition 7.24 applied to the word $W = A_1 A_2 A_1^{-1} A_2^{-1}$, there exist neighborhoods U_1 and U_2 of the elements a and b such that for a pair of elements $\tilde{a}, \tilde{b}$ of the group G for which $J(\tilde{a}) \in U_1$, $J(\tilde{b}) \in U_2$, we also have $J\left(\tilde{a}\tilde{b}\tilde{a}^{-1}\tilde{b}^{-1}\right) \in V$. The elements a and b belong to the group $\overline{G}(S)$; hence there exist elements $\tilde{a}$ and $\tilde{b}$ of the group G for which these relations hold. For such pair of elements $\tilde{a}, \tilde{b}$, the element $\tilde{a}\tilde{b}\tilde{a}^{-1}\tilde{b}^{-1}$ lies in the group G_1, since the quotient group G/G_1 is abelian. Thus we have found an element belonging to the image $J(G_1)$ of the group G_1 in the given neighborhood V of the element $aba^{-1}b^{-1}$. This completes the proof of the theorem. □

Theorem 7.26 *For every almost homomorphism $J : G \to T$ of the group G near the group S and every subgroup G_1 of the group G having finite index k in the group G, the induced closures $\overline{G}_1(S), \overline{G}(S)$ of the groups G_1, G in the group S with respect to the homomorphism J satisfy the following condition: the group $\overline{G}_1(S)$ is a subgroup of finite index in the group $\overline{G}(S)$; moreover, this index is less than or equal to k.*

Proof Let $R_1, \dots, R_k$ be right cosets of the subgroup G_1 in the group G. Let P_i, $i = 1, \dots, k$, denote the intersection of the group S with the closure of the image

$J(R_i)$ of the coset R_i under the map J. We will show that every right coset of the subgroup $\overline{G}_1(S)$ in the group $\overline{G}(S)$ coincides with one of the sets $P_1, \dots, P_k$. This implies the statement of the theorem immediately, since this means that there are no more than k right cosets.

We now show that the union of the sets $P_1, \dots, P_k$ coincides with the group $\overline{G}(S)$. Indeed, the group G is the union of the right cosets $R_1, \dots, R_k$; therefore, $J(G) = \bigcup_{i=1}^k J(R_i)$. This implies that $\overline{J}(G) = \bigcup_{i=1}^k \overline{J}(R_i)$. Intersecting both parts of this equality with the group S, we obtain $\overline{G}(S) = \bigcup_{i=1}^k P_i$.

Let a be a point in the group $\overline{G}(S)$, and let $a\overline{G}_1(S)$ be the right coset containing this point. By the above, the point a lies in one of the subsets P_i. Let us show that this set P_i includes the entire coset $a\overline{G}_1(S)$. Indeed, every point of this coset has the form $b = ac$, where $c \in \overline{G}_1(S)$. All three points a, b, c belong to the group S. By definition of an almost homomorphism, for every neighborhood V of the point b there exist neighborhoods U_1 and U_2 of the points a and c such that if $J(\tilde{a}) \in U_1$ and $J(\tilde{c}) \in U_2$, then $J(\tilde{a}\tilde{c}) \in V$. The points $\tilde{a}$ and $\tilde{c}$ satisfying these relations can be chosen respectively in the coset R_i and the group G_1, since $a \in \overline{J}(R_i)$, $c \in \overline{G}_1(S)$. For this choice of points $\tilde{a}$ and $\tilde{c}$, the point $\tilde{a}\tilde{c}$ belongs to the coset R_i. Therefore, the point b lies in the set P_i, as desired.

Suppose that the set P_i is nonempty, and $a \in P_i$ is one of its points. We will prove that the set P_i is contained in the right coset $a\overline{G}_1(S)$. Let b be any point in P_i. Consider the element $c = a^{-1}b$. Let us show that $c \in \overline{G}_1(S)$. For this, we need to prove that every neighborhood V of the element c intersects the image $J(G_1)$ of the group G_1. All three points a, b, c belong to the group S. By definition of an almost homomorphism, for every neighborhood V of the point c, there exist neighborhoods U_1 and U_2 of the points a and b such that if $J(\tilde{a}) \in U_1$ and $J(\tilde{b}) \in U_2$, then $J(\tilde{a}^{-1}\tilde{b}) \in V$. The points $\tilde{a}, \tilde{b}$ satisfying these relations can be chosen in the coset R_i, since $a, b \in P_i$. For such choice of the points $\tilde{a}, \tilde{b}$, the point $\tilde{a}^{-1}\tilde{b}$ lies in the group G_1. Therefore, the point c lies in the group $\overline{G}_1(S)$, as desired, which completes the proof of the theorem. □

7.3.3 *Induced Closure of a Group Acting on a Set in the Transformation Group of a Subset*

We now describe a principal example of an almost homomorphism J near a group.

The topological space $T = T(L, X)$ Let X be an arbitrary set, and $L \subseteq X$ any subset of it. Consider the space $T = T(L, X)$ of all maps from the set L to the set X, equipped with the following topology. For every map $f : L \to X$ and every finite subset $K \subset L$, let $U_K(f)$ denote the set of maps from L to X that coincide with the map f on the set K. By definition, a basis of neighborhoods of an element f in the space $T(L, X)$ consists of the sets $\{U_K(f)\}$, where the index K runs through all finite subsets of L. In other words, the topology in the space $T(L, X)$ can be described as the topology of pointwise convergence of maps from

L to X with respect to the discrete topology on X. For infinite sets L, the topology on $T(L, X)$ is nontrivial and will be useful in what follows.[4]

The group $S = S(L) \subset T(L, X)$ The group $S(L)$ consisting of all one-to-one transformations of the set L can be naturally embedded in the space $T(L, X)$: every one-to-one transformation $F : L \to L$ can be regarded as a map $f : L \to X$, since $L \subset X$.

The group G and the map $J : G \to T(L, X)$ As the group G, we take any subgroup of the group $S(X)$ of one-to-one transformations of the set X. As a map $J : G \to T(L, X)$ we consider the map taking every transformation $f : X \to X$ from the group G to its restriction to the set L, i.e., $J(f) = f|_L$.

Theorem 7.27 *In the situation described by the space T and groups S and G described above, the map $J : G \to T(L, X)$ is an almost homomorphism near the group $S = S(L)$.*

Proof The restriction of the identity self-map of X to the subset L is the identity self-map of L. Therefore, the map J takes the identity element of the group G to the identity element of the group $S(L)$.

Suppose that $a \in S(L)$. Fix a finite subset $K \subset L$, and consider the neighborhood $V = U_K(a^{-1})$ of the element a^{-1} in the space $T(L, X)$. Let K_1 denote the image of the set K under the map $a : L \to L$. Let $\tilde{a}$ be a transformation from $S(X)$, and $J(\tilde{a})$ its restriction to L, $J(\tilde{a}) = \tilde{a}|_L$. Consider the neighborhood $U_1 = U_{K_1}(a)$ of the element a. If $J(\tilde{a}) \in U_1$, then $J(\tilde{a}^{-1})|_K = a|_K$.

Take $a, b \in S(L)$, and let $ab \in S(L)$ be their composition $b \circ a$. Fix any finite set $K \subset L$ and consider the neighborhood $V = U_K(ab)$ of the element ab in the space $T(L, X)$. Let K_1 denote the image of the set K under the map $a : L \to L$. Let $\tilde{a}, \tilde{b}$ be transformations from $S(X)$, and $J(\tilde{a}), J(\tilde{b})$ their restrictions to the set L, $J(\tilde{a}) = \tilde{a}|_L$, $J(\tilde{b}) = \tilde{b}|_L$.

Consider the neighborhoods $U_1 = U_K(a)$ and $U_2 = U_{K_1}(b)$ of the elements a and b. If $J(\tilde{a}) \in U_1$ and $J(\tilde{b}) \in U_2$, then $J(\tilde{a}\tilde{b}) \in V$. Indeed, if $\tilde{b}|_{K_1} = b|_{K_1}$ and $\tilde{a}|_K = a|_K$, then $\tilde{a}\tilde{b}|_K = ab|_K$. □

7.3.4 The Monodromy Groups of Induced Functions

To every single-valued analytic function f, we can associate its jet extension $\mathscr{F}$ that takes every point x to the germ of f at that point. Similarly, we can talk about the jet extension $\mathscr{F}$ of a multivalued analytic function f: this is a multivalued function taking values in the set of all germs of the function f and mapping a point x to all regular germs of the function f at that point.

[4]Note that the topology of the space $T(L, X)$ is the same as the direct product topology on X^L.

Let f be a multivalued $\mathcal{S}$-function on the space $\mathbb{C}^n$, and f_a some distinguished germ of the function f at the point a. A continuous map $\pi : (Y, y_0) \to (\mathbb{C}^n, a)$ from a path-connected topological space Y with a marked point y_0 to the space $\mathbb{C}^n$ such that $\pi(y_0) = a$ is said to be *admissible for the germ* f_a if the germ f_a admits an analytic continuation along the image of every path in the space Y originating at the marked point y_0.

Remark 7.28 Typical examples of spaces Y with which we will need to deal are given by such locally non-simply-connected spaces as complements on the complex line $\mathbb{C}$ to countable dense subsets $A \subset \mathbb{C}$, i.e., $Y = \mathbb{C} \setminus A$.

To a map $\pi : (Y, y_0) \to (\mathbb{C}^n, a)$ admissible for a germ f_a, we associate the *multivalued function* $\pi^*\mathcal{F}$ *on the space* Y taking values in the set of all germs of the multivalued function f at points of the space $\mathbb{C}^n$. Namely, every value of the multivalued function $\pi^*\mathcal{F}$ at a point $y \in Y$ is a germ of a function f at the point $\pi(y) \in \mathbb{C}^n$ obtained by analytic continuation of the germ f_a along some path of the form $\pi \circ \gamma : [0, 1] \to \mathbb{C}^n$, where $\gamma : [0, 1] \to \mathbb{C}^n$ is a path in the space Y originating at the point y_0 and terminating at the point y, i.e., $\gamma(0) = y_0$, $\gamma(1) = y$. For every multivalued function $\pi^*\mathcal{F}$ on the space Y, the monodromy group (which may well have cardinality of the continuum) and the monodromy pair of a germ f_a of this function $\pi^*\mathcal{F}$ at a point y_0 are defined in the same way as for an $\mathcal{S}$-function.

Let L_a denote the collection of all germs of the function f at the point a that are the values of the multivalued function $\pi^*\mathcal{F}$ at the point y_0. The monodromy group M of the function $\pi^*\mathcal{F}$ is a transitive group of one-to-one transformations of the set L_a. We need to relate the pair M_0, M, in which M_0 is the stabilizer of the germ f_a in the group M, to the monodromy pair of the $\mathcal{S}$-function f. To this end, we need the identifications described below.

Let O be the set of singular points of the function f, and $x \notin O$ any nonsingular point in the space $\mathbb{C}^n$. Let X denote the set of all germs of the function f at the point x. Fix a path $\delta : [0, 1] \to \mathbb{C}^n$ connecting the points a and x, $\delta(0) = a$, $\delta(1) = x$, and intersecting the set of singular points of the function f at most at the initial moment, i.e., $\delta(t) \notin O$ for $t > 0$. The existence of such a path δ follows from the lemma on releasing a path from a meager set (Lemma 7.19). Each germ of the function f lying in the set L_a admits an analytic continuation along the path δ. Indeed, the Taylor series of every germ in the set L_a converges at points $\delta(t)$ of the path δ for sufficiently small t, $0 \le t \le t_0$. For $t \ge t_0$, there are no obstructions to continuation of the germ, since by our assumption, for $t > 0$, the points $\gamma(t)$ do not lie in the set O.

Identify every germ $f_{i,a}$ in the set L_a with the germ of the function $f_{i,x}$ at the point x obtained by continuation of the germ $f_{i,a}$ along the path δ. In this way, the set L_a can be identified with some subset L_x of the set X, the distinguished germ f_a can be identified with some distinguished germ $f_x \in X$, the monodromy group M of the function $\pi^*\mathcal{F}$ can be identified with some transitive group $M(x)$ of transformations of the set L_x, and the stabilizer of the germ f_a in M can be identified with the stabilizer $M_0(x)$ of the germ f_x in the group $M(x)$. Let G denote the monodromy group of the function f, regarded as a group of transitive one-to-one transformations of the set X. Let G_0 denote the stabilizer of the germ f_x in the group G.

The groups G_0, G will be viewed as groups of transformations of the set X including the subset $L_x \subset X$.

Theorem 7.29 *The induced closure $\overline{G}(S)$ of the monodromy group G of the function f in the group $S = S(L_x)$ of one-to-one transformations of the set L_x includes the monodromy group $M(x)$ of the function $\pi^*\mathcal{F}$. Moreover, the stabilizer $M_0(x)$ is equal to the intersection of the group $M(x)$ with the induced closure $\overline{G}_0(S)$ of the stabilizer G_0 of the germ f_x in the group G.*

Proof If a germ g of an analytic function admits an analytic continuation along some path $\gamma : [0,1] \to \mathbb{C}^n$, then it admits an analytic continuation along every path $\tilde{\gamma}$ with the same endpoints, $\gamma(0) = \tilde{\gamma}(0)$, $\gamma(1) = \tilde{\gamma}(1)$, sufficiently close to the path γ. Moreover, the continuations of the germ g along the paths γ and $\tilde{\gamma}$ yield the same result. The proof of the theorem is based on this analytic fact. Using the identifications we have discussed, every transformation $m : L_x \to L_x$ in the group $M(x)$ is obtained by simultaneous analytic continuation of the germs from the set L_x along some path γ of the form $\delta\gamma_1\delta^{-1}$, where δ^{-1} is the path δ traversed in the opposite direction, and the path γ_1 is the pushforward under the map π of some path $\gamma_2 : [0,1] \to Y$ in the space Y originating and terminating at the point y_0. The endpoints of the path γ coincide with the point $x \in \mathbb{C}^n$, but the path γ may intersect the singular point set O of the function f. Fix any finite set K of germs in L_x. Perturb the path γ slightly with fixed endpoints so as not to change analytic continuations of the finite set K of germs along the path and to make the perturbed path $\tilde{\gamma}$ not intersect O. This is possible due to the analytic fact given at the beginning of the proof and due to the lemma on releasing a path from a meager set.

All germs in the set X admit analytic continuations along the path $\tilde{\gamma}$, since the path $\tilde{\gamma}$ does not intersect the set O. The transformation $g : X \to X$ corresponding to the path $\tilde{\gamma}$ belongs to the monodromy group G of the function f. Thus, for a neighborhood U_K of the transformation $m : L_x \to L_x$ lying in the group $M(x)$, we have defined a transformation $g : X \to X$ from the group G such that $m|_K = g|_K$. Therefore, $M(x) \subset \overline{G}(S)$.

Furthermore, the subgroup $M_0(x)$ consists of transformations from the group $M(x)$ mapping the point f_x to itself. For finite sets $K \subset L_x$ containing the point f_x, every transformation $g : X \to X$ whose restriction to the set K coincides with that of some transformation $m : L \to L$, where $m \in M_0$, also maps the point f_x to itself. Therefore, $M_0 = M \cap \overline{G}_0(S)$, completing the proof of the theorem. □

7.3.5 *Classes of Group Pairs*

In the one-dimensional version of topological Galois theory, it is described how the monodromy pairs of functions of a single variable change under composition, differentiation, integration, and so on. To this end, some notions are used that concern group pairs (see Sects. 5.5.5, and 5.7.1). For functions of several variables, Theorem 7.29 forces us to modify these notions slightly. We now recall the definitions and perform the necessary modifications.

A *group pair* is always a pair consisting of a group and some subgroup of it. Moreover, a group is identified with the group pair consisting of that group and its trivial subgroup (containing only the identity element).

Definition 7.30 A collection $\mathscr{L}$ of group pairs is called an *almost complete class of group pairs* if for every group pair $[\Gamma, \Gamma_0] \in \mathscr{L}$, where $\Gamma_0 \subseteq \Gamma$, the following hold:

1. For every homomorphism $\tau : \Gamma \to G$, where G is some group, the group pair $[\tau\Gamma, \tau\Gamma_0]$ is also an element of $\mathscr{L}$.
2. For every homomorphism $\tau : G \to \Gamma$, where G is some group, the group pair $[\tau^{-1}\Gamma, \tau^{-1}\Gamma_0]$ is also an element of $\mathscr{L}$.
3. For every group G equipped with a T_2-topology and containing the group Γ, $\Gamma \subseteq G$, the group pair $\left[\overline{\Gamma}, \overline{\Gamma}_0\right]$ is also an element of $\mathscr{L}$, where $\overline{\Gamma}$, $\overline{\Gamma}_0$ are the closures of the groups Γ, Γ_0 in the group G.

Definition 7.31 An almost complete class of group pairs $\mathscr{M}$ is said to be *complete* if the following hold:

1. For every group pair $[\Gamma, \Gamma_0] \in \mathscr{M}$ and a group Γ_1 such that $\Gamma_0 \subseteq \Gamma_1 \subseteq \Gamma$, the group pair $[\Gamma, \Gamma_1]$ is also an element of $\mathscr{M}$.
2. For every two group pairs $[\Gamma, \Gamma_1]$, $[\Gamma_1, \Gamma_2] \in \mathscr{M}$ such that $\Gamma_2 \subseteq \Gamma_1 \subseteq \Gamma$, the group pair $[\Gamma, \Gamma_2]$ is also an element of $\mathscr{M}$.

The minimal complete and almost complete classes of group pairs containing a given set $\mathscr{B}$ of group pairs will be denoted, respectively, by $\mathscr{L}\langle\mathscr{B}\rangle$ and $\mathscr{M}\langle\mathscr{B}\rangle$.

Let $\mathscr{K}$ be the class of all finite groups, $\mathscr{A}$ the class of all abelian groups, $S(k)$ the permutation group on k elements. The minimal complete classes of group pairs $\mathscr{M}\langle\mathscr{A}, \mathscr{K}\rangle$, $\mathscr{M}\langle\mathscr{A}, S(k)\rangle$, and $\mathscr{M}\langle\mathscr{A}\rangle$ are called, respectively, *almost solvable, k-solvable* and *solvable* classes of group pairs.

These classes of group pairs are important for Galois theory. They admit the following explicit description. A chain of subgroups Γ_i, $i = 1, \dots, m$, $\Gamma = \Gamma_1 \supseteq \dots \supseteq \Gamma_m \subseteq \Gamma_0$ is called a *normal tower* of the group pair $[\Gamma, \Gamma_0]$ if the group Γ_{i+1} is a normal subgroup of the group Γ_i for every $i = 1, \dots, m-1$. The collection of quotient groups Γ_i/Γ_{i+1} is called the collection of *quotients of the normal tower.*

Theorem 7.32 (On classes of pairs $\mathscr{M}\langle\mathscr{A}, \mathscr{K}\rangle$, $\mathscr{M}\langle\mathscr{A}, S(n)\rangle$, and $\mathscr{M}\langle\mathscr{A}\rangle$ (cf. Theorem 5.46))

1. *A group pair is almost solvable if and only if it has a normal tower in which every quotient is a finite or abelian group.*
2. *A group pair is k-solvable if and only if it has a normal tower in which every quotient is either a subgroup of the group $S(k)$ or an abelian group.*
3. *A group pair is solvable if and only if the monodromy group of this pair is solvable (the monodromy group of a group pair $[\Gamma, \Gamma_0]$ is by definition the*

quotient group of the group Γ by the largest normal subgroup lying in the group Γ_0).

We can now state a stronger version of property 3 from the definition of an almost complete class of group pairs:

3′. For every almost homomorphism $J : \Gamma \to T$ near the group S, the group pair $[\overline{\Gamma}(S), \overline{\Gamma}_0(S)]$ is also an element of $\mathscr{L}$, where $\overline{\Gamma}(S)$, $\overline{\Gamma}_0(S)$ are the induced closures of the groups Γ, Γ_0 in the group S with respect to the almost homomorphism J.

Definition 7.33 An I-almost complete class of group pairs, an I-complete class of group pairs, and the classes $I\mathscr{L}\langle\mathscr{B}\rangle$ and $I\mathscr{M}\langle\mathscr{B}\rangle$ are defined in the same way as an almost complete class of group pairs, a complete class of group pairs, and the classes $\mathscr{L}\langle\mathscr{B}\rangle$ and $\mathscr{M}\langle\mathscr{B}\rangle$; we have only to replace property 3 by the stronger property 3′ in all definitions.

Proposition 7.34 *The following equalities hold:*

$$I\mathscr{M}\langle\mathscr{A},\mathscr{K}\rangle = \mathscr{M}\langle\mathscr{A},\mathscr{K}\rangle,$$
$$I\mathscr{M}\langle\mathscr{A},S(k)\rangle = \mathscr{M}\langle\mathscr{A},S(k)\rangle,$$
$$I\mathscr{M}\langle\mathscr{A}\rangle = \mathscr{M}\langle\mathscr{A}\rangle.$$

Proof The statement follows immediately from the explicit description of the classes $\mathscr{M}\langle\mathscr{A},\mathscr{K}\rangle$, $\mathscr{M}\langle\mathscr{A},S(k)\rangle$, and $\mathscr{M}\langle\mathscr{A}\rangle$ and from Theorems 7.25 and 7.26 about induced closures. □

Theorem 7.35 *The monodromy pair of every multivalued analytic function induced by some continuous map from an $\mathscr{S}$-function f belongs to the minimal I-almost complete class of group pairs containing the monodromy pair of f. In particular, if an $\mathscr{S}$-function f has a solvable monodromy group (almost solvable monodromy pair, k-solvable monodromy pair), then the monodromy group (the monodromy pair) of every multivalued function induced by some continuous map from the function f has the same property.*

Proof The statement follows from Theorem 7.29 and from the stability of the classes $\mathscr{M}\langle\mathscr{A},\mathscr{K}\rangle$, $\mathscr{M}\langle\mathscr{A},S(k)\rangle$, and $\mathscr{M}\langle\mathscr{A}\rangle$ under the operation of induced closure. □

7.4 Multidimensional Results on Nonrepresentability of Functions by Quadratures

In this section, we describe topological obstructions to representability of multivariate functions by quadratures. Similar results for functions of one variable are described in Chap. 5.

In Sect. 5.4, we constructed a wide class of infinite-valued functions of one variable for which the monodromy group is defined. Is there a sufficiently wide class of germs of infinite-valued multivariate functions (containing germs of functions representable by generalized quadratures and germs of entire multivariate functions and stable under natural operations such as composition) with a similar property? In this section, we define the class of $\mathcal{SC}$-germs that gives an affirmative answer to this question. The proof uses results on continuation of multivalued analytic functions along their ramification sets (see Sect. 7.2).

The main theorem (see Sect. 7.4.5) describes how the monodromy groups of $\mathcal{SC}$-germs change as we apply natural operations to the germs. This theorem is very close to the corresponding one-dimensional theorem (see Sect. 5.6), but it also uses new results of an analytic (see Sect. 7.2) and group-theoretic (see Sect. 7.3) nature. As a corollary, we obtain topological results on the unsolvability of equations in explicit form that are stronger than the classical results.

7.4.1 Formulas, Their Multigerms, Analytic Continuations, and Riemann Surfaces

We consider Liouvillian classes of multivariate analytic functions representable by explicit formulas (see Sects. 7.1.2 and 7.1.3). For every formula, one can define a multigerm containing the germs of all functions appearing in this formula (see Sect. 7.4.3).

We can talk about the analytic continuation of a multigerm of a formula along a path (which is, in essence, the analytic continuation of the germs appearing in this formula along various paths related to each other by the formula). We can define the notion of the Riemann surface of a formula,[5] the $\mathcal{S}$-property of a formula, etc. We will discuss these definitions in detail for the case of a simple formula $y = f \circ G$. For simplicity, we will not give similar definitions for more complicated formulas. The main ideas are clear from the example that we work out below. A germ of an analytic function (which may be multivalued) will sometimes be denoted by the same letter as the function itself without any specification in the notation as to which point and which germ we mean, provided that all this is clear from the context.

Consider the composition of a germ of an analytic map G from a connected analytic manifold M to $\mathbb{C}^n$ and a germ of an analytic function $f : \mathbb{C}^n \to \mathbb{C}$. A *multigerm of the formula* $y = f \circ G$ is defined as a triple $\{y_b \mid G_b, f_a\}$, where y_b and G_b are germs at the point $b \in M$ of the analytic function y and the analytic map $G : (M, b) \to (\mathbb{C}^n, b)$, and f_a is the germ of the analytic function f at the point $a \in \mathbb{C}^n$ for which the following formula holds: $y_b = f_a \circ G_b$.

[5]In the multivariable case, the Riemann surface of a formula is not a surface but rather a higher-dimensional complex analytic manifold.

Let $\gamma : [0, 1] \to M, \gamma(0) = b$, be a continuous path in the manifold M. Consider the path $G_{\gamma(\cdot)} \circ \gamma : [0, 1] \to \mathbb{C}^n$ in the space $\mathbb{C}^n$ taking the point t, $1 \leq t \leq 1$, to the point $(G_{\gamma(t)} \circ \gamma)(t)$, where $G_{\gamma(t)}$ is the analytic continuation of the germ G_b along the path $\gamma : [0, t] \to M$. The *analytic continuation of a multigerm* $\{y_{b_1} \mid G_{b_1}, f_{a_1}\}$ *of the formula* $y = f \circ G$ along the path $\gamma : [0, 1] \to M, \gamma(0) = b_1, \gamma(1) = b_2$, is defined as the triple $\{y_{b_2} \mid G_{b_2}, f_{a_2}\}$, where y_{b_2} and G_{b_2} are germs obtained by analytic continuation of the germs y_{b_1} and G_{b_1} along the path γ, and f_{a_2} is the germ obtained by analytic continuation of the germ f_{a_1} along the path $G_{\gamma(\cdot)} \circ \gamma : [0, 1] \to \mathbb{C}^n$. It is clear that these germs are related by the equality $y_{b_2} = f_{a_2} \circ G_{b_2}$.

We say that two multigerms of the formula $y = f \circ G$ are equivalent if one of them can be obtained from the other by analytic continuation along some path. As a set of points, the *Riemann surface* R of the formula $y = f \circ G$ is the collection of all multigerms equivalent to a given multigerm $\{y_b \mid G_b, f_a\}$. We can define the *natural projection* $\pi : R \to M$ from the Riemann surface of the formula $y = f \circ G$ to the manifold M that maps the multigerm $\{y_{b_1} \mid G_{b_1}, f_{a_1}\}$ to the point $b_1 \in M$. Given a small neighborhood U of the point b_1 in the manifold M, we can define a neighborhood $\tilde{U}$ of the multigerm $\tilde{b}_1 = \{y_{b_1} \mid G_{b_1}, f_{a_1}\}$ on the Riemann surface R. For this, we need to require that the neighborhood U belong to some coordinate neighborhood of the point b_1 in the manifold M, that the Taylor series of the germ $G_{b_1} : M \to \mathbb{C}^n$ converge in the domain U to some map $\tilde{G} : U \to \mathbb{C}^n$, and that the image $\tilde{G}(U) \subset \mathbb{C}^n$ of the neighborhood U under the map $\tilde{G}$ lie in the domain of convergence of the Taylor series for the germ f_{a_1}. If these requirements are fulfilled, then the neighborhood $\tilde{U}$ of the multigerm $\tilde{b}_1$ in the Riemann surface R is defined as the set of multigerms $\{y_{b_2} \mid G_{b_2}, f_{a_2}\}$, where $b_2 \in U$, G_{b_2} is a germ at the point b_2 of the map $\tilde{G}$, f_{a_2} is a germ at the point $a_2 = \tilde{G}(b_2)$ of the function $\tilde{f}$ equal to the sum of the Taylor series of the germ f_{b_1}, and $y_{b_2} = f_{a_2} \circ G_{b_2}$.

Neighborhoods $\tilde{U}$ of this form define a topology on the Riemann surface R. In this topology, the natural projection $\pi : R \to M$ is a local diffeomorphism from R to M. With the help of the local homeomorphism π, we can equip R with the structure of a complex analytic manifold, which exists by definition on the manifold M.

The Riemann surface R of a formula $y = f \circ G$ plays exactly the same role as the Riemann surface of an analytic function. Namely, a multigerm $\{y^*_{\tilde{b}} \mid G^*_{\tilde{b}}, f_a\}$ of the formula $y^* = f \circ G^*$, where $G^* = \pi^* G$, has a unique analytic extension to the entire Riemann surface R, and the Riemann surface R is the maximal manifold for which this condition is satisfied (this means that if $\pi_1 : R_1 \to M$ is another manifold R_1 with a local homeomorphism π_1 to M for which the above-mentioned fact holds, then there exists an analytic embedding $j : R_1 \to R$ compatible with the projections, i.e., $\pi_1 = \pi \circ j$).

A point $b_2 \in M$ is called a *singular point* of a multigerm $\{y_{b_1} \mid G_{b_1}, f_{a_1}\}$ of a formula $y = f \circ G$ if there exists a path $\gamma : [0, 1] \to M, \gamma(0) = b_1, \gamma(1) = b_2$, such that the multigerm has no analytic continuation along this path but for every t, $0 \leq t < 1$, it has a regular continuation along the truncated path $\gamma : [0, t] \to M$. Equivalent multigerms have the same sets of singular points. We say that a formula

$y = f \circ G$ has the $\mathcal{S}$-*property* if the set of singular points of each of its multigerms is meager (see Sect. 7.3.1).

Along with the set of singular points, it is convenient to consider some other sets outside of which every multigerm of a formula admits analytic continuations. A meager set A is called a *forbidden set* for a multigerm of a formula if the multigerm has a regular continuation along every path $\gamma(t)$, $\gamma(0) = a$, intersecting the set A at most at the initial moment. The following theorem can be proved in the same way as the corresponding theorem on $\mathcal{S}$-functions in the one-dimensional case (see Theorem 5.21).

Theorem 7.36 (On forbidden sets) *A meager set is a forbidden set for some multigerm of a formula if and only if it contains the set of singular points of that formula. In particular, a multigerm of a formula has a forbidden set if and only if the formula has the $\mathcal{S}$-property.*

7.4.2 The Class of $\mathcal{S}\mathcal{C}$-Germs, Its Stability Under the Natural Operations

The following definition plays a key role in what follows.

Definition 7.37 A germ f_a of an analytic function f at a point a in the space $\mathbb{C}^n$ is an $\mathcal{S}\mathcal{C}$-*germ* if the following condition is satisfied. For every connected complex analytic manifold M, every analytic map $G : M \to \mathbb{C}^n$, and every preimage b of the point a, $G(b) = a$, there exists a meager set $A \subset M$ such that for every path $\gamma : [0,1] \to M$ originating at the point b, $\gamma(0) = b$, and intersecting the set A at most at the initial moment, $\gamma(t) \notin A$ for $t > 0$, the germ f_a admits an analytic continuation along the path $G \circ \gamma : [0,1] \to \mathbb{C}^n$.

In other words, a germ f_a at a point $a \in \mathbb{C}^n$ is an $\mathcal{S}\mathcal{C}$-germ if for every analytic map $G : M \to \mathbb{C}^n$ and every point $b \in M$ such that $G(b) = a$, the multigerm $\{y_b \mid G_b, f_a\}$ of the formula $y = f \circ G$ has the $\mathcal{S}$-property on the manifold M.

Proposition 7.38 *Every germ of an $\mathcal{S}$-function f of one variable is an $\mathcal{S}\mathcal{C}$-germ.*

Proof If the map $G : M \to \mathbb{C}^1$ is constant, then the function $f \circ G$ is also constant. If the map G not constant, then as a meager set A, it suffices to take the set $G^{-1}(O)$, where O is the set of singular points of the function f. □

Proposition 7.39 *If $f_1, \dots, f_m$ are $\mathcal{S}\mathcal{C}$-germs at a point $a \in \mathbb{C}^n$, and g an $\mathcal{S}\mathcal{C}$-germ at the point $(f_1(a), \dots, f_m(a))$ of the space $\mathbb{C}^m$, then $g(f_1, \dots, f_m)$ is an $\mathcal{S}\mathcal{C}$-germ at the point a.*

Proof Let $G : M \to \mathbb{C}^n$ be an analytic map of a connected complex manifold M into $\mathbb{C}^n$, and let $b \in M$ be a point such that $G(b) = a$. Since the germs $f_1, \dots, f_m$ at the point $a \in \mathbb{C}^n$ are $\mathcal{S}\mathcal{C}$-germs, it follows that for every $i = 1, \dots, m$, there exists a meager set $A_i \subset M$ forbidden for the multigerm of the formula $y_i = f_i \circ G$. As a forbidden set of the multigerm of the formula $\mathbf{z} = \mathbf{f} \circ G$, where $\mathbf{f} = (f_1, \dots, f_m)$ is the germ of the vector function at the point $a \in \mathbb{C}^n$, it suffices to take the set $A = \bigcup_{i=1}^m A_i$.

Let $\pi : R \to M$ denote the natural projection from the Riemann surface R of the formula $\mathbf{z} = \mathbf{f} \circ G$ to M, and let $\tilde{b}$ denote the point of the Riemann surface R corresponding to the multigerm $\{\mathbf{z}_b \mid G_b, \mathbf{f}_a\}$. The germ of the function g at the point $c = (f_1(a), \dots, f_m(a))$ of the space $\mathbb{C}^m$ is an $\mathscr{SC}$-germ. Therefore, the space R includes a meager set $B \subset R$ forbidden for the multigerm $\{w_{\tilde{b}} \mid (\mathbf{f} \circ G \circ \pi)_{\tilde{b}}, g_c\}$ of the formula $w = g \circ (\mathbf{f} \circ G \circ \pi)$. As a forbidden set for the multigerm $\{u_b \mid (\mathbf{f} \circ G)_b, g_c\}$ of the formula $u = g \circ (\mathbf{f} \circ G)$, it suffices to take the meager set $A \cup \pi(B)$. □

Definition 7.40 An operation $\aleph$ taking a germ of an analytic vector function $\mathbf{f}$ at a point $a \in \mathbb{C}^n$ to a germ of an analytic function $\varphi = \aleph(f)$ at the same point a is called an *operation with controlled singularities* if the natural projection $\pi : R \to M$, the germ $\mathbf{f}$, the Riemann surface R, and the germ $\pi^*\varphi$ have a closed forbidden analytic subset $A \subset R$ (i.e., the germ $\pi^*\varphi$ admits an analytic continuation along every path $\gamma : [0, 1] \to R$, $\gamma(0) = \tilde{a}$, where $\tilde{a}$ is a point in R corresponding to the germ $\mathbf{f}$ intersecting the set A at most at the initial moment, $\gamma(t) \notin A$ for $0 < t \le 1$).

Proposition 7.41 *The following properties hold:*

1. *For every $i = 1, \dots, n$, the operation of differentiation mapping a germ of an analytic function f at a point $a \in \mathbb{C}^n$ to the germ of the function $\partial f / \partial x_i$ at the same point is an operation with controlled singularities.*
2. *The operation of integration mapping a germ of a vector function $\mathbf{f} = (f_1, \dots, f_n)$ at a point $a \in \mathbb{C}^n$ to the germ of an analytic function φ at the same point, for which $d\varphi = f_1\,dx_1 + \cdots + f_n\,dx_n$, is an operation with controlled singularities.*

Proof If a germ of a function f (respectively of a one-form $\alpha = f_1\,dx_1 + \cdots + f_n\,dx_n$) admits an analytic continuation along some path in $\mathbb{C}^n$, then the partial derivatives of the germ f (respectively, the antiderivative of a form α) admit an analytic continuation along the same path. Therefore, a partial derivative (respectively the antiderivative) has no singular points on the Riemann surface of the germ f (respectively of the germ of the vector function $\mathbf{f}$). □

Proposition 7.42 *The operation of solving an algebraic equation that maps a germ of a vector function $\mathbf{f} = (f_0, \dots, f_k)$ at a point $a \in \mathbb{C}^n$ where f_0 is not identically zero to a germ y at the same point $a \in \mathbb{C}^n$ satisfying the equation $f_0 y^k + \cdots + f_k = 0$ is an operation with controlled singularities.*

Proof Consider the field K generated by the germs $f_0, \dots, f_k$ over the field of complex numbers $\mathbb{C}$. By definition, the germ y satisfies the algebraic equation $f_0 y^k + \cdots + f_k = 0$ over the field K. However, this equation may be reducible. Choose an irreducible equation

$$Q_0 y^l + \cdots + Q_l = 0 \tag{7.1}$$

over the field K that holds at the germ y. We may assume that the coefficients $Q_0, \dots, Q_l$ of this equation belong to the algebra over the field $\mathbb{C}$ generated by the germs $f_0, \dots, f_k$ (if this is not the case, then it suffices to multiply all coefficients of this equation by the common denominator to arrange this). The coefficients $Q_0, \dots, Q_l$ extend to single-valued functions on the Riemann surface R of the germ of the vector function **f**.

Let $D(Q_0, \dots, Q_l)$ denote the discriminant of (7.1). The discriminant does not vanish identically on R, since (7.1) is irreducible. Let $\Sigma_D \subset R$ be the analytic set on which the discriminant $D(Q_0, \dots, Q_l)$ vanishes. Let $\Sigma_0 \subset R$ be the analytic set on which the coefficient Q_0 vanishes. As the set A from the definition of an operation with controlled singularities, it suffices to take $\Sigma = \Sigma_0 \cup \Sigma_D$. □

Recall that a system of N linear partial differential equations

$$L_j(y) = \sum a^j_{i_1,\dots,i_n} \frac{\partial^{i_1+\dots+i_n} y}{\partial x_1^{i_1} \cdots \partial x_n^{i_n}} = 0, \quad j = 1, \dots, N, \tag{7.2}$$

on an unknown function y whose coefficients $a^j_{i_1,\dots,i_n}$ are analytic functions of n complex variables $x_1, \dots, x_n$ is said to be *holonomic* if the affine space of all solution germs at every point of the space $\mathbb{C}^n$ is finite-dimensional.

Definition 7.43 The operation of solving a holonomic system of equations is defined as the operation of mapping a germ of a vector function $\mathbf{a} = (a^j_{i_1,\dots,i_n})$ at a point a whose components are coefficients, arranged in arbitrary order, of the holonomic system of Eqs. (7.2) to a germ y at the point a of some solution of this system.

Proposition 7.44 *The operation of solving a holonomic system of equations is an operation with controlled singularities.*

This statement follows from general theorems on holonomic systems. (Note that even a local theory of nonholonomic linear systems is rather complicated (see [61]).

Theorem 7.45 *Let* **f** *be a germ of an analytic vector function at a point* $a \in \mathbb{C}^n$, $\mathbf{f} = (f_1, \dots, f_N)$, *whose components* $f_1, \dots, f_N$ *are* $\mathcal{SC}$*-germs. Suppose that a germ* φ *at the point* $a \in \mathbb{C}^n$ *is obtained from the germ* **f** *by an operation with controlled singularities. Then the germ* φ *is an* $\mathcal{SC}$*-germ.*

Proof Let $\pi : R \to \mathbb{C}^n$ be the natural projection of the Riemann surface R of the germ **f** to the space $\mathbb{C}^n$, and let $\tilde{a} \in R$ be the marked point in R corresponding to this germ, $\pi(\tilde{a}) = a$. By definition, the germ $\pi^*\varphi$ at the point $\tilde{a} \in R$ admits an analytic continuation along every path on the manifold R intersecting some analytic set $\Sigma \subset R$ at most at the initial moment. Fix a Whitney stratification for the pair (R, Σ) such that the closure of every stratum is a closed complex analytic set. We are interested only in strata whose closures contain the marked point $\tilde{a}$ of the Riemann surface R. Let $\overline{\Sigma}_1$ be the closure of one such stratum Σ_1, and Σ_1^0 the union of all strata except for the stratum Σ_1 lying in $\overline{\Sigma}_1$. By Theorem 7.18, the germ $\pi^*\varphi$ admits an analytic continuation along every path $\gamma : [0, 1] \to \Sigma_1$, $\gamma(0) = \hat{a}$, intersecting the set Σ_1^0 at most at the initial moment. The theorem now follows.

Indeed, let $G : M \to \mathbb{C}^n$ be an analytic map from a connected complex manifold M to $\mathbb{C}^n$, and let $b \in M$ be a point such that $G(b) = a$. Since all components of the germ of the vector function $\mathbf{f}$ are $\mathcal{SC}$-germs, there exists a meager set $A \subset M$ forbidden for the multigerm $\{\mathbf{y}_b \mid G_b, \mathbf{f}_a\}$ of the formula $\mathbf{y} = \mathbf{f} \circ G$. Let $\pi_1 : R_1 \to M$ be the natural projection of the Riemann surface R_1 of this formula, and $\tilde{b} \in R_1$ the marked point of M_1 corresponding to this germ. On the Riemann surface R_1, there is an analytic extension of the germ $\pi^{-1} G \pi_1$ at the point $\tilde{b} \in R_1$ of the map from R_1 to R taking the point $\tilde{b}$ to the point $\tilde{a}$. The map obtained by this analytic extension will be denoted by $\tilde{G} : R_1 \to R$. Let $\overline{\Sigma}_1$ be the smallest closure of a stratum in the Whitney stratification of the pair (R, Σ) containing the image $\tilde{G}(R_1)$ of the manifold R_1. Let Σ_1^0 be the union of all strata, except for the stratum Σ_1, that lie in $\overline{\Sigma}_1$. The set $B \subset R_1$, where $B = \tilde{G}^{-1}(\Sigma_1^0)$, is a proper analytic subset in R_1. According to [8], a germ φ_a admits an analytic continuation along the pushforward $G \circ \pi_1 \circ \gamma : [0, 1] \to \mathbb{C}^n$ of every path $\gamma : [0, 1] \to M_1$, $\gamma(0) = \tilde{b}$, intersecting the set B at most at the initial moment. Therefore, the set $A \cup \pi_1(B) \subset M$ is a forbidden set for the multigerm $\{y_c \mid G_c, \varphi_a\}$ of the formula $y = \varphi \circ G$. The theorem is proved. □

Corollary 7.46 *Suppose that the set of singular points of a multivalued analytic function on $\mathbb{C}^n$ is a closed analytic set. Then every germ of this function is an $\mathcal{SC}$-germ.*

Proof By definition, a germ of such a function at a point $a \in \mathbb{C}^n$ can be regarded as the result of an operation with controlled singularities applied to the germ of the vector function $\mathbf{x} = x_1, \dots, x_n$ at the point a whose components are the coordinate functions. □

Theorem 7.47 (On stability of the class of $\mathcal{SC}$-germs) *The class of $\mathcal{SC}$-germs contains all germs of $\mathcal{S}$-functions of one variable and all germs of $\mathcal{S}$-functions of several variables with an analytic set of singular points.*

The class of $\mathcal{SC}$-germs on $\mathbb{C}^n$ is stable under the operation of taking the composition with $\mathcal{SC}$-germs of m-variable functions and the operations of differentiation, integration, solving algebraic equations, and solving holonomic systems of linear partial differential equations.

Proof The germs of $\mathcal{S}$-functions that are mentioned in the statement of the theorem belong to the class of $\mathcal{SC}$-germs by Proposition 7.38 and Corollary 7.46. The class of $\mathcal{SC}$-germs is stable under composition by Proposition 7.39. The stability of the class of $\mathcal{SC}$-germs under the other operations follows from Theorem 7.45 due to Propositions 7.41, 7.42, and 7.44. □

Corollary 7.48 *If a germ of a function f can be obtained from germs of $\mathcal{S}$-functions having analytic sets of singular points and from germs of $\mathcal{S}$-functions of one variable with the help of integration, differentiation, arithmetic operations, superpositions, solving algebraic equations, and solving holonomic systems of linear differential equations, then the germ f is an $\mathcal{SC}$-germ. In particular, a germ that is not an $\mathcal{SC}$-germ cannot be represented by generalized quadratures.*

7.4.3 The Class of Formula Multigerms with the $\mathcal{SC}$-Property

Suppose that a class $\mathcal{A}$ of analytic function germs is given by a list of basic germs $\mathcal{B}$ and a list of admissible operations $\mathcal{D}$. Suppose that the list $\mathcal{D}$ contains only operations discussed in the introduction to this chapter. By definition, every germ of the class $\mathcal{A}$ can be expressed in terms of the basic germs by formulas containing admissible operations only. We will now define multigerms of formulas of this kind by induction.

The *multigerm* of a formula Ω expressing the membership of a germ φ in the set of basic germs consists of germs φ and g, where g is an element of the set $\mathcal{B}$, and the equality $\varphi = g$, i.e., $\Omega = \{\varphi|g|\varphi = g\}$. Suppose that a germ φ at a point $a \in \mathbb{C}^n$ can be expressed through the germs of the functions $f_1, \dots, f_m$ at the point a with the help of one of operations 1–8 from Sects. 7.1.2 and 7.1.3 or by solving a system of holonomic equations. Let $\Omega_1, \dots, \Omega_m$ be multigerms of the formulas expressing the functions $f_1, \dots, f_m$ through the set of basic germs. Then the *multigerm of the formula expressing the germ* φ is the collection consisting of the germ φ, the multigerms of all formulas $\Omega_1, \dots, \Omega_m$, and the equality describing the considered operation. For example, if φ can be obtained from $f_1, \dots, f_m$ by solving an algebraic equation $\varphi^m + f_1\varphi^{m-1} + \cdots + f_m = 0$, then $\Omega = \{\varphi|\Omega_1, \dots, \Omega_m|\varphi^m + f_1\varphi^{m-1} + \cdots + f_m = 0\}$.

If a germ φ at a point $a \in \mathbb{C}^n$ can be expressed through the germs of functions $f_1, \dots, f_m$ at the point a and through the germ of a function g at the point $b = (f_1(a), \dots, f_m(a)) \in \mathbb{C}^m$ by composition, then the multigerm Ω of the formula expressing φ is by definition $\Omega = \{\varphi|\Omega_1, \dots, \Omega_m, \Omega_0|\varphi = g(f_1, \dots, f_m)\}$, where for $i = 1, \dots, m$, the multigerm Ω_i is the multigerm of the formula for the germ of the function f_i at the point a, and Ω_0 is the multigerm of the formula for the germ of the function g at the point b. (Because of the composition operation, formula multigerms may contain germs of functions on different spaces.)

For the multigerm of the formula Ω expressing the germ of φ at a point $a \in \mathbb{C}^n$, the notions of *analytic continuation* and *Riemann surface* are defined in the same way as was done in Sect. 7.4.1 for the formula $y = f \circ G$. Note that the Riemann surface R of the formula Ω projects to the space $\mathbb{C}^n$ (i.e., there is a natural projection $\pi : R \to \mathbb{C}^n$), although germs of functions of a different number of variables may appear in the construction of R. If $n > 2$, then the Riemann surface of a formula is not a surface but rather a complex analytic manifold of dimension n.

Repeating Definition 7.37, we say that the multigerm Ω of the formula expressing the germ of a function φ at a point $a \in \mathbb{C}^n$ through the basic germs has the *$\mathcal{SC}$-property* if the following condition holds: for every connected complex analytic manifold M, every analytic map $G : M \to \mathbb{C}^n$, and every preimage b of the point a, $G(b) = a$, there exists a meager set $A \subset M$ such that for every path $\gamma : [0, 1] \to M$ originating at the point b, $\gamma(0) = b$, and intersecting the set A at most at the initial moment, $\gamma(t) \notin A$ for $t > 0$, the multigerm of the formula Ω admits an analytic continuation along the path $G \circ \gamma : [0, 1] \to \mathbb{C}^n$.

Theorem 7.49 *The following properties hold:*

1. *Let a class $\mathscr{A}$ of germs be defined by a list $\mathscr{B}$ of basic germs containing only $\mathscr{S}\mathscr{C}$-germs and a list of admissible operations $\mathscr{D}$ containing only operations listed in Sect. 7.3.1, except possibly for the operation of solving a holonomic system of equations. Then for every germ of the given class, every formula expressing this germ through the basic functions with the help of admissible operations has the $\mathscr{S}\mathscr{C}$-property.*
2. *If additionally, the set $\mathscr{B}$ of basic germs is stable under the operation of analytic continuation, then for every germ $\varphi_a \in \mathscr{A}$ defined at a point a of the space $\mathbb{C}^n$, there exists a forbidden set $A \subset \mathbb{C}^n$ such that at every point $b \notin A$, every germ φ_b equivalent to the germ φ_a is also in the class $\mathscr{A}$ (and can be expressed through the basic germs by the same formula, in a natural sense, as the germ φ).*

Proof To prove claim 1, it suffices to repeat the arguments from Sect. 7.4.2 (replacing germs of functions with multigerms of formulas). Let us prove claim 2. By claim 1, the multigerm of a formula Ω expressing the germ φ_a through the germs of basic functions has the $\mathscr{S}\mathscr{C}$-property and, in particular, admits a meager forbidden set A. Suppose that a germ φ_b is obtained from the germ φ_a by analytic continuation along a path γ. We can assume that $\gamma(t)$ does not belong to the set A for $0 < t \leq 1$ (see Lemma 7.19, on releasing a path from a meager set). The analytic continuation of the multigerm of the formula Ω is the multigerm of a formula expressing the germ φ_b through basic germs by admissible operations, since the set of basic germs is stable under analytic continuation. □

Under the assumptions of claim 2 from the statement of the theorem, we have the following dichotomy. For every multivalued analytic function φ, either no germ of it belongs to the class $\mathscr{A}$, or all the germs outside of some meager set belong to this class (and can be expressed through the basic germs by the "same" formula). In the first case, we say that the function φ cannot be expressed through the basic germs by admissible operations, and in the second case, we say that such an expression exists. In particular, representability of multivalued analytic functions by quadratures, k-quadratures, or generalized quadratures is well defined.

7.4.4 Topological Obstructions to Representability of Functions by Quadratures

Fix some nonempty I-almost complete class of group pairs IM (see Sect. 7.3). Let $\widehat{IM}$ denote the class of $\mathscr{S}\mathscr{C}$-germs of analytic functions (at points of all spaces $\mathbb{C}^n$, $n \geq 1$, simultaneously) whose monodromy pairs belong to IM.

Theorem 7.50 (Main theorem) *The class $\widehat{IM}$ of germs contains $\mathscr{S}\mathscr{C}$-germs of all single-valued functions and is stable under composition and differentiation. Moreover:*

1. *If the class IM contains the additive group $\mathbb{C}$ of complex numbers, then the class $\widehat{IM}$ is stable under integration.*
2. *If the class IM contains the permutation group $S(k)$ on k elements, then the class $\widehat{IM}$ is stable under solving algebraic equations of degree at most k.*

Proof To prove this theorem, we analyze what happens with the monodromy pairs of the function germs under the operations listed in the theorem. The arguments are similar to those from the theorem on $\mathscr{S}$-functions of one variable (see Sect. 5.6). For this reason, we just list the differences between the two proofs.

Firstly, the theorem on stability of the class of $\mathscr{S}\mathscr{C}$-germs (see Sect. 7.4.2) is more involved than the corresponding one-dimensional theorem. It is based on the results of Sect. 7.2. Secondly, the operation of composition in the multidimensional setting is connected to the new operation on the group pairs, the operation of induced closure. This circle of ideas is described in Sect. 7.3. □

Proposition 7.51 (Result on quadratures) *The monodromy group of a function germ f representable by quadratures is solvable. Moreover, the same conclusion holds for every function germ f representable through the germs of single-valued $\mathscr{S}$-functions with analytic set of singular points and the germs of single-valued $\mathscr{S}$-functions of one variable by integrations, differentiations, and compositions.*

Proposition 7.52 (Result on k-quadratures) *The monodromy group of a function germ f representable by k-quadratures is k-solvable (see Chap. 5). Moreover, the same conclusion holds for every function germ f representable through the germs of single-valued $\mathscr{S}$-functions with analytic set of singular points and the germs of single-valued $\mathscr{S}$-functions of one variable by integrations, differentiations, compositions, and solving algebraic equations of degree $\leq k$.*

Proposition 7.53 (Result on generalized quadratures) *The monodromy group of a function germ f representable by generalized quadratures is almost solvable (see Chap. 5). Moreover, the same conclusion holds for every function germ f representable through the germs of single-valued $\mathscr{S}$-functions with analytic set of singular points and the germs of single-valued $\mathscr{S}$-functions of one variable by integrations, differentiations, compositions, and solving algebraic equations.*

Proof The results stated above follow from Theorem 7.50, since the germs mentioned there are $\mathscr{S}\mathscr{C}$-germs (see Sect. 7.4.2), and classes of group pairs having respectively solvable, k-solvable, or almost solvable monodromy groups contain the additive group $\mathbb{C}$. The last two classes of group pairs contain, in addition, the group $S(k)$ and, respectively, all groups $S(m)$ for $0 < m < \infty$ (see Chap. 5). □

The following theorem follows easily from Galois theory.

Theorem 7.54 *A solution of an algebraic equation $y^m + r_1 y^{m-1} + \cdots + r_m = 0$ in which the r_i are rational functions of n variables is expressible by radicals (by radicals and solving algebraic equations of degree at most k) if and only if its monodromy group is solvable (k-solvable).*

Our results allow us to make the negative results of this theorem stronger.

Corollary 7.55 *If the monodromy group of an algebraic equation*

$$y^k + r_1 y^{k-1} + \cdots + r_k = 0,$$

in which the r_i are rational functions of n variables, is not solvable, then every germ of this solution is not only impossible to express by radicals but also impossible to express through germs of single-valued $\mathcal{S}$-functions with analytic set of singular points by integrations, differentiations, and compositions.

The following version of a classical theorem of Abel is stronger than other known results in this direction.

Theorem 7.56 (cf. [1, 46]) *For $n \geq 5$, no germ of a solution of the general algebraic equation $y^n + x_1 y^{n-1} + \cdots + x_n = 0$, in which $x_1, \ldots, x_n$ are independent variables, cannot be expressed through the germs of elementary functions, the germs of single-valued $\mathcal{S}$-functions with analytic set of singular points, and the germs of single-valued $\mathcal{S}$-functions of one-variable by composition, integrations, differentiations, and solving algebraic equations of degree less than n.*

7.4.5 Monodromy Groups of Holonomic Systems of Linear Differential Equations

Consider a holonomic system of N linear differential equations $L_j(y) = 0$, $j = 1, \ldots, N$,

$$L_j(y) = \sum a^j_{i_1,\ldots,i_n} \frac{\partial^{i_1+\cdots+i_n} y}{\partial x_1^{i_1} \cdots \partial x_n^{i_n}} = 0,$$

on an unknown function y whose coefficients $a^j_{i_1,\ldots,i_n}$ are rational functions of n complex variables $x_1, \ldots, x_n$.

It is known that there exists an algebraic hypersurface Σ in the space $\mathbb{C}^n$, the *singular hypersurface* of the holonomic system, with the following property: every solution of the system admits an analytic continuation along every path avoiding the hypersurface Σ. Let V be the finite-dimensional solution space of the holonomic system in a neighborhood of a point x_0 not belonging to the hypersurface Σ. Take an arbitrary path $\gamma(t)$ in the space $\mathbb{C}^n$ originating and terminating at x_0 and avoiding the hypersurface Σ. Solutions of this system admit analytic continuations along the path γ, which are also solutions of the system. Therefore, every such path γ gives rise to a linear map M_γ of the solution space V to itself. The collection of linear transformations M_γ corresponding to all paths γ form a group, which is called the *monodromy group of the holonomic system.*

Kolchin obtained a generalization of Picard–Vessiot theory to the case of holonomic systems of differential equations. We now state some corollaries from Kolchin's theory that relate to solvability of regular holonomic systems of PDEs by quadratures. As in the one-dimensional case, a holonomic system is said to be *regular* if near the singular set Σ and at infinity, the solutions of the system grow at most polynomially.

Theorem 7.57 *A regular holonomic system of linear differential equations is solvable by quadratures (or by generalized quadratures) if and only if its monodromy group is solvable (or almost solvable).*

Thus Kolchin's theory proves two results.

1. If the monodromy group of a regular holonomic system of linear differential equations is solvable (almost solvable), then this system of equations is solvable by quadratures (by generalized quadratures).
2. If the monodromy group of a regular holonomic system of linear differential equations is not solvable (not almost solvable), then this system of equations is unsolvable by quadratures (by generalized quadratures).

Our theory allows us to make the negative result two stronger.

Theorem 7.58 *If the monodromy group of a holonomic system of linear differential equations is not solvable (not almost solvable), then every germ of almost every solution of this system cannot be expressed through the germs of single-valued $\mathcal{S}$-functions with analytic set of singular points by compositions, meromorphic operations, integrations, differentiations (by compositions, meromorphic operations, integrations, differentiations, and solving algebraic equations).*

7.4.6 Holonomic Systems of Linear Differential Equations with Small Coefficients

Consider a completely integrable system of linear differential equations of the form

$$dy = Ay, \tag{7.3}$$

where $y = y_1, \dots, y_n$ is an unknown vector function, and A is an $n \times n$ matrix consisting of differential one-forms with rational coefficients on the space $\mathbb{C}^n$ satisfying the condition of complete integrability $dA + A \wedge A = 0$ and having the following form:

$$A = \sum_{i=1}^{k} A_i \frac{dl_i}{l_i},$$

where A_i are constant matrices, and l_i are linear (not necessarily homogeneous) functions on $\mathbb{C}^n$.

If the matrices A_i can be simultaneously reduced to triangular form, then system (7.3), like every completely integrable triangular system, is solvable by quadratures. Of course, there exist solvable systems that are not triangular. However, if the matrices A_i are sufficiently small, then there are no such systems. Namely, the following theorem holds.

Theorem 7.59 *A system of the form* (7.3) *that cannot be reduced to triangular form and is such that the matrices A_i have sufficiently small norm is strongly unsolvable, i.e., it cannot be solved even using the germs of all single-valued $\mathcal{S}$-functions with analytic set of singular points, compositions, meromorphic operations, integrations, differentiations, and solving algebraic equations.*

The proof of this theorem is similar to the proof of Corollary 6.38 (see also Sect. 6.2.3). We have only to replace the reference to the (one-dimensional) Lappo-Danilevsky theory with a reference to the multidimensional version of it from [71].

Appendix A
Straightedge and Compass Constructions

The ancient Greeks solved many beautiful problems about geometric constructions using straightedge and compass, and they raised some questions that became widely known because of numerous unsuccessful attempts to solve them. These problems remained open for many centuries. The ancient Greeks constructed regular n-gons for $n = 2^k3$, 2^k4, 2^k5, and 2^k15, where k is any nonnegative integer. It was unknown how to construct a regular n-gon for any other value of n until Gauss constructed the regular 17-gon and gave a characterization of all numbers n for which such a construction is possible. Gauss obtained this remarkable result before Galois theory had been discovered. His amazing result had a huge impact on the development of several branches of mathematics. Below, we discuss this problem and other problems on constructibility.

Problems about constructions using straightedge and compass are the oldest problems on "solvability in finite terms." We treat them just as we have treated other problems of this type. We distinguish three classes of constructions. In problems of the first, simplest, class, only three operations are allowed: constructing a line through two given points, a circle of given radius with a given center, and points of intersection of given lines and circles. In the third, hardest, class we allow all these constructions, and in addition, we also allow ourselves to choose an arbitrary point, but we require the result to be independent of the choice of arbitrary points. In the second, intermediate, class we do not allow arbitrary choice, but we allow two additional constructions: construction of the center of a given circle and construction of the line orthogonal to a given line passing through a given point not lying on the given line.

Logically, constructions in the third class are different from the constructions in other classes and from everything we saw before in problems on solvability in finite terms: intermediate objects that we get in the process can be dependent on arbitrary

---

This appendix was published originally as A. Khovanskii [57].

A. Khovanskii, *Topological Galois Theory*, Springer Monographs in Mathematics,
DOI 10.1007/978-3-642-38871-2

choices we have made. In this case, we do not consider them to be constructions made using straightedge and compass. We try to avoid, when possible, the operation of choice of an arbitrary point. We prove that in most problems, this operation does not give anything new (everything that is constructible using this operation can be constructed without it). Some of the classical problems can be rather satisfyingly formulated and proved inside the first class of constructions.

The problem of trisection of an angle requires the operation of choice of an arbitrary point: given two lines passing through a given point, nothing new can be constructed using operations of the first class. We show that if we add to these given lines an arbitrary point on one of them, then using operations of the first class only, we can construct everything that can be constructed from two given lines using operations from the third class.

Operations of the second class allow us to extend initial data. For example, operations from the first class of constructions do not allow us to construct anything if the initial datum consists of several nonintersecting circles. But if we mark the centers of the circles, then operations of the first class can be used to construct anything constructible from the initial datum using operations from the third class (except when all the given circles are concentric).

In the first section, we discuss problems of solvability of algebraic equations using square roots that are later needed for problems of constructibility. We do it in greater generality than we need (we do not assume that the base field is perfect and that it is of characteristic not equal to 2). This problem is interesting in its own right, and additional generality does not cause much trouble. The second section is devoted to problems of constructibility using straightedge and compass.

A.1 Solvability of Equations by Square Roots

In this section, we discuss the following question about solvability of equations in finite terms: when is an irreducible algebraic equation over a field K solvable using arithmetic operations and the operation of taking the square root? Galois theory answers this question when the field K is perfect and its characteristic is not equal to 2. But some simple additional observations help to get rid of any assumptions on the structure of the field K.

The material is structured as follows: In Sect. A.1.1 we give necessary background material. In Sect. A.1.2 we prove necessary and sufficient conditions for an equation to be solvable using square roots if the characteristic of the field K is not equal to 2. In Sect. A.1.3, the same question is solved when the characteristic of the field is equal to 2. In Sects. A.1.4 and A.1.5, Gauss's results on roots of unity of degree n are presented (we use them in the problem of constructibility of the regular n-gon).

A.1.1 Background Material

We begin by recalling some elementary facts about fields. If the field K is a subfield of the field F, then F is a vector space over K. If $\dim_K F < \infty$, then F is said to be a *finite extension* of the field K. The *degree* of the extension is defined to be $\dim_K F$ and is denoted by $[F : K]$.

Theorem A.1 *If $K \subset F$ and $F \subset M$ are finite field extensions, then:*

1. *$K \subset M$ is a finite extension.*
2. *$[M : K] = [M : F][F : K]$.*

Proof Let $u_1, \dots, u_n$ be a basis of F over K and let $v_1, \dots, v_m$ be a basis of M over F. It is easily shown that elements $u_i v_j$ with $1 \le i \le n$, $1 \le j \le m$ form a basis of M over K. □

Proposition A.2 1. *If $[F : K] = n$ and $a \in F$, then there exists a polynomial Q over the field K of degree not greater than n such that $Q(a) = 0$.*
2. *If Q is an irreducible polynomial over the field K and $Q(a) = 0$, then $[K(a) : K] = \deg Q$.*

Proof 1. Since the $\dim_K F = n$ elements $1, a, \dots, a^n$ are linearly dependent, there exist $\lambda_i \in K$ such that $\lambda_n a^n + \lambda_{n-1} a^{n-1} + \dots + \lambda_0 = 0$ and for some $i > 0$, the coefficient λ_i is nonzero.
2. The field $K(a)$ is isomorphic to the field $K[x]/I$, where I is an ideal generated by the polynomial Q of degree n.

□

Proposition A.3 *If the degree of an irreducible polynomial over a field of characteristic $p > 0$ is not divisible by p, then it does not have multiple roots.*

Proof A multiple root of the polynomial Q is also a root of the derivative Q'. An irreducible polynomial Q cannot have a common root with a nonzero polynomial of smaller degree. Thus if Q has a multiple root, then $Q' \equiv 0$, i.e., $Q(x) = R(x^p)$ for some polynomial R, in which case, $\deg Q = p \cdot \deg R$, and so $\deg Q$ is divisible by p. □

A.1.2 Extensions by 2-Radicals

We now return to the question of solvability of equations by square roots.

Definition A.4 An extension $K \subset F$ is said to be an *extension by* 2-*radicals* if there exists a tower of fields $K = F_0 \subset F_1 \subset \dots \subset F_n$ such that $F \subset F_n$ and for $1 \le i \le n$, $F_i = F_{i-1}(a_i)$ with $a_i^2 \in F_{i-1}$ and $a_i \notin F_{i-1}$.

Theorem A.5 *If $K \subset F$ is a 2-radical extension, then $[F : K] = 2^k$.*

Proof For $K = F_0 \subset F_1 \subset \cdots \subset F_n$, we have $[F_n : K] = [F_n : F_{n-1}] \cdots [F_1 : F_0] = 2^n$. If $K \subset F \subset F_n$, then $[F : K] \cdot [F_n : F] = 2^n$. Therefore, $[F : K]$ is a power of 2. □

Corollary A.6 *If a polynomial P is irreducible over the field K and has a root in some extension of K by 2-radicals, then* $\deg P = 2^k$.

Proof If $P(a) = 0$, then $[K(a) : K] = \deg P$. □

Corollary A.7 *Let K be a field of characteristic not equal to 2. A cubic equation $P = 0$ over the field K is solvable in square roots if and only if one of its roots lies in K.*

Proof If $a \in K$ and $P(a) = 0$, then $P = (x-a)Q$ for some $Q \in K[x]$. A quadratic equation $Q = 0$ is solvable in square roots, because the characteristic of the field is not equal to 2. If the cubic polynomial does not have a root in K, then it is irreducible, and we can use Corollary A.6. □

Remark A.8 For $K = \mathbb{Q}$, Corollary A.7 has an effective form: the rational roots of a rational polynomial $P \in \mathbb{Q}[x]$ can be explicitly found (in particular, we can check whether there are none). If a root a is found, then the quadratic equation $0 = Q(x) = P/(x - a)$ is explicitly solvable.

Let E_P denote the splitting field of the polynomial P over the field K.

Corollary A.9 *If a polynomial P is irreducible over K and has a root in some extension of the field K by 2-radicals, then $[E_P : K] = 2^m$.*

Proof The extension $K \subset E_P$ in the assumptions of Corollary A.9 is 2-radical. □

Corollary A.9 has the following partial converse.

Theorem A.10 *Let K be a field of characteristic not equal to 2. If the equality $[E_P : K] = 2^m$ holds for a polynomial $P \in K[x]$, then the extension $K \subset E_P$ is 2-radical.*

Proof The degree of the extension $[E_P : K]$ is divisible by the degree of the polynomial P. Therefore, $\deg P = 2^k$. Hence by Proposition A.3, the equation $P = 0$ is separable, and Galois theory is applicable. The order of the Galois group G of the field E_P over the field K is equal to $[E_P : K] = 2^m$. Since the order of the group G is a power of 2, there exists a normal tower of subgroups $G = G_0 \supset G_1 \supset \cdots \supset G_m = e$ such that $G_i/G_{i-1} = \mathbb{Z}_2$. For the tower of fields $K = K_0 \subset K_1 \subset \cdots \subset K_m = E_P$ corresponding to this tower of subgroups, we have $[K_i : K_{i-1}] = 2$. Since the characteristic of the field K (and so of the fields K_i) is not equal to 2, the field K_i is obtained from the field K_{i-1} by adjoining a square root. □

A.1.3 2-Radical Extensions of a Field of Characteristic 2

In this subsection, K is some field of characteristic 2, and $\overline{K}$ is its algebraic closure. We are interested only in algebraic elements over the field K and algebraic extensions of K. These elements and extensions lie in the field $\overline{K}$.

Lemma A.11 *The subset of $\overline{K}$ containing the elements $y \in \overline{K}$ such that $y^2 \in K$ is a field.*

Proof The result follows from the equality $(a + b)^2 = a^2 + b^2$, which holds in fields of characteristic 2. □

We define a chain of subfields $K = K_0 \subset K_1 \subset \cdots \subset K_n \subset \cdots$ of the field $\overline{K}$ by the rule $y \in K_{i+1}$ if $y^2 \in K_i$. We will call the field $\tilde{K} = \bigcup K_i$ the *perfect closure* of the field K. It is easy to check that the field $\tilde{K}$ is the minimal perfect field that contains the field K.

Theorem A.12 *A finite extension $K \subset M$ of the field K is 2-radical if and only if $M \subset \tilde{K}$.*

Proof If for the tower of fields $K = F_0 \subset F_1 \subset \cdots \subset F_n$, we have $F_i = F_{i-1}(a_i)$, where $a_i^2 \in F_{i-1}$, then $F_i \subset K_i$. □

A polynomial $P \in K[x]$ is called the *minimal polynomial* of an algebraic element a over K if $P(a) = 0$ and the polynomial P is monic and irreducible.

Theorem A.13 *A polynomial P is a minimal polynomial of some element $a \in K_n \setminus K_{n-1}$ if and only if $P(x) = x^{2^n} - b$, where $b \in K$ and $b \neq c^2$ for all $c \in K$.*

Proof The element $a \in K_n$ is the only (multiple) root of the polynomial $x^{2^n} - b$, where $b = a^{2^n} \in K$. Therefore, the minimal polynomial P of a has the unique root a. The numbers m such that $a^m \in K$ form an additive subgroup in $\mathbb{Z}$. If $a \in K_n \setminus K_{n-1}$, then $a^m \in K$ only if m is divisible by 2^n. Hence $P(x) = x^{2^n} - b$. □

Corollary A.14 *If $P \in K[x]$ is monic and irreducible over K, then the equation $P(x) = 0$ is solvable in square roots if and only if $P(x) = x^{2^n} - b$, where $b \in K$ and $b \neq c^2$ for all $c \in K$. In particular, every irreducible equation of degree greater than 1 over a perfect field K of characteristic 2 is not solvable in square roots.*

A.1.4 Roots of Unity

Here we recall some classical results that were obtained by Gauss before Galois theory has been discovered.

Let $\Omega_n \subset \mathbb{C}$ be the set of numbers x such that $x^n = 1$, and let Ω_n^* be the set of all *primitive roots of unity* of order n, i.e., the set of numbers $a \in \Omega_n$ such that $a^m \neq 1$ for $0 < m < n$. If $\omega \in \Omega_n^*$, then $a \in \Omega_n$ if and only if $a = \omega^k$ for some integer k;

and $a \in \Omega^*$ if and only if $a = \omega^k$, where k is relatively prime to n, i.e., the residue k modulo n lies in the multiplicative group $U(n)$ of invertible elements over the ring $\mathbb{Z}/n\mathbb{Z}$. Denote by Φ_n the nth cyclotomic polynomial $\Phi_n(x) = \prod_{a \in \Omega_n^*}(x - a)$.

Lemma A.15 *The equality $x^n - 1 = \prod_{d|n} \Phi_d(x)$ holds, where the product is taken over all divisors d of the number n.*

Proof The result follows from the relations $\Omega_n = \bigcup_{d|n} \Omega_d^*$ and $\Omega_{d_1}^* \bigcap \Omega_{d_2}^* = \varnothing$ for $d_1 \neq d_2$. □

Corollary A.16 *The polynomial Φ_n is monic and has integer coefficients.*

Proof If $P, Q \in \mathbb{Z}[x]$ are monic, then (1) the polynomial PQ is monic, and $PQ \in \mathbb{Z}[x]$. Furthermore, (2) if $T = P/Q$ is a polynomial, then the polynomial T is monic, and $T \in \mathbb{Z}[x]$. We use these two facts to prove the corollary inductively. For $n = 1$, the corollary is true because $\Phi_1(x) = x - 1$. Set $\Psi_n = \prod \Phi_{d'}$, where the product is taken over all divisors d' of n that are less than n. If the corollary is true for $d' < n$, then according to fact (1), the polynomial Ψ_n is monic, and $\Psi_n \in \mathbb{Z}[x]$. By Lemma A.15, $\Phi_n(x) = (x^n - 1)/\Psi_n(x)$. According to (2), the corollary is true for $d = n$. □

A polynomial $f \in \mathbb{Z}[x]$ is said to be *primitive* if its coefficients do not have a common divisor. Gauss proved that the product of primitive polynomials is a primitive polynomial. From this, we automatically get the following *integrality property*: if for $f_1 = f_2 f_3$, where f_1, f_2 are monic, $f_1 \in \mathbb{Z}[x]$ and $f_2, f_3 \in \mathbb{Q}[x]$, then the polynomials f_2, f_3 have integer coefficients, and f_3 is monic.

Recall also that if p is relatively prime to n, then the polynomial $x^n - 1 \in \mathbb{Z}_p[x]$ does not have multiple roots (since $n \not\equiv 0 \pmod p$ and the derivative nx^{n-1} of the polynomial $x^n - 1$ does not have nonzero roots).

Theorem A.17 (Gauss) *The polynomial Φ_n is irreducible over $\mathbb{Z}$.*

Lemma A.18 (Gauss) *Let $\omega \in \Omega_n^*$, let f be the minimal polynomial of ω, and suppose p is a prime number relatively prime to n. Then $f(\omega^p) = 0$.*

Proof The number ω is a root of the polynomial Φ_n. Therefore, $\Phi_n = fg$, where $g \in \mathbb{Q}[x]$. According to the integrality property, g is in $\mathbb{Z}[x]$ and is monic. Assume that $f(\omega^p) \neq 0$. Since $0 = \Phi_n(\omega^p) = f(\omega^p)g(\omega^p)$, we get that $g(\omega^p) = 0$. In this case, ω is a root of the polynomial $g(x^p)$, and thus $g(x^p) = fh$, where $h \in \mathbb{Q}[x]$. By the integrality property, h is in $\mathbb{Z}[x]$ and is monic.

Let $\pi : \mathbb{Z}[x] \to \mathbb{Z}_p[x]$ be a homomorphism extending the natural map $\mathbb{Z} \to \mathbb{Z}_p$ to the ring of polynomials. We have $(\pi g)(x^p) = (\pi f)(\pi h)$. In the ring $\mathbb{Z}_p[x]$, for every polynomial φ, we have the identity $\varphi(x^p) = \varphi^p(x)$ (it follows from the formulas $a^p = a$ in $\mathbb{Z}_p$ and $(\phi + \psi)^p = \phi^p + \psi^p$ in $\mathbb{Z}_p[x]$). Therefore, the polynomial $\pi(f)\pi(g)$ has a multiple root. But the polynomial $\pi(\Phi_n) = \pi(f)\pi(g)$ is a divisor of the polynomial $\pi(x^n - 1) = (x^n - 1) \in \mathbb{Z}_p[x]$, which does not have multiple roots. This contradiction completes the proof of Gauss's lemma.

Proof (of Gauss's theorem) Let (ω, f, p) be the same as in Gauss's lemma. We can apply Gauss's lemma to the triple (ω_1, f, p_2), where $\omega_1 = \omega^{p_1}$, $p_1 = p$, and p_2 is any number relatively prime to n. Indeed, $\omega_1 \in \Omega_n^*$ and $f(\omega_1) = 0$. Similarly, $f(\omega^{p_1 \cdots p_m}) = 0$ for every set of prime numbers relatively prime to n.

Every element $\alpha \in \Omega_n^*$ can be written in the form $\alpha = \omega^m$, where m is the product of prime numbers that are relatively prime to n. The polynomial Φ_n has the same roots as the polynomial f, and both are monic. Therefore, $\Phi_n = f$ and Φ_n is irreducible.

Corollary A.19 *Let E_n be the splitting field of the polynomial $x^n - 1$ over the field $\mathbb{Q}$. Then its Galois group G is isomorphic to the multiplicative group $U(n)$ of the ring $\mathbb{Z}/n\mathbb{Z}$.*

Proof It is easy to prove (see Sect. 2.8.1) that the Galois group G is a subgroup of the group $U(n)$. The roots of the irreducible polynomial Φ_n lie in the splitting field of the polynomial $x^n - 1 = 0$. Therefore, $\#G \geq \deg \Phi_n$. But $\deg \Phi_n = \#U(n)$, so the group G coincides with the group $U(n)$. □

Remark A.20 Let $\mathbb{Q} \subset E$ be a Galois extension and let $E \subset E_n$. The Galois group G of the extension $\mathbb{Q} \subset E$ is abelian, since G is a factor group of the group $U(n)$. The famous Kronecker–Weber theorem states that the converse is also true: if the Galois group of the extension $\mathbb{Q} \subset E$ is abelian, then E is contained in the field E_n for some n.

A.1.5 Solvability of the Equation $x^n - 1 = 0$ by 2-Radicals

Here we describe numbers n for which the extension $\mathbb{Q} \subset E_n$ is 2-radical.

Proposition A.21 *Let $n = p_1^{k_1} \cdots p_m^{k_m}$ be the prime factorization of n. Then $U(n) = U(p_1)^{k_1} \times \cdots \times U(p_m)^{k_m}$ and $\#U(n) = \prod (p_i^k - p^{k_i - 1})$.*

Proof The result follows from the Chinese remainder theorem and from the fact that in the ring $\mathbb{Z}/p^k\mathbb{Z}$, there are exactly p^{k-1} noninvertible elements. □

The numbers $F_n = 2^{(2^n)} + 1$ are called *Fermat numbers*. They are named for the seventeenth-century French mathematician Pierre Fermat, who was the first to study prime numbers of the form $p = 2^n + 1$. A prime Fermat number is called a *Fermat prime*. Since for odd m, the number $2^{km} + 1$ is divisible by $2^k + 1$ and is therefore composite, the Fermat numbers are the only numbers of the form $2^n + 1$ that can be prime. The first five Fermat numbers are $3, 5, 17, 257, 65{,}537$, and they are all prime. On that basis, Fermat conjectured that all the numbers now known as Fermat numbers are prime. But in 1732, Leonhard Euler showed that $F_5 = 2^{2^5} + 1 = 2^{32} + 1 = 4{,}294{,}967{,}297 = 641 \times 6{,}700{,}417$. It is not known whether there exist any Fermat primes greater than F_4.

Let us call an integer n a *Gauss number* if $n = 2^k p_1 \cdots p_m$, where $k \geq 0$, and $p_1, \ldots, p_m$ are distinct Fermat primes.

Theorem A.22 *The extension* $\mathbb{Q} \subset E_n$ *is* 2*-radical if and only if* n *is a Gauss number.*

Proof Since $\deg \Phi(n) = \#U(n)$, we see from Proposition A.21 that $\#U(n) = 2^k$ if and only if n is a Gauss number.

Example A.23 Let us solve the equation $\Phi_5(x) = 0$. We have

$$\Phi_5(x) = (x^5 - 1)/(x - 1) = x^4 + x^3 + x^2 + x + 1,$$

so

$$x^{-2}\Phi_5(x) = x^2 + x + 1 + x^{-1} + x^{-2} = u^2 + u - 1,$$

where $u = x + x^{-1}$. To find x, it is enough to solve the quadratic equation $u^2 + u - 1 = 0$ and then the quadratic equation $xu = x^2 + 1$.

An explicit solution of the equation $\Phi_{17}(x) = 0$ was found by Gauss. It was the starting point of many remarkable discoveries. Even now with the knowledge of Galois theory, it is not easy to solve this equation. My students Y. Burda and L. Kadets did this in [21].

A.2 What Can Be Constructed Using Straightedge and Compass?

This section is devoted to the question of solvability and unsolvability of problems related to constructions using straightedge and compass.

In Sects. A.2.1 and A.2.2, we describe the class of points, lines, and circles that can be constructed using operations from the first class, where a finite number of points constitute the initial data. We also give necessary (Sect. A.2.1) and sufficient (Sect. A.2.2) conditions for determining whether an object belongs to this class. In Sect. A.2.3, we discuss a number of classical problems regarding constructibility (including the problem of constructing a regular n-gon) that can be solved using material from Sects. A.2.1 and A.2.2.

In Sect. A.2.4, we distinguish two constructions that use a choice of arbitrary point, which later will be considered as two new operations. In Sect. A.2.6, we describe what can be constructed from any (apart from a few exceptional cases) initial data using the operation of arbitrary choice of a point. It turns out that everything that can be constructed using this operation is constructible without it using two new operations and constructions from Sects. A.2.1 and A.2.2. In Sect. A.2.7, we describe what can be constructed from one specific initial datum that arises in the problem of trisecting an angle, and we discuss the question of solvability of this problem in detail.

In Sect. A.2.8, we prove a theorem from real affine geometry that is related to the implementability of arithmetic operations over the field of real numbers using geometric constructions.

A.2.1 *The Unsolvability of Some Straightedge and Compass Construction Problems*

Before we prove that a particular construction is impossible, we need to define explicitly what this means. Let M be the set of all points, lines, and circles in the plane (these are the only objects that can be constructed using straightedge and compass). We can define an *admissible class* $\mathscr{M} \subset M$ and say that a point, line, or circle can be constructed if it belongs to the class $\mathscr{M}$. Such a class $\mathscr{M}$ can be defined by giving initial data and admissible operations.

List of Admissible Operations

1. *Constructing a line:* given two distinct points, we construct a line passing through them.
2. *Constructing a circle:* given points P, Q, O, where $P \neq Q$, construct a circle with center O and radius equal to $[P, Q]$.
3. *Intersection:* given a pair of distinct intersecting curves Γ_1, Γ_2, where Γ_i is a line or a circle, construct their points of intersection (of which there can be one or two).

Definition A.24 The class $\mathscr{M}(\mathscr{D}) \subset M$ consists of points, lines, and circles that are constructible from the initial data $\mathscr{D} \subset M$. It is the minimal class that contains $\mathscr{D}$ and is closed under the three admissible operations listed above.

The fact that the class $\mathscr{M}(\mathscr{D})$ is closed under intersection means that if $\Gamma_1, \Gamma_2 \in \mathscr{M}(\mathscr{D})$, the curves Γ_1, Γ_2 are distinct, and $P \in \Gamma_1 \bigcap \Gamma_2$, then $P \in \mathscr{M}(\mathscr{D})$. For the class to be closed under other operations is defined analogously. Theorems about the impossibility of one or another construction are based on a simple algebraic fact that is formulated below as Theorem A.26.

Definition A.25 Given a real field T, we define $\mathscr{M}_T$ to be the class of all points, lines, and circles in the coordinate plane defined over T (a point is defined over T if both its coordinates lie in T; a line or a circle is defined over T if it can be given by an equation $ax + by + c = 0$ or $(x-a)^2 + (y-b)^2 + c = 0$, where $a, b, c \in T$).

If a real field $T \subset \mathbb{R}$ is closed under the operation of taking the square root, i.e., if $a \in \mathbb{R}$ and $a^2 \in T$ implies $a \in T$, then the class $\mathscr{M}_T$ contains the center of every circle in the class, and the distance between points $P, Q \in \mathscr{M}_T$ is a real number in T.

Theorem A.26 *If a real field $T \subset \mathbb{R}$ is closed under the operation of taking square roots, then the class $\mathscr{M}_T$ is closed under the three admissible operations given above.*

Proof In the coordinate plane $\mathbb{R}^2$, the three operations reduce to finding the real solutions of linear and quadratic equations. Every solution of such an equation lies in the field T, since T is closed under taking square roots. □

Let $\mathscr{D}_0$ be some set in the plane that contains at least two points. Euclidean transformations and homotheties map lines to lines and circles with marked center to circles with marked center. Such transformations are compatible with the three construction operations.

Definition A.27 The *field corresponding to* $\mathscr{D}_0$ is the smallest real field $T(\mathscr{D}_0)$ that is closed under taking square roots and that contains the ratios of lengths of segments $[P, Q]$ with $P, Q \in \mathscr{D}_0$.

Definition A.28 Choose two distinct points $O, E \in \mathscr{D}_0$ and choose a normalization such that the length of the segment $[O, E]$ is equal to 1. We say that an orthonormal system of coordinates is *compatible with* $\mathscr{D}_0$ if $O = (0, 0)$ and $E = (0, 1)$.

Theorem A.29 *If a coordinate system is compatible with $\mathscr{D}_0$, then the inclusion $\mathscr{M}(\mathscr{D}_0) \subset \mathscr{M}_T$ holds, where $T = T(\mathscr{D}_0)$. In other words, if a point, line, or circle is not defined over the field T, then it cannot be constructed using the three operations from the initial data set $\mathscr{D}_0$.*

Proof From the assumptions of the theorem, the coordinates of the points in $\mathscr{D}_0$ lie in the field $T(\mathscr{D}_0)$. The result now follows from Theorem A.26. □

A.2.2 Some Explicit Constructions

To carry out a construction, we need certain building blocks that the reader may have encountered in high school during the study of construction using straightedge and compass. We recall them here.

1. *Problem:* Given two distinct points A, B, construct the midpoint P of segment $[A, B]$. *Solution:* Let Q, R be the points of intersection of the circles with centers A, B and radii equal to the length $|AB|$ of the segment $[A, B]$. Then the points Q, R lie outside of AB, and the desired point P is the point of intersection of AB and QR.
2. *Problem:* Construct a perpendicular to a line ℓ at point $P \in \ell$. *Solution:* Choose a point $A \in \ell \setminus \{P\}$. We shall construct points Q, R such that lines $AP = \ell$ and QR are orthogonal and $P \in QR$. Let $B \neq A$ be the point of intersection (in addition to the point A) of the line ℓ with the circle with center P and radius equal to $|PA|$. Now take A, B and Q, R to be the same points as in the previous construction.

3. *Problem:* Construct a perpendicular to a line ℓ from a point $P \notin \ell$. *Solution:* If we choose distinct points $A, B \in \ell$, it will suffice to construct a point $Q \neq A, B, P$ such that the lines AB and PQ are perpendicular. We can take Q to be the point of intersection $Q \neq P$ of circles with centers A and B that pass through the point P.
4. *Problem:* Construct a line ℓ_1 parallel to a given line ℓ and passing through a point $P \notin \ell$. *Solution:* It is enough to construct a perpendicular ℓ_2 from the point P to the line ℓ and then a perpendicular ℓ_1 to the line ℓ_2 passing through P.

Let $O \neq E$ be two points. Consider a system of coordinates in the plane that is compatible with the set $\mathscr{D}_0 = \{O, E\}$ (see Definition A.28). Denote by $\ell_0 = OE$ the first coordinate line. We identify a point on the line ℓ_0 with the number equal to its coordinate. Then O and E are identified with 0 and 1.

Lemma A.30 *Let $a, b \in \ell_0 \bigcap \mathscr{M}(\mathscr{D}_0)$. Then the following hold:*

1. $-a$, a^{-1}, $a + b$, ab *belong to* $\ell_0 \bigcap \mathscr{M}(M_0)$.
2. *If* $ab > 0$, *then* $(ab)^{1/2} \in \ell_0 \bigcap \mathscr{M}(\mathscr{D}_0)$.

Remark A.31 When we constructed the line parallel to a given line and passing through a given point, we used straightedge and compass. This construction can be viewed as a single operation, and this operation suffices to construct, given points $0, 1$, points $-a$, a^{-1}, $a + b$, and ab from points a, b on the coordinate line (see Figs. A.1–A.4). This fact has a beautiful application in affine geometry (see Sect. A.2.8).

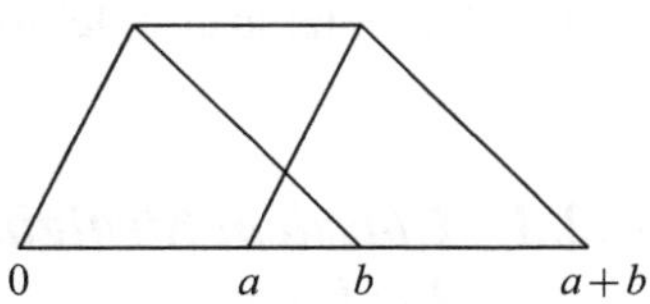

Fig. A.1 Construction of $a + b$ given a and b

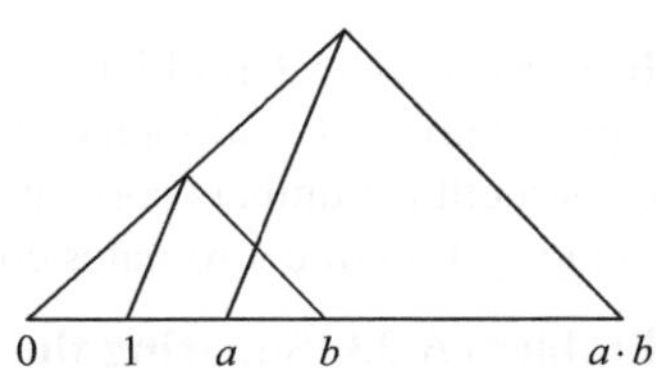

Fig. A.2 Construction of $a \cdot b$ given 1, a, and b

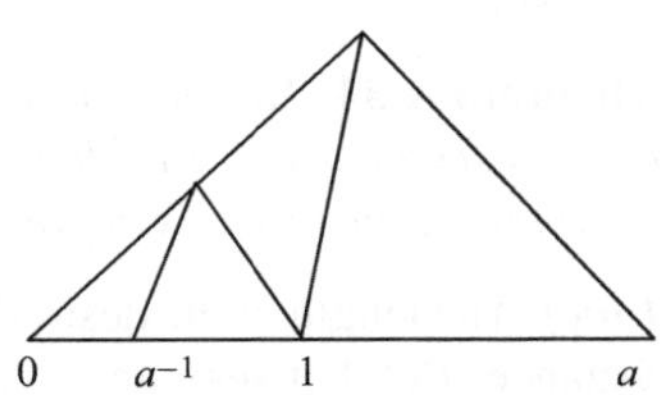

Fig. A.3 Construction of a^{-1} given 1 and a

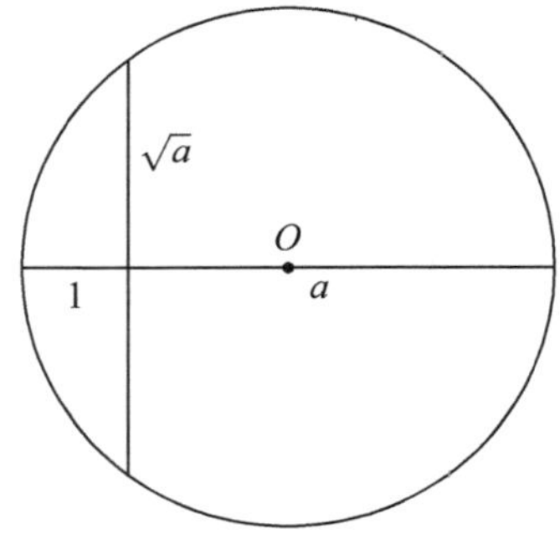

Fig. A.4 Construction of $\sqrt{a}$ given 1 and a

Theorem A.32 *Under the conditions of Theorem A.29, the equality $\mathscr{M}(\mathscr{D}_0) = \mathscr{M}_T$ holds.*

Proof It is enough to show that $\mathscr{M}(\mathscr{D}_0) \supset \mathscr{M}_T$, i.e., that every element from $\mathscr{M}_T$ can be constructed. If $P, Q \in \mathscr{D}_0$, then the set $\mathscr{M}(\mathscr{D}_0)$ contains the point $\rho \in \ell_0$, where ρ is the ratio of the lengths of the segments $[P, Q]$ and $[O, E]$. This is one of the points of intersection of the line ℓ_0 with the circle with center O and radius equal to $|PQ|$. According to Lemma A.30, every point $(a, 0)$ with $a \in T$ lies in $\mathscr{M}(\mathscr{D}_0)$. Every point $(0, b)$ with $b \in T$ also lies in $\mathscr{M}(\mathscr{D}_0)$: it can be constructed by intersecting the y-axis with the circle centered at O and passing through the point $(b, 0)$. Constructing the lines perpendicular to the axes, we see that when $a, b \in T$, the point (a, b) is in $\mathscr{M}(\mathscr{D}_0)$. A line ℓ defined over T has a pair of points defined over T. Therefore, $\ell \in \mathscr{M}(D_0)$. A circle S defined over T contains a point defined over T. Its center is also defined over T, and thus $S \in \mathscr{M}(D_0)$. □

A.2.3 Classical Straightedge and Compass Constructibility Problems

In some classical problems on construction, the initial datum is a segment, or equivalently, its endpoints O, E. In this section, we denote by T the field of constructible numbers corresponding to the set $\mathscr{D}_0 = \{O, E\}$. From Theorem A.32, in the system of coordinates compatible with $\mathscr{D}_0$ we have $\mathscr{M}(\mathscr{D}_0) = \mathscr{M}_T$.

Problem A.33 (Squaring the circle) Given the points O, E, construct a segment I such that the area of the circle with radius OE is equal to the area of the square with side I.

Theorem A.34 *For no points $P, Q \in \mathscr{M}_T$ is the segment $[P, Q]$ equal to the desired segment I in Problem A.33. In other words, in the class $\mathscr{M}_T$, the operation of squaring the circle is impossible.*

Proof The length of the desired segment I is the number $|OE| \cdot \pi^{1/2} = \pi^{1/2}$. But the distance $|PQ|$ between any two points $P, Q \in \mathscr{M}_T$ is a constructible number, and therefore algebraic, while the length of the desired segment I is the transcendental number $\pi^{1/2}$. □

Problem A.35 (Doubling the cube) Given points O, E, construct a segment J such that the volume of a cube of side J is twice the volume of a cube with side length $|OE| = 1$.

Theorem A.36 *For no points $P, Q \in \mathscr{M}_T$ is the segment $[P, Q]$ equal to the desired segment J in Problem A.35. In other words, in the class $\mathscr{M}_T$, the operation of doubling the cube is impossible.*

Proof The distance between points $P, Q \in \mathscr{M}_T$ is a constructible number, but the length of the desired segment J is equal to $2^{1/3}$. The equation $x^3 - 2 = 0$ is irreducible over $\mathbb{Q}$ and is not solvable using square roots.

Problem A.37 Construction of a regular n-gon. Construct a regular n-gon with given side OE.

Theorem A.38 (Gauss) *The regular n-gon can be constructed, that is, all its vertices lie in the class $\mathscr{M}_T$, if and only if n is a Gauss number.*

Proof It is not hard to see that the problem is equivalent to the following: construct all the vertices of the regular n-gon with center O with one of the vertices at the point E. When we identify the plane with the complex line, the vertices of this n-gon are roots of the equation $z^n = 1$. This equation is solvable in square roots (in the field of complex numbers) if and only if n is a Gauss number. Now we note that a complex number can be expressed in square roots over the field $\mathbb{Q}$ if and only if its real and imaginary parts are constructible numbers. □

A.2.4 Two Specific Constructions

Below, we present two simple constructions that use the choice of an arbitrary point from a continuous set. These constructions cannot be performed using the three operations given above. Instead, they can be considered new operations (as we shall do later).

Problem A.39 *Given a line ℓ and a point $P \notin \ell$, find the point $E \in \ell$ such that the lines ℓ and EP are perpendicular.*

Solution The class $\mathscr{M}(\mathscr{D})$, where the set of initial data $\mathscr{D}$ consists of the line ℓ and a point $P \notin \ell$, coincides with the set $\mathscr{D}$: application of the admissible operations does not enlarge the set $\mathscr{D}$. However, if we arbitrarily choose two distinct points A, B on the line ℓ, then this construction is easy to perform (see Sect. A.2.2). We obtain the perpendicular ℓ_P and a point $E = \ell \cap \ell_P$ that does not depend on the arbitrary choice.

Using the points $O = P$ and E, one can construct all the objects of the class $\mathcal{M}_T$. The result of each of these constructions does not depend on the arbitrarily chosen points $A, B \in \ell$, $A \neq B$, that were used in the construction.

Problem A.40 Construct the center of a given circle S.

Solution The class $\mathcal{M}(\mathcal{D})$, where $\mathcal{D} = \{S\}$, coincides with the set $\mathcal{D}$. However, if we choose two distinct points $A, B \in S$, then the perpendicular bisector of the segment $[A, B]$ passes through the center of the circle. Finding the midpoint of the constructed diameter, we find the center O of the circle S.

To include such constructions in our considerations, we have to allow an arbitrary choice of points lying in one of the sets (strata) into which the plane is divided by the points, lines, and circles we have already constructed. However, an object is considered to be constructed only if it does not depend on the arbitrary choices that were made. Later, we shall show that such an extended interpretation of the process of constructing using straightedge and compass does not change the results we obtained previously and allows us to consider other problems, in particular the problem of trisecting an angle.

We begin with consideration of the stratification of the plane induced by a finite subset M of points, lines, and circles.

A.2.5 Stratification of the Plane

Let $V \subset M$ be a finite subset. There is a stratification Σ_V of the plane that is induced by V, i.e., its division into strata $S_\alpha \in \Sigma_V$ of different dimensions. (The stratification Σ_V is very natural. If the finite set V of points, lines, and circles is drawn, the corresponding stratification Σ_V becomes clearly visible.)

A *zero-dimensional stratum* in Σ_V is any point of the set V_0 of all points of intersection of the lines and circles of V and all of the one-point subsets contained in V.

A *one-dimensional stratum* in Σ_V is any connected component of the set $\Gamma_i \setminus (\Gamma_i \bigcap V_0)$, where Γ_i is any line or circle from V.

A *two-dimensional stratum* in Σ_V is any connected component of the complement in the plane of the union of all points, lines, and circles in V. Let T be a real field closed under the operation of taking square roots. From Theorem A.26, we get the following corollary.

Corollary A.41 *If $V \subset \mathcal{M}_T$ and P is a zero-dimensional stratum in Σ_V, then $P \subset \mathcal{M}_T$.*

Proposition A.42 *The points defined over T are dense in the plane, on every line defined over T, and on every circle defined over T.*

Proof That the points are dense in the plane and every line defined over T is trivial. Lines of the form $y = c$, where $c \in T$, are dense in $\mathbb{R}^2$. They intersect every circle defined over T at points defined over T. The set of such points is everywhere dense on the circle. □

Corollary A.43 *If* $V \subset \mathscr{M}_T$, *then the points defined over* T *are dense in every stratum of positive dimension of the stratification* Σ_V.

A.2.6 Classes of Constructions That Allow Arbitrary Choice

The result of applying the operation of intersection depends on the choice of one of the points of intersection of two curves. Define operation (4) depending not only on the choice of an element from a finite set, but also on the choice of an element from a set of cardinality of the continuum. Using this operation, we can make two simple constructions (see Sect. A.2.4) that can be viewed as new operations (5) and (6).

Extension of the List of Admissible Operations

4. *Operation of choice of an arbitrary point:* given a finite set of points $V \subset M$, choose a stratum $S_\alpha \in \Sigma_V$ of positive dimension and a point P from this stratum.
5. *Operations of constructing the foot of a perpendicular:* given a line ℓ and a point $P \notin \ell$, construct the point $E \in \ell$ such that lines EP and ℓ are perpendicular.
6. *Operation of restoring the center:* given a circle, find its center.

Define the class $\mathscr{M}_G(\mathscr{D})$ of elements that can be constructed in the generalized sense from a finite set $\mathscr{D}$. We say that $v \in \mathscr{M}_G(\mathscr{D})$ if there exists a finite algorithm (i.e., a rule that describes all discrete choices) whose kth step is passage from the finite set $V_k \subset M$ to the next set $V_{k+1} \subset M$ such that (1) $V_1 = \mathscr{D}$; (2)$V_{k+1} = V_k \bigcup \{a\}$, where either a is obtained by applying one of the operations (1)–(3) to some elements from the set V_k (see Sect. A.2.1), or $a = P$ and P is the point obtained using the operation of the choice from the set V_k; (3) the element v is contained in some of the sets V_N and is independent of the continuous choices that occurred on the previous steps.

Theorem A.44 *Let* T *be a real field closed under the operation of taking square roots, and let* $\mathscr{D} \subset \mathscr{M}_T$ *be a finite set. Then* $\mathscr{M}_G(\mathscr{D}) \subset \mathscr{M}_T$.

Proof If $v \in \mathscr{M}_G(\mathscr{D})$, then the continuous choice that occurs in the construction of v can be made arbitrarily. From the conditions of the theorem, we have $V_1 = \mathscr{D} \subset \mathscr{M}_T$ and $V_2 = V_1 \bigcup \{a\}$. If the first step is the operation of adding the point a from a stratum of positive dimension in the stratification Σ_{V_1}, then we choose the point a defined over T. It follows from Corollary A.43 that this is possible. With this choice, $V_2 \subset \mathscr{M}_T$. If the point, line, or circle a is obtained by applying to some elements of the set V_1 one of the operations (1)–(3), then $V_2 \subset \mathscr{M}_T$ by Theorem A.26. Now every time we encounter the operation of choosing an arbitrary point, we choose it to be defined over T. With this rule of choice, we have $V_k \subset \mathscr{M}_T$ for every $k > 0$. □

Corollary A.45 *If $\mathscr{D}_0$ is a finite set of points that contains at least two points, then $\mathscr{M}_G(\mathscr{D}_0) = \mathscr{M}(\mathscr{D}_0)$. In particular, operation (4) does not help us solve the problems of squaring the circle and doubling the cube. With its help, we can construct only the regular n-gons that can be constructed without it.*

Definition A.46 The minimal class $\mathscr{M}_r(\mathscr{D})$ that contains $\mathscr{D}$ and is closed under operations (1)–(3), (5), and (6) is called the class of objects that are *constructible in the generalized sense from the initial data* $\mathscr{D}$.

1. We say that D is an *exceptional set of type* R_1 if D consists of a single point.
2. We say that D is an *exceptional set of type* R_2 if D consists of a single line.
3. We say that D is an *exceptional set of type* R_3 if D consists of $k > 1$ parallel lines.
4. We say that D is an *exceptional set of type* R_4 if D consists of $k > 0$ lines passing through a point O.
5. We say that D is an *exceptional set of type* R_5 if D consists of $k > 0$ lines passing through a point O together with the point O.
6. We say that D is an *exceptional set of type* R_6 if D consists of $k > 0$ circles with a common center O.
7. We say that D is an *exceptional set of type* R_7 if D consists of $k > 0$ circles with center O and the point O.

Proposition A.47 *For a finite unexceptional set $\mathscr{D}$, there exists a finite set $\mathscr{D}_0 \subset \mathscr{M}_r(\mathscr{D})$ that contains only points and for which $\mathscr{D} \subset \mathscr{M}(\mathscr{D}_0)$ (moreover, for a given $\mathscr{D}$, the sets $\mathscr{D}_0$ can be found explicitly).*

For example, for $\mathscr{D} = \{S, \ell\}$, where S is a circle with center O, ℓ is a line, and $O \notin \ell$, it is enough to take $\mathscr{D}_0 = \{O, E, P\}$, where $E \in \ell$ is the foot of the perpendicular dropped from O to ℓ, and $P \in S \cap \ell$. For other unexceptional sets D, the set $\mathscr{D}_0$ can be found just as easily.

Corollary A.48 *For a finite unexceptional set of initial data $\mathscr{D} \subset M$, the following equalities hold: $\mathscr{M}_G(\mathscr{D}) = \mathscr{M}_r(\mathscr{D}) = \mathscr{M}(\mathscr{D}_0) = \mathscr{M}_T$, where T is the field compatible with $\mathscr{D}_0$.*

Proof According to the claim, $\mathscr{D} \subset \mathscr{M}(\mathscr{D}_0)$. But $\mathscr{M}(\mathscr{D}_0) = \mathscr{M}_T$ (see Theorem A.32) and $\mathscr{M}_G(\mathscr{D}) \subset \mathscr{M}_T$ (see Theorem A.44). We have the inclusions $\mathscr{M}_G(\mathscr{D}) \supset \mathscr{M}_r(\mathscr{D}) \supset \mathscr{M}(\mathscr{D}_0)$, which completes the proof. □

We have described the class $\mathscr{M}_G(\mathscr{D})$ for unexceptional $\mathscr{D}$ and shown that operation (4) is not needed to construct its objects: $\mathscr{M}_G(\mathscr{D}) = \mathscr{M}_r(\mathscr{D})$.

A.2.7 *Trisection of an Angle*

The next classical problem that we shall consider is a problem related to the class $\mathscr{M}_G(\mathscr{D})$ for the exceptional set $\mathscr{D}$ of type R_4.

Problem A.49 (Trisecting an angle) Divide a given angle into three equal parts.

Let us describe the class $\mathcal{M}_G(\mathcal{D})$ for the set $\mathcal{D}$ of type R_4 (the classes $\mathcal{M}_G(\mathcal{D})$ for the exceptional sets $\mathcal{D}$ of other types can also be completely described). Let $\mathcal{D}$ consist of $k > 0$ lines passing through the point O. Fix any circle S with center O. We use the following notation: $\mathcal{D}'$ is the set of points equal to $\bigcup_{\ell\in\mathcal{D}}(S \cap \ell)$; T is a field compatible with $\mathcal{D}'$; $\ell_0 \in \mathcal{D}$ is a fixed line.

Theorem A.50 *The class $\mathcal{M}_G(\mathcal{D})$ consists of the point O and of all lines ℓ passing through the point O such that $|\cos(\ell, \ell_0)| \in T$ (here (ℓ, ℓ_0) is either of the two angles formed by the lines ℓ and ℓ_0).*

Proof Choose any point $E \in \ell_0 \setminus O$ using operation (4). Construct the circle S with center O and radius OE. Consider the set $\mathcal{D}'$ associated with S. The class $\mathcal{M}(\mathcal{D}') = \mathcal{M}_T$ contains all lines ℓ passing through the point O such that $|\cos(\ell, \ell_0)| \in T$, and it does not contain any other lines passing through O.

The objects of the set $\mathcal{D}$ are invariant under the group G_O of homotheties with center O, and therefore all objects of the class $\mathcal{M}_G(\mathcal{D})$ should also be invariant under G_O. Indeed, under homothety, every construction is mapped to a homothetic construction if the arbitrary points in the construction are chosen to be homothetic to the points of the initial construction. But objects of the class $\mathcal{M}_G(\mathcal{D})$ do not depend on arbitrary choices that were made in the process of their construction, i.e., they are invariant under the group G_O. Only the point O and lines passing through O are invariant under this group. □

The solvability of the problem of trisecting an angle depends crucially on the magnitude of the angle (see Corollaries A.51–A.54).

Corollary A.51 *If $\mathcal{D} = \{\ell_0, \ell_1\}$, where ℓ_0, ℓ_1 are lines passing through O, and $a = |\cos(\ell_0, \ell_1)|$, then the class $\mathcal{M}_G(\mathcal{D})$ consists of the point O and lines ℓ such that $O \in \ell$ and $|\cos(\ell, \ell_0)| \in T$, where T is the minimal real field that contains a and is closed under the operation of taking square roots.*

Corollary A.52 *With the assumptions of Corollary A.51, we can construct lines that divide the angle (ℓ_0, ℓ_1) into n equal parts if and only if the equation $P_n(x) = a$, where P_n is the Chebyshev polynomial of degree n, is solvable in 2-radicals over T.*

We note that if a is a transcendental number, then the equation $P_n(x) = a$ is irreducible over the field $\mathbb{Q}(a)$. This is so because the field $\mathbb{Q}(a)$ is isomorphic to the field of rational functions $\mathbb{Q}(t)$ over $\mathbb{Q}$, and the equation $P_n(x) = t$ is irreducible even over the field $\mathbb{C}(t)$ (the Riemann surface of the algebraic function $x(t)$ defined by this equation is the Riemann sphere).

Corollary A.53 *If in the assumptions of Corollary A.51, a is transcendental, then an angle can be divided into n equal parts if and only if $n = 2^k$. In particular, trisection of this angle is impossible.*

Indeed, if an irreducible equation is solvable in 2-radicals, then its degree is equal to 2^k. On the other hand, we can divide an angle into 2^k parts by successively constructing its bisectors.

Corollary A.54 *If in the assumptions of Corollary A.51, the number a is rational, then trisection of an angle is possible if and only if the equation* $4x^3 - 3x = a$ *has a rational root.*

Corollary A.54 gives an explicit criterion for solvability of the angle trisection problem for angles with a rational cosine. In particular, it is easy to see that trisection of a 60° angle is impossible.

A.2.8 A Theorem from Affine Geometry

The theorem formulated below shows that in the definition of an affine automorphism of the real plane, only the preservation of points and lines is important, since the continuity of the automorphism comes for free and could be dropped from the definition. Its proof is based on the possibility of performing arithmetic operations using parallel lines (see Sect. A.2.2).

Theorem A.55 *Let* $F : \mathbb{R}^2 \to \mathbb{R}^2$ *be a bijection that takes lines to lines. Then* F *is an affine transformation.*

Lemma A.56 *If* $\varphi : \mathbb{R} \to \mathbb{R}$ *is an automorphism of the field* $\mathbb{R}$, *then* $\varphi(x) = x$ *for* $x \in \mathbb{R}$.

Proof Every automorphism of the field $\mathbb{Q}$ of rational numbers is obviously the identity map, so if $x \in \mathbb{Q}$, then $\varphi(x) = x$. If $x \in \mathbb{R}$ and $x \geq 0$, then $x = a^2$ and $\varphi(x) = \varphi(a^2) \geq 0$, i.e., φ is monotone. Thus $\varphi(x) = x$ for $x \in \mathbb{R}$. □

Lemma A.57 *If in the assumptions of the theorem,* $F(O) = O$ *and* $F(E) = E$, $O \neq E$, *then the restriction of* F *to the line* OE *is the identity map.*

Proof Set coordinates on the line OE identifying O with 0 and E with 1. Then F maps nonintersecting lines to nonintersecting lines, i.e., it preserves the relation of parallelism between lines. Using parallel lines and the points $O = 0$ and $E = 1$, from given points $a, b \in OE$ we can construct the points $-a$, a^{-1}, $a + b$, and ab (see Lemma A.30). Therefore, restriction of F to OE gives an automorphism of the real line. Now apply Lemma A.56.

Proof (of Theorem A.55) Maps that satisfy the assumptions of the theorem form a group G that contains the group of affine transformations. The subgroup G_0 that fixes noncollinear points A, B, C is trivial. Indeed, if $\Psi \in G_0$, then by Lemma A.57, the restriction of Ψ on continuation of the sides of the triangle ABC is the identity map (since Ψ fixes the vertices of the triangle). From every point P, we can draw a line ℓ_P that intersects the sides of triangle ABC in two different points (that Ψ fixes). Applying Lemma A.57 to the line ℓ_P, we get that $\Psi(P) = P$, i.e., the group G_0 is trivial. Therefore, the group G cannot contain more than one map that takes points A, B, C to noncollinear points A', B', C'. But there is an affine map of this kind. This completes the proof of the theorem. □

Appendix B
Chebyshev Polynomials and Their Inverses

The Chebyshev polynomial of degree n is defined by the formula

$$T_n(x) = \cos n \arccos x.$$

These polynomials were discovered by Pafnuty Chebyshev (1821–1894) when he was considering the problem of the best approximation of a given function by polynomials of degree $\leq n$. They play an important role in approximation theory. Rather surprising is the fact that these polynomials became useful in algebra: the problem from which they originally appeared is far from algebra, and even their definition uses transcendental functions.

Nevertheless, in some algebraic problems, the series T_n of Chebyshev polynomials appears along with the polynomials $P(x) = x^n$. From a "philosophical" point of view, these two classes result from the existence of two families of finite groups of projective transformations of the space $\mathbb{C}P^1$: cyclic groups C_n and dihedral groups D_n.

In complex analysis, the class of polynomials x^n extends to the family of multivalued analytic functions x^α, $\alpha \in \mathbb{R}$, which contains, along with the polynomials x^n, their inverses $x^{1/n}$ and satisfies the composition relation $(x^\alpha)^\beta = x^{\alpha\beta}$.

In a similar manner, we extend the class of Chebyshev polynomials T_n to the family of multivalued analytic functions T_α, $\alpha \in \mathbb{R}$, which contains, along with the polynomials T_n, their inverses $T_{1/n}$ and satisfies the composition relation $T_\beta \circ T_\alpha = T_{\alpha\beta}$.

A multivalued function can be defined without the notion of analytic continuation, just by giving its set of values at each point. This sometimes helps us in carrying over the definition of the multivalued function to an arbitrary field (where the operation of analytic continuation is not defined). For example, for positive

This appendix was published originally as A. Khovanskii [58].

A. Khovanskii, *Topological Galois Theory*, Springer Monographs in Mathematics,
DOI 10.1007/978-3-642-38871-2

integers n, the function $x^{1/n}$ is defined over every field $\mathbf{k}$: it is a multivalued function that assigns to every $x \in \mathbf{k}$, the set of elements z from the algebraic closure of $\mathbf{k}$ such that $z^n = x$.

It is easier to work with a germ of a single-valued function than with a multivalued function. This can be done when all values of a multivalued function come from the analytic continuation of a single-valued germ.

In Sect. B.1.1, the multivalued Chebyshev function T_α, $\alpha \in \mathbb{R}$, is defined as a function of a complex variable x by means of its set of values. In Sect. B.1.2, we define a series at the point $x = 1$ whose analytic continuation is T_α (see Sect. B.1.3).

In Sect. B.2.1, we give an algebraic definition of Chebyshev polynomials and their inverses over an arbitrary field of characteristic not equal to 2. In addition, if the characteristic of the field is not equal to 3, then these functions are used to construct solutions in radicals of equations of degree 3 and 4 over this field (see Sects. B.2.2 and B.2.3).

In Sects. B.3.1–B.3.3, we discuss three classical problems whose solution involves the families of polynomials x^n and T_n. In Sect. B.3.1, we discuss the problem of describing all complex polynomials that can be inverted in radicals. This problem was solved by Joseph Ritt. In Sect. B.3.2, we discuss Schur's problem, which was solved by Michael Fried, of describing all polynomials $P \in \mathbb{Q}[x]$ for which the maps $P : \mathbb{Z}_p \to \mathbb{Z}_p$ are invertible for infinitely many prime numbers p. In Sect. B.3.3, we formulate a result of Julia, Fatou, and Ritt on the affine classification of *integrable* polynomial maps from the complex line to itself.

B.1 Chebyshev Functions over the Complex Numbers

B.1.1 Multivalued Chebyshev Functions

The *Chebyshev function of degree* $\alpha \in \mathbb{R}$ is the multivalued function T_α of a complex variable x that is defined by the relation

$$T_\alpha(x) = \frac{u^\alpha(x) + u^{-\alpha}(x)}{2}, \tag{B.1}$$

where u is the two-valued function defined by relation

$$x = \frac{u(x) + u^{-1}(x)}{2}. \tag{B.2}$$

In formula (B.1), we mean that every value $f(x)$ of the multivalued function $u^\alpha(x)$ is summed with the value $(f(x))^{-1}$ of the function $u^{-\alpha}(x)$ (and not with any other of its values). According to formula (B.2), the function $u(x)$ satisfies the equation $u^2(x) - 2xu(x) + 1 = 0$. Its roots $u_1(x)$, $u_2(x)$ satisfy $u_1(x)u_2(x) = 1$, so it doesn't matter which of the two roots we use in formula (B.1). (Note that these roots can be explicitly calculated: $u_{1,2}(x) = x \pm \sqrt{x^2 - 1}$.) The choice of the other root only permutes the summands $u^{\alpha}(x)$ and $u^{-\alpha}(x)$ and does not change the sum.

Theorem B.1 *The functions T_α can be defined by the relations*

$$x = \cos z(x), \quad T_\alpha(x) = \cos \alpha z(x).$$

Proof If $x = \cos z_0$, then $z(x) = \pm(z_0 + 2k\pi)$ and

$$\cos(\alpha z(x)) = \frac{\exp(i\alpha z(x)) + \exp(-i\alpha z(x))}{2}.$$

We also have $u_{1,2}(x) = \exp(\pm i z(x))$ and $u_{1,2}^{\pm\alpha}(x) = \exp i\alpha(\pm z(x))$. The theorem follows. □

Proposition B.2 *The function T_n, for positive integers n, is the polynomial of degree n with integer coefficients that satisfies the following formula:*

$$T_n(x) = \sum_{0 \le k \le [n/2]} \binom{n}{2k} x^{n-2k}(x^2 - 1)^k.$$

Proof The relation $T_n(x) = (u^n(x) + u^{-n}(x))/2$ combined with the equalities

$$u^n(x) = \left(x + \sqrt{x^2 - 1}\right)^n \quad \text{and} \quad u^{-n}(x) = \left(x - \sqrt{x^2 - 1}\right)^n$$

and Newton's binomial theorem gives the formula for $T_n(x)$. □

Definition B.3 The function T_n is called the *Chebyshev polynomial of degree n*.

The Chebyshev polynomials satisfy the identity $T_n(\cos z) = \cos nz$ (see Theorem B.1). They can be defined using this identity (and that is how Chebyshev defined them). The polynomial T_n is an even function for even n, and an odd function for odd n. The leading coefficient of the polynomial T_n is equal to 2^n. Later, we will need the formula $T_3(x) = 4x^3 - 3x$.

Corollary B.4 *The equation $T_n(x) = a$ can be explicitly solved by radicals. Its roots are the values $T_{1/n}(a)$ of the multivalued function $T_{1/n}$ at the point a.*

Proof If $\cos z = a$ and $x = \cos(z/n)$, then $x = T_{1/n}(a)$ and $T_n(x) = a$. □

This "trigonometric" computation, when carried over to algebra, gives a solution of the equation $T_n(x) = a$, where a is an element of a field with characteristic not equal to 2 (see Corollary B.9). Note that $T_{1/n}$ is an n-valued function: a choice of a value of the function $u(a)$ does not change the values $T_\alpha(a)$, but the function $u^{1/n}(a)$ assumes n values.

B.1.2 Germs of a Chebyshev Function at the Point $x = 1$

The multivalued function $T_\alpha(x)$, like the function x^α, has a special germ at the point $x = 1$, with value equal to 1. It is easier to work with single-valued germs than with their multivalued analytic continuations. From now on, by x^α we denote the germ

$$1 + \sum_{k=1}^{\infty} \frac{\alpha \cdots (\alpha - k + 1)}{k!}(x-1)^k.$$

Properties of the Germs of Power Functions at the Point $x = 1$

A power function enjoys the following properties.

1. *Composition property:* if $f = x^\alpha$ and $g = x^\beta$, then $f \circ g = x^{\alpha\beta}$; in other words, $(x^\beta)^\alpha = x^{\alpha\beta}$.
2. *Multiplicative property:* $x^\alpha x^\beta = x^{\alpha+\beta}$.
3. *Algebraicity property:* for $\alpha = 1/n$, where n is a positive integer, the germ $z = x^\alpha$ satisfies the algebraic equation $z^n = x$.

Analytic Germs Invariant Under Involution

The involution τ of the complex line $\tau(u) = u^{-1}$ maps the point $u = 1$ to itself. It is easy to describe all germs f of analytic functions at this point that are invariant under the involution τ, i.e., such that $f = f \circ \tau$.

Proposition B.5 *The equality $f = f \circ \tau$ holds if and only if $f(u) = \varphi(x)$, where $x = (u + u^{-1})/2$ and φ is a germ of an analytic function at the point $x = 1$.*

Proof Let $u(x)$ be one of the two branches of the function defined by the equation

$$\frac{u(x) + u^{-1}(x)}{2} = x.$$

If $f = f(\tau)$, then the function $\varphi(x) = f(u(x))$ does not depend on the choice of branch and is analytic in a punctured neighborhood of the point $x = 1$. By the theorem on removable singularities, it is analytic at this point as well. □

Germs of analytic functions of a variable u that are not invariant under the involution τ give *two-valued Puiseux germs* of the variable x.

The *germ of the Chebyshev function T_α at the point $x = 1$* is the germ of the analytic function of the variable x such that the germ of the function $(u^\alpha + u^{-\alpha})/2$ (which is invariant under the involution τ) is equal to $T_\alpha(x(u))$, where $x(u) = (u + u^{-1})/2$. In this section, the germ of the Chebyshev function is denoted by T_α, the same symbol as was used for the multivalued function itself. The germs T_α inherit the properties of germs of power functions.

Properties of the Germs of Chebyshev Functions at the Point $x = 1$

1. *Composition property:* $T_\alpha \circ T_\beta = T_{\beta\alpha}$.
2. *Multiplicative property:* $T_\alpha T_\beta = (T_{\alpha+\beta} + T_{\alpha-\beta})/2$.
3. *Algebraicity property:* for $\alpha = n$, where n is a natural number, the germ T_α is the germ of the Chebyshev polynomial T_n. The germ $T_{1/n}$ satisfies the algebraic equation $T_n(T_{1/n}(x)) = x$.
4. *Trigonometric property:* $T_\alpha(\cos z) = \cos\alpha z$, in the sense that the germs of functions of the variable z at the point $z = 0$ are equal. The composition $T_\alpha(\cos z)$ is well defined, since $\cos 0 = 1$.

Proposition B.6 *The family of germs of Chebyshev functions satisfies properties 1–4 above.*

Proof Property 4 follows from Theorem B.1. This property completely characterizes the germ T_α. Indeed, the function $\cos z$ is even. By the implicit function theorem, the germ of the function z^2 at zero is an analytic function of the germ at $z = 1$ of the function $\cos z$. The function $\cos\alpha z$ is an analytic function of z^2. Properties 1–3 are simple properties of the function cos: to prove property 1, if $\cos v = \cos\beta z = T_\beta(\cos z)$, then $\cos\alpha v = T_\alpha(\cos v)$ and $T_\alpha T_\beta \cos z = \cos\alpha\beta z$. Property 2 follows from the identity $\cos\alpha z\cos\beta z = [\cos((\alpha+\beta)z) + \cos((\alpha-\beta)z)]/2$. Property 3 is proved for $\alpha = n$ in Proposition B.2; for $\alpha = 1/n$, it follows from the composition property. □

B.1.3 Analytic Continuation of Germs

In this section, we show that the set of values of the multivalued function generated by the germ T_α is consistent with the definition from Sect. B.1.1. The compositional inverse of the germ at 0 of the function $\cos z$ is a two-valued Puiseux germ at the point $x = 1$. Its values differ by a sign. Let $\pi^{-1}(x)$ be one of the two inverses (differing by sign) of the function $\cos z = x$ that has this Puiseux germ at the point $x = 1$. Consider the even function $\Phi_\alpha(z) = \cos\alpha z$ of the variable z. By definition, $T_\alpha = \Phi_\alpha \circ \pi^{-1}$.

The function $\cos z$ has simple critical points $z = k\pi$ and two critical values $x = \pm 1$. We say that the curve $x(t)$ that goes from point 1 to point x_0, i.e., $x(0) = 1$, $x(1) = x_0$, is *admissible* if $x(t) \neq \pm 1$ for $0 \le t \le 1$. The Puiseux germ of the function π^{-1} at the point $x = 1$ can be continued along the admissible curve $x(t)$ that goes from $x = 1$ to x_0 in the following sense: either of the two branches of the germ can be continued analytically along $x(t)$ up to $t = 1$ if $x_0 \neq \pm 1$, and up to any $t < 1$ if $x_0 = \pm 1$. In the second case, the continuation up to $t = 1$ is a two-valued Puiseux germ at the point $x_0 = \pm 1$ (whose branches at x_0 coincide).

In the same sense, the germ $T_\alpha = \Phi_\alpha \circ \pi^{-1}$ can be continued along any admissible curve $x(t)$. The germ T_α is regular and single-valued (not two-valued, like π^{-1}); therefore, it has a unique continuation along an admissible curve. For some admissible curves that go from $x = 1$ to the point $x = \pm 1$, the result of continuation may also turn out to be an analytic germ (and not a two-valued Puiseux germ).

Let us show that formulas (B.1) and (B.2) describe all values of the multivalued function that is obtained by continuation of the germ T_α. Let x_0 and $a = T_\alpha(x_0)$ be any numbers that satisfy (B.1) and (B.2).

Proposition B.7 *There exists an admissible curve $x(t)$ that goes from the point $x = 1$ to the point x_0 such that the analytic germ (or the Puiseux germ) that is obtained by continuation of the germ T_α along $x(t)$ takes the value a at the point x_0, where a, x_0 are as defined above.*

Proof Choose z_0 such that $\exp i z_0 = u(x_0), \exp(\alpha i z_0) = u^\alpha(x_0)$. Let $z(t)$ be a curve with $z(0) = 0$, $z(1) = z_0$ such that $z(t)$ does not pass through the points $z = k\pi$ for $0 < t < 1$. Then the curve $x(t) = \cos z(t)$ is admissible; it goes from the point $x = 1$ to the point x_0, and the analytic continuation along this curve of the germ $T_\alpha = \cos\alpha(\cos^{-1})$ gives the germs that take the value a at the point x_0. □

Of special importance to us are the Chebyshev polynomials T_n and their inverses $T_{(1/n)}$. Proposition B.7 provides a description of the set of values of the function $T_{1/n}$ at a point a. Let u_1, u_2 be the roots of the equation $\left(u + u^{-1}\right)/2 = a$ (it is enough to take one of these roots). Let $\{v_{i,j}\}$ be the roots of the equation $v^n = u_i$, where $i = 1, 2, 1 \leq j \leq n$. The set $T_{1/n}(a)$ of all values of the function at the point a is equal to the set

$$\left\{\frac{v_{1,j} + v_{i,j}^{-1}}{2}\right\}$$

and to the set

$$\left\{\frac{v_{2,j} + v_{2,j}^{-1}}{2}\right\}.$$

B.2 Chebyshev Functions over Fields

B.2.1 Algebraic Definition

The Chebyshev polynomials $T_n \in \mathbb{Z}[x]$ are defined over every field $\mathbf{k}$. If the characteristic of the field is zero, then $\mathbb{Z} \subset \mathbf{k}$ and $T_n \in \mathbf{k}[x]$. If the field has characteristic $p > 0$, then $\mathbb{Z}_p \subset \mathbf{k}$, and the polynomial obtained from T_n by reduction of the coefficients modulo p (which we denote by the same symbol T_n) lies in $\mathbf{k}[x]$. If $p \neq 2$, then $\deg T_n = n$, since the leading coefficient of the polynomial T_n is equal to 2^{n-1}.

Proposition B.8 *If the characteristic of the field* $\mathbf{k}$ *is not equal to* 2, *then the following identity holds in the field of rational functions* $\mathbf{k}(x)$:

$$T_n\left(\frac{x+x^{-1}}{2}\right)=\frac{x^n+x^{-n}}{2}. \tag{B.3}$$

Proof The result follows from the formulas (B.1) and (B.2). □

Corollary B.9 *If the characteristic* p *of the field* $\mathbf{k}$ *is not equal to* 2, *then the equation* $T_n(x)=a$ *with* $a\in\mathbf{k}$ *is explicitly solvable in radicals over the field* $\mathbf{k}$.

Proof Plugging $x=(v+v^{-1})/2$ into the identity (B.3), we obtain $(v^n+v^{-n})/2=a$. Then we solve the quadratic equation $u^2-2au+1=0$ for $u=v^n$. Let u_1,u_2 be its roots and $\{v_{1,j}\}$ the set of all roots of u_1 of degree n. Then the elements $v_{2,j}=v_{1,j}^{-1}$ form the set of all roots of degree n of u_2, since $u_1u_2=1$. All roots of the equation $T_n(x)=a$ can be expressed in the form

$$x=\frac{\left(v_{1,j}+v_{1,j}^{-1}\right)}{2}\quad\text{or}\quad x=\frac{\left(v_{2,j}+v_{2,j}^{-1}\right)}{2}.$$

□

The proof of Corollary B.9 shows that the equation $T_n(x)=a$ over a field $\mathbf{k}$ of characteristic not equal to 2 is solvable explicitly using the formula $x=T_{1/n}(a)$, which makes sense over $\mathbf{k}$.

B.2.2 Equations of Degree Three

Let F be a polynomial of degree n over a field $\mathbf{k}$ with characteristic equal to zero or greater than n. Define $Q(y)=aF(\lambda y+x_0)$, where $a,\lambda\neq 0$, and x_0 is an element of the field $\mathbf{k}$ or some finite extension. Under the assumptions about the characteristic of $\mathbf{k}$, we have

$$Q(y)=\sum\frac{a\lambda^k F^{(k)}(x_0)}{k!}y^k.$$

The linear function $Q^{(n-1)}$ takes the value 0 at some point q. Assume that when $x_0=q$, the coefficient of Q at y^{n-1} vanishes. By varying a and λ, we can make any two nonzero coefficients of Q equal to any two given nonzero numbers.

Using this transformation, we can reduce the polynomial $F(x)=a_3x^3+a_2x^2+a_1x+a_0$ to the form y^3+c or to the form $4y^3-3y+c$. Indeed, the polynomial F'' vanishes at the point $x_0=-a_2/3a_3$. There are two possible cases:

1. $F'(x_0)=0$. In this case, the polynomial F reduces to the form y^3+c via the transformation $aF(y+x_0)$, where $a=a_3^{-1}$, and we obtain $c=F(x_0)a$.

2. $F'(x_0) \neq 0$. In this case, the polynomial F reduces to the form $4y^3 - 3y + c$ via the transformation $aF(\lambda y + x_0)$ with $\lambda = (-4F'(x_0)/3a_3)^{1/2}$, $a = -3(\lambda F'(x_0))^{-1}$. Here $c = F(x_0)a$. (We choose any sign of λ, since we are looking for one transformation that has the properties we need rather than a description of all such transformations.)

Corollary B.10 *A cubic equation $F(x) = a_3x^3 + a_2x^2 + a_1x + a_0$ over a field* **k** *of characteristic not equal to* 2 *or* 3 *is solvable in radicals in the following way. Let $x_0 = -a_2/3a_3$ be the root of the polynomial F''.*

1. *If $F'(x_0) = 0$, then $x = x_0 + (-F(x_0)/a_3)^{1/3}$.*
2. *If $F'(x_0) \neq 0$, then $x = x_0 + \lambda T_{1/3}(-c)$, where λ and c are as defined above.*

B.2.3 Equations of Degree Four

An equation of degree four can be reduced to an equation of degree three (which is solvable using the function $T_{1/3}$) by considering a pencil of planar quadrics [12].

Let $Q : V \to \mathbf{k}$ be a quadratic form and $\dim_{\mathbf{k}} V = n$. A quadratic form in the plane or on the line can be decomposed as a product of linear factors (possibly not over the original field **k**, but over a quadratic extension K). Let K be an extension of the field **k**. Let V_K and Q_K denote the space and the form that correspond to V and Q under the extension $\mathbf{k} \subset K$.

Lemma B.11 *If Q_K can be factored, then* $\dim_{\mathbf{k}} \ker Q \geq n - 2$. *If this inequality holds, then we can explicitly find a factorization $Q_K = L_1L_2$ over a quadratic extension K of* **k**.

Proof If $Q_K = L_1L_2$, then $\ker Q_K \supset \cap_{i=1,2}\{L_i = 0\}$ and $\dim_K \ker Q_K \geq n-2$. The form Q is defined over k, and therefore, $\dim_{\mathbf{k}} \ker Q \geq n - 2$. If the inequality holds, then V can be expressed in the form $V = \ker Q \oplus W$, where $\dim_{\mathbf{k}} W \leq 2$. Let $\pi : V \to W$ be the projection along $\ker Q$, and $\tilde{Q}$ the restriction of the form Q to W. On W, we have the factorization $\tilde{Q} = \tilde{L}_1\tilde{L}_2$, and therefore $Q = (\pi^*\tilde{L}_1)(\pi^*\tilde{L}_1)$. □

Proposition B.12 *Let P, Q be quadratic polynomials of two variables. The coordinates x, y of the points of intersection of two planar quadrics $\mathscr{P} = 0$ and $\mathscr{R} = 0$ can be found by solving one cubic equation and a number of quadratic and linear equations.*

Proof All quadrics of the pencil $0 = \mathscr{Q}_\lambda = \mathscr{P} + \lambda\mathscr{R}$, where λ is a parameter, pass through the desired points. For some λ, some quadric $\mathscr{Q}_\lambda = 0$ splits into a union of two lines, i.e., $\mathscr{Q}_\lambda = \mathscr{L}_1\mathscr{L}_2$, where $\mathscr{L}_1, \mathscr{L}_2$ are polynomials of degree 1. These λ satisfy the cubic equation $\det(Q_\lambda) = 0$, where $Q_\lambda = P + \lambda Q$ is the 3×3 matrix of quadratic forms that corresponds to the equation of the quadric in homogeneous coordinates. Indeed, for this λ, the form Q_λ has nontrivial kernel, and therefore, $Q_\lambda = L_1L_2$, where L_1, L_2 can be found by solving one quadratic equation and a number of linear equations. Returning to the coordinates x, y, from L_1, L_2 we will obtain the polynomials $\mathscr{L}_1, \mathscr{L}_2$. Now we need to solve quadratic equations to find the points of intersection of the quadric $\mathscr{P} = 0$ and lines $\mathscr{L}_1 = 0$ and $\mathscr{L}_2 = 0$. □

Corollary B.13 *The roots of a polynomial $a_0x^4 + a_1x^3 + a_2x^2 + a_3x + a_4$ can be found by solving one cubic and a number of quadratic and linear equations.*

Proof To find the roots of this polynomial, we need to project the points of the intersection of the quadrics $y = x^2$ and $a_0y^2 + a_1xy + a_2y + a_3x + a_4 = 0$ to the x-axis. □

A polynomial F is said to be *decomposable (in the sense of composition)* if it can be written in the form $F = P(Q)$, where P and Q are polynomials of degree greater than 1.

Proposition B.14 *A polynomial F of degree 4 is decomposable if and only if it has one of the following equivalent properties:*

1. *$F(x - x_0) \equiv F(x_0 - x)$ for some x_0.*
2. *$F'(x_0) = 0$, where x_0 is the point where $F^{(3)}(x_0) = 0$.*

Proof If the first assertion holds, then F is a polynomial of degree 2 in the variable y^2, where $y = x - x_0$. By Taylor's formula, this property is equivalent to the equalities $F'(x_0) = F^{(3)}(x_0) = 0$. Let $F = Q(P)$. Then since the polynomial P can be represented in the form $P = a(x - x_0)^2 + b$, we obtain $F(x - x_0) \equiv F(x_0 - x)$. □

B.3 Three Classical Problems

B.3.1 Inversion of Mappings in Radicals

When is a polynomial map $P : \mathbb{C} \to \mathbb{C}$ invertible in radicals? We start by providing several examples.

Example B.15 If P is the power polynomial x^n, then the inverse map $x = z^{1/n}$ by definition can be expressed in radicals. If $n = km$ is a composite number, then the map x^n can be expressed as the composition $x^n = (x^m)^k$. For prime n, the polynomial x^n is indecomposable.

Example B.16 If $P = T_n$ is the nth Chebyshev polynomial, then the inverse map $T_{1/n}$ can be expressed in radicals. If $n = km$ is a composite number, then the map T_n can be expressed as the composition $T_n = T_k(T_m)$. For prime n, the polynomial T_n is indecomposable.

Example B.17 If P is a polynomial of degree 4, then the inverse map can be expressed in radicals (since equations of degree 4 are solvable in radicals). Polynomials of degree 4 are generally indecomposable. Exceptions are described in Corollary B.14.

Theorem B.18 *If $P = P_1 \circ \cdots \circ P_k$, where for $1 \leq i \leq k$, the polynomial P_i is either a linear polynomial, an indecomposable polynomial of degree four, x^n with prime n, or T_n with prime $n > 2$, then the map $P : \mathbb{C} \to \mathbb{C}$ is invertible in radicals.*

Proof The result follows from Examples B.15–B.17. □

Ritt [84] proved the converse statement (see also [19,55]).

Theorem B.19 (J. Ritt) *If a map $P : \mathbb{C} \to \mathbb{C}$ is invertible in radicals, then the polynomial P can be expressed in the form described in Theorem B.18.*

One can ask also the following question (see [22]): when is a polynomial map $P : \mathbb{C} \to \mathbb{C}$ invertible in k-radicals? Here is the answer: A polynomial that is invertible in radicals and solutions of equations of degree at most k is a composition of power polynomials, Chebyshev polynomials, polynomials of degree at most k, linear polynomials, and if $k \leq 14$, certain polynomials with exceptional monodromy groups.

The proofs in [22] rely on the classification of monodromy groups of primitive polynomials obtained by Muller [80] based on group-theoretic results of Feit [31] and on previous work on primitive polynomials with exceptional monodromy groups by many authors (references to their works and descriptions of these exceptional polynomials can be found in [22]). I have to add that we failed to obtain an exhaustive description of the exceptional polynomials of degree 15.

For $k \geq 15$, the answer is much simpler: for $k \geq 15$, a polynomial is invertible in k-radicals if and only if it is a composition of power polynomials, Chebyshev polynomials, and polynomials of degree at most k.

There is an interesting question related to Theorem B.19 and to the paper [22]. To what extent is the decomposition of the polynomial P as

$$P = P_1 \circ \cdots \circ P_k, \tag{B.4}$$

where for $1 \leq i \leq k$, the polynomials P_i are indecomposable, unique? Ritt gave a complete answer to this question ([85]; see also [103]). There are several relations of the form

$$A \circ B = C \circ D, \tag{B.5}$$

where A, B, C, D are polynomials. For example, we know that $T_m \circ T_n = T_n \circ T_m$. There is also the following generalization of the equality $(x^m)^n = (x^n)^m$: for every polynomial H, the equality (B.5) holds for $A(x) = x^n$, $B(x) = x^m H(x^n)$, $C(x) = x^m H^n(x)$, $D(x) = x^n$. Ritt proved that modulo these relations and composition relations with linear functions, the representation in the form (B.4) is unique.

Ritt completely described the polynomials that are invertible in radicals. Families of power polynomials and Chebyshev polynomials play a fundamental role in this description. Ritt also completely described rational functions $R : \mathbb{C} \to \mathbb{C}$ of prime degree p that are invertible in radicals [84]. Functions related to the division of the argument of an elliptic function appear in his description (just as the polynomial T_n is related to division of the argument of the function cos). Appendix C contains a more detailed description of these mappings (see also [19]).

B.3.2 Inversion of Mappings of Finite Fields

A polynomial $P \in \mathbb{Q}[x]$ can be defined over $\mathbb{Z}_p$ if the prime number p does not divide denominators of its coefficients. For which P is the map $P : \mathbb{Z}_p \to \mathbb{Z}_p$ invertible (i.e., bijective) for infinitely many prime numbers p? This question was formulated by Schur [91]. He found a conjectural answer and obtained some results in this direction. Fried proved Schur's conjecture in even greater generality [32]. Instead of the field $\mathbb{Q}$, he considered a finite extension K. Here we consider only the case $K = \mathbb{Q}$. We will use the notation $\mathbf{k}$ to denote the quadratic extension of the field $\mathbb{Z}_p$ with p^2 elements.

Example B.20 For $p > 2$, an even polynomial $P \in \mathbb{Z}[x]$ (for example, x^{2n} or T_{2n}) gives a noninvertible map $P : \mathbb{Z}_p \to \mathbb{Z}_p$, since $P(x) = P(-x)$ and the number of values of the polynomial is not greater than $(p-1)/2 + 1 < p$.

Example B.21 For a linear polynomial

$$P(x) = \frac{a_1}{b1}x + \frac{a_2}{b_2},$$

the map $P : \mathbb{Z}_p \to \mathbb{Z}_p$ is defined and invertible if b_1, b_2 are not divisible by p.

Example B.22 The map $P : K \to K$ for $P(x) = x^q$, where $q \neq 2$ is a prime number and K is a finite field, is invertible if $\#K \not\equiv 1 \pmod q$. For $K = \mathbb{Z}_p$, the condition $p \not\equiv 1 \pmod q$ holds in particular if $p \equiv 2 \pmod q$. For the quadratic extension $\mathbf{k}$ of the field $\mathbb{Z}_p$, the condition $p \not\equiv \pm 1 \pmod q$ for $q > 3$ holds in particular if $p \equiv 2 \pmod q$.

Proposition B.23 *Let $q, p > 2$ be prime numbers and $p \not\equiv \pm 1 \pmod q$. Then the map $T_q : \mathbb{Z}_p \to \mathbb{Z}_p$ is invertible.*

Proof Let us prove that for every $a \in \mathbb{Z}_p$, the equation $T_q(x) = a$ has a solution in $\mathbb{Z}_p$. Let $\mathbf{k}$ be an extension of degree 2 of the field $\mathbb{Z}_p$. The equation $v^2 - av + 1 = 0$ has solutions $v_1, v_2 \in \mathbf{k}$. Since $p \not\equiv \pm 1 \pmod q$, there is a unique solution $u_1 \in \mathbf{k}$ of the equation $u^q = v_1$, where v_1 is one of the solutions v_1, v_2. Let g be the nontrivial element of the Galois group of $\mathbf{k}$ over $\mathbb{Z}_p$. Denote $g(u_1)$ by u_2. Since $g(v_1) = v_2$, we have $u_2^q = v_2$. From the equality $(u_1u_2)^q = v_1v_2 = 1$, we get that $u_1u_2 = 1$. It follows that $x = (u_1 + u_2)/2$ is a solution of the equation $T_q(x) = a$. Since $g(x) = x$, we have $x \in \mathbb{Z}_p$. We have proved that the map $T_q : \mathbb{Z}_p \to \mathbb{Z}_p$ is onto. Since the field $\mathbb{Z}_p$ is finite, the map is invertible. □

Remark B.24 For T_3, Proposition B.23 says only that the map $T_3 : \mathbb{Z}_3 \to \mathbb{Z}_3$ is invertible (which is trivial). One can check that the map $T_3 : \mathbb{Z}_p \to \mathbb{Z}_p$ is not invertible for $p > 3$.

Theorem B.25 *Let $P = P_1 \circ \cdots \circ P_k$, where for $1 \leq i \leq k$, the polynomial $P_i \in \mathbb{Q}[x]$ is either linear, x^q with $q > 2$ a prime number, or T_q with $q > 3$ a prime number. Then the map $P : \mathbb{Z}_p \to \mathbb{Z}_p$ is invertible for infinitely many prime numbers.*

Proof Denote by E the finite set of prime numbers p for which linear polynomials from the decomposition of P are not defined over $\mathbb{Z}_p$. Let $M = \{q_i\}$ be the set of different degrees of the polynomials T_{q_i} and x^{q_i} from the decomposition of P and $m = \prod_{q_i \in M} q_i$. Let S be the set of natural numbers congruent to 2 modulo m. If $a \in S$ and $q_i \in M$, then $a \equiv 2 \pmod{q_i}$. By Dirichlet's theorem on primes in arithmetic progressions, in the arithmetic progression S there are infinitely many prime numbers p that are not in the finite set E. For each of these prime numbers p, every map $P_i : \mathbb{Z}_p \to \mathbb{Z}_p$ from the decomposition of P is invertible (see Examples B.21 and B.22, and Proposition B.23). □

Theorem B.26 (Fried) *Assume that for $P \in \mathbb{Q}[x]$, the map $P : \mathbb{Z}_p \to \mathbb{Z}_p$ is invertible for infinitely many prime numbers p. Then P can be expressed in the form $P = P_1 \circ \cdots \circ P_k$, where for $1 \leq i \leq k$, the polynomial P_i is linear, x^q, or T_q.*

Fried's paper [32] contains beautiful results about complex polynomials that are close to Ritt's Theorem B.19. It also uses relations between number theory and algebraic geometry (in particular, some results of André Weil).

B.3.3 Integrable Mappings

Iterations of a polynomial map $P : \mathbb{C} \to \mathbb{C}$ on the complex line exhibit very unusual behavior for the polynomials x^n and T_n. Their dynamics look like the behavior of completely integrable systems in Hamiltonian mechanics.

Example B.27 Iterations of the map $x \mapsto x^n$ can be explicitly described: the kth iteration is the map $x \mapsto x^{n^k}$. As $k \to \infty$, then $x_0^{n^k} \to 0$ for $|x_0| < 1$ and $x_0^{n^k} \to \infty$ for $|x_0| > 1$. The projection $x = \exp it$ of the line $\mathbb{R}$ to the circle $|x| = 1$ conjugates the dilatation $t \mapsto nt$ with the map $x \mapsto x^n$. The segment $|t - t_0| \leq \varepsilon$ under the kth iteration of the dilatation goes to the segment $|t - n^k t_0| \leq \varepsilon n^k$. For $k \gg 0$, every point in the circle has about $(\varepsilon/\pi)n^k$ preimages in this segment. Points $\exp 2\pi i n^{-k}$ after the kth iteration are mapped to the point 1 and stay at that point under all subsequent iterations. Despite the fact that the iterations of the map can be explicitly described, its dynamics on the circle $|u| = 1$ are chaotic.

Example B.28 Iterations of the map $x \mapsto T_n(x)$ can be described explicitly: the kth iteration is the map $x \mapsto T_{n^k}$. As $k \to \infty$, then $T_{n^k}(x_0) \to \infty$ for $x_0 \notin I$, where $I \subset \mathbb{R}$ is the segment defined by the inequality $|x| \leq 1$. The projection

$x = (u + u^{-1})/2$ of the circle $|u| = 1$ to the segment I conjugates the map $u \mapsto u^n$ with the map $x \mapsto T_n(x)$. On the segment I, the dynamics of the map T_n are as chaotic as the dynamics of the map u^n on the circle $|u| = 1$.

Definition B.29 A polynomial map $P : \mathbb{C} \to \mathbb{C}$ is *integrable* (see [99]) if there exists a polynomial map $G : \mathbb{C} \to \mathbb{C}$ such that $P \circ G = G \circ P$ and (1) $\deg P > 1$, $\deg G > 1$; and (2) the kth iteration of the polynomial P does not coincide with the qth iteration of the polynomial G for any natural numbers k, q.

The map $x \to x^n$ is integrable, since it commutes with all power functions $x \mapsto x^m$. If $m \neq n^{k/q}$, where $k, q \in \mathbb{Z}$, then the iterations of these maps do not coincide. The map $x \mapsto T_n(x)$ is integrable, since it commutes with all maps of the form $x \mapsto T_m(x)$. If $m \neq n^{k/q}$, where $k, q \in \mathbb{Z}$, then the iterations of these maps do not coincide.

Polynomials P and G are *equivalent* if there exists a polynomial $H(x) = ax+b$, $a \neq 0$, such that $P \circ H = H \circ G$. Ritt, Julia, and Fatou described all integrable polynomial mappings up to this equivalence relation. Below, we formulate their remarkable results (see [30, 42, 85]).

Theorem B.30 *The map $P : \mathbb{C} \to \mathbb{C}$ is integrable if and only if the polynomial P is equivalent to one of the polynomials x^n, T_{2m}, T_{2m+1}, $-T_{2m+1}$.*

Julia and Fatou proved this theorem using methods of dynamics. Ritt's proof is quite different (see Sect. B.3.1). Earlier, Lattès gave examples of integrable (in the same sense) rational mappings of $\mathbb{C}P^1$ to itself [70, 79]. Ritt proved that there are no integrable rational mappings other than Lattès maps. No one was able to prove Ritt's theorem using the dynamical methods that go back to Julia and Fatou until it was done by Eremenko [29].

It is interesting that Lattès maps are invertible in radicals. Ritt described a wonderful class of rational maps that are invertible in radicals (see [19, 84] and Appendix C). This class is quite large. For example, it contains all Lattès maps and all rational maps of prime degree invertible in radicals.

Multidimensional examples of integrable polynomials and rational maps are known (they can be found in the references given in Milnor's survey [79]).

Appendix C
Signatures of Branched Coverings and Solvability in Quadratures

Yuri Burda and Askold Khovanskii

In this appendix, we deal with branched coverings over the complement in the Riemann sphere of finitely many exceptional points having the property that the local monodromy around each of the branch points is of finite order. To such a covering we assign its *signature*, i.e., the set of its exceptional and branch points together with the orders of local monodromy operators around the branch points.

What can be said about the monodromy group of a branched covering if its signature is known? It seems at first that the answer is nothing or next to nothing. Indeed, generically, such is the case. However, there is a (small) list of signatures of *elliptic* and *parabolic* types for which the monodromy group can be described completely, or at least determined up to an abelian factor. This appendix is devoted to an investigation of these signatures. For all these signatures (with one exception), the corresponding monodromy groups turn out to be solvable.

Linear differential equations of Fuchsian type related to these signatures are solvable in quadratures (in the case of elliptic signatures in algebraic functions). A well-known example of this type is provided by Euler differential equations, which can be reduced to linear differential equations with constant coefficients.

The algebraic functions related to all (except one) of these signatures are expressible in radicals. A simple example of this kind is provided by the possibility of expressing the inverse of a Chebyshev polynomial in radicals. Another example of this kind is provided by functions related to division theorems for the argument of elliptic functions. Such functions play a central role in Ritt's work [84].

A talk based on the contents of this appendix was given at the Sixth European Congress of Mathematics, Krakow, Poland, July 27, 2012, at the minisymposium "Differential Algebra and Galois Theory." A slightly updated exposition can be found in [23].

A. Khovanskii, *Topological Galois Theory*, Springer Monographs in Mathematics,
DOI 10.1007/978-3-642-38871-2

C.1 Coverings with a Given Signature

C.1.1 Definitions and Examples

The mapping $\pi : Y \to S$ of a connected Riemann surface Y to the Riemann sphere S is said to be *admissible* if the following conditions hold:

1. $\pi(Y) = S \setminus B$, where $B = \{b_1, \dots, b_k\}$ is the *exceptional* set.
2. $\pi : Y \to S \setminus B$ is a branched covering with branch locus $A = \{a_1, \dots, a_n\}$.
3. For $1 \leq j \leq n$, the order of the local monodromy operator at the point a_j is a finite number $r_j > 1$ (the local monodromy operator at the point x is the element of the monodromy group, defined up to conjugation, that corresponds to a small path going around the point x). We do not assume anything about the order of local monodromy operators at points b_j (i.e., points $b_j \in B$ can be branch points of infinite order).

Definition C.1 The *signature* of an admissible mapping $\pi : Y \to S$ is the triple (A, B, R), where $R = \{r_1, \dots, r_n, \infty, \dots, \infty\}$ is the *set of orders*. If $B = \varnothing$, we do not mention B in the signature. We call an admissible mapping with a given signature a *covering with a given signature.*

We assume that the inequality $n + k \geq 2$ holds for the signature (A, B, R). We also assume that for the signature (A, R) with $\#A = 2$ and $R = (k, n)$, the equality $k = n$ holds. If a signature does not satisfy these conditions, then every covering with such a signature is either trivial or does not exist.

Example C.2 Consider an algebraic function with the branch locus $A = \{a_1, \dots, a_n\}$. Suppose that the local monodromy operator at the point $a_i \in A$ has order r_i. Then the Riemann surface of this function is a covering with signature (A, R), where $R = \{r_1, \dots, r_n\}$.

Example C.3 Consider a linear differential equation of Fuchsian type with set of singular points $A \cup B$, where $A = \{a_1, \dots, a_n\}$, $B = \{b_1, \dots, b_k\}$. Suppose that the local monodromy operator has finite order r_i at each of the points $a_i \in A$ and infinite order at each of the points $b_j \in B$. Then the Riemann surface of a generic solution of this differential equation is a covering with signature (A, B, R), where $R = \{r_1, \dots, r_n, \infty, \dots, \infty\}$.

We will see below that for all but one of the exceptional signatures, the set $A \cup B$ contains two or three points.

Claim If $\#A \cup B \leq 3$, then up to an automorphism of the sphere S, the signature (A, B, R) is defined by the set of orders R.

Proof There exists an automorphism of the sphere that takes every triple of points to any other triple. □

C.1.2 Classification

The covering $\pi : Z \to S$ with signature (A, B, R) is said to be *universal* if (1) the surface Z is simply connected, (2) the multiplicity of the mapping π at points $c_j \in \pi^{-1}(a_k)$ is r_k.

The universal covering $\pi : Z \to S$ with signature (A, B, R) has the following universal property.

Theorem C.4 *Let $\pi_1 : Y \to S$ be a covering with signature (A, B, R) and let $z_0 \in Z$, $y_0 \in Y$ be points with $\pi(z_0) = \pi_1(y_0) = x_0 \notin A$. Then there exists a mapping $\pi_2 : Z \to Y$ such that $\pi = \pi_1 \circ \pi_2$ and $\pi_2(z_0) = y_0$.*

Proof Let $C = \pi^{-1}(A) \subset Z$. Since the surface Z is simply connected, the fundamental group of the complement $Z \setminus C$ is generated by the curves τ_j going around the points $c_j \in C$. Suppose $\pi(c_j) = a_k$. By definition, the mapping π has multiplicity r_k at the point c_j. Hence the image of the curve τ_j under the projection π_1 goes around the point a_k exactly r_k times. By the definition of signature, the lift of the curve $\pi(\gamma)$ to the surface Y based at the point y_0 is a closed curve. The theorem follows. □

Let $\pi_1 : Y \to S$ be a covering with signature (A, B, R). Fix a point $x_0 \in S \setminus (A \cup B)$. A branched covering $\pi : Y \to S \setminus A$ corresponds to a conjugacy class of subgroups of the fundamental group of the set $S \setminus (A \cup B)$ with base point x_0. To the intersection of these subgroups corresponds a branched covering $\pi_{\text{nor}} : Y_{\text{nor}} \to S \setminus A$. This covering will be called the *normalization* of the original covering.

The following theorem obviously holds.

Theorem C.5 *The normalization of a covering with a given signature (A, B, R) is a covering with the same signature and isomorphic monodromy group. If $\pi_{\text{nor}}(c_j) = a_k$, then the multiplicity of the mapping π_{nor} at the point c_j is r_k.*

The following theorem provides an explicit construction of the universal covering with a given signature if some covering with this signature is given.

Theorem C.6 *Let $\pi_{\text{nor}} : Y_{\text{nor}} \to S$ be the normalization of the covering $\pi_1 : Y \to S$ with signature (A, B, R) and let $\pi : Z \to Y$ be the universal covering of Y. Then $\pi_{\text{nor}} \circ \pi : Z \to S$ is the universal covering with signature (A, B, R).*

Proof By construction, the surface Z is simply connected. If $\pi \circ \pi_{\text{nor}}(z) = a_k$, then the multiplicity of the mapping $\pi \circ \pi_{\text{nor}}$ at the point z is equal to r_k. Indeed, the mapping π is a local diffeomorphism at the point z, while the mapping π_{nor} has multiplicity r_k at the point $\pi(z)$. □

Theorems C.4–C.6 provide a way to classify all the coverings with a given signature (A, B, R) by considering the universal covering with the given signature and its group of deck transformations.

Let $\pi : Z \to S$ be the universal covering with signature (A, B, R). The group G of deck transformations of π acts on Z. The quotient space of Z by the action of G is isomorphic to $S \setminus B$. The set of orbits on which G acts freely is isomorphic to $S \setminus (A \cup B)$. If a point $c \in Z$ is mapped to the point $a_k \in A$ in the quotient space, then the stabilizer of the point c has r_k elements.

We say that $H \subset G$ is a *free normal subgroup* of the group G if H acts freely on Z and H is a normal subgroup of G. We say that the subgroup $F \subset G$ is *admissible* if the intersection $H = \bigcap F_i$ of all the subgroups F_i conjugate to F is a free normal subgroup of G.

Corollary C.7 *Every covering with signature* (A, B, R) *is isomorphic to a quotient of* Z *by an admissible subgroup* $F \subset G$. *Conjugate subgroups* F_i *correspond to equivalent coverings. The monodromy group of the covering is isomorphic to the quotient* G/H, *where* $H = \bigcap F_j$. *A normal covering with signature* (A, B, R) *corresponds to a free normal subgroup* H. *Its group of deck transformations is isomorphic to the monodromy group* G/H.

Admissible mappings can be divided into three natural classes.

Definition C.8 The signature of a covering is *elliptic*, *parabolic*, or *hyperbolic* if the universal covering $\pi : Z \to S$ with this signature has total space Z isomorphic respectively to the Riemann sphere, the line $\mathbb{C}$, or the open unit disk.

In Sects. C.2 and C.3, we discuss coverings with elliptic and parabolic signatures. We turn now to a geometric construction of a large class of branched coverings.

C.1.3 Coverings and Classical Geometries

Using the geometry of a sphere and Euclidean and hyperbolic planes, one can construct universal coverings with many signatures. In this section, we use realizations of each of these geometries on a subset E of the Riemann sphere $\mathbb{C} \cup \{\infty\}$: the sphere is identified with the set $\mathbb{C} \cup \{\infty\}$ by means of stereographic projection; the Euclidean plane is identified with the line $\mathbb{C}$; and the hyperbolic plane is identified with its Poincaré model in the unit disk $|z| < 1$.

We consider polygons in E that may have "vertices at infinity" lying in $\overline{E}$. For the plane $\mathbb{C}$, such a vertex is the point ∞ at which two parallel sides meet. For the hyperbolic plane, such a vertex is a point on the circle $|z| = 1$ at which two neighboring sides meet. The angle at a vertex at infinity is equal to zero.

Let E be the sphere, plane, or hyperbolic plane, and let $\Delta \subset \overline{E}$ be an $(n+k)$-gon with finite vertices $A' = \{a'_1, \dots, a'_n\}$ and vertices at infinity $B' = \{b'_1, \dots, b'_k\}$. Let $R = (r_1, \dots, r_{n+k})$, where $r_i > 1$ are natural numbers for $1 \le i \le n$ and $r_i = \infty$ for $n < i \le n + k$.

Definition C.9 The polygon $\Delta \subset \overline{E}$ has *signature* (A', B', R), if its angle at each vertex a_i' is π/r_i and its angle at each vertex b_j' is 0.

It is clear that the signature (A', B', R) with $\#A' \cup B' \leq 2$ can be a signature of a polygon only if $R = (k, k)$ or $R = (\infty, \infty)$. We assume that when $n + k \leq 2$, this condition on the set R holds.

Definition C.10 The *characteristic* of the signature $R = (r_1, \dots, r_{n+k})$ is

$$\chi(R) = \sum_{1 \leq i \leq n+k} \left(1 - \frac{1}{r_i}\right).$$

Definition C.11 We say that the set R is *elliptic*, *parabolic*, or *hyperbolic* if respectively $\chi(R) < 2$, $\chi(R) = 2$, or $\chi(R) > 2$.

Claim Suppose that the polygon $\Delta \subset \overline{E}$ has signature (A', B', R). The set R is elliptic, parabolic, or hyperbolic if and only if E is respectively the sphere, Euclidean plane, or the hyperbolic plane.

Proof On the sphere, the sum of the external angles of a polygon is less than 2π, on the plane it is equal to 2π, and on the hyperbolic plane, it is greater than 2π. For Δ, this sum is equal to

$$\pi \sum_{1 \leq i \leq n+k} \left(1 - \frac{1}{k_i}\right) = \pi\chi(R).$$

□

Definition C.12 Given a polygon $\Delta \subset \overline{E}$ with signature (A', B', R), define $\tilde{G}_\Delta$ to be the group of isometries of the space E generated by reflections in the sides of the polygon. Define the group G_Δ to be the index-two subgroup of the group $\tilde{G}_\Delta$ consisting of orientation-preserving isometries.

The condition on the angles of the polygon guarantees that the images $g(\Delta)$ of the polygon Δ under the action of the group $\tilde{G}_\Delta$ cover the space E without overlaps. Divide the polygons $g(\Delta)$, $g \in \tilde{G}_\Delta$, into two classes: white if $g \in G_\Delta$, and black otherwise. Let g_ℓ be the reflection in the side ℓ of the polygon Δ. Define a (possibly nonconvex) polygon $\diamondsuit$ as the union of polygons Δ and $g_\ell(\Delta)$ sharing the side ℓ. It can be seen from the construction that the polygon $\diamondsuit$ is a fundamental domain for the action of the group G_Δ. The polygon $\diamondsuit$ contains ℓ, and ℓ is not one of its sides. The transformation g_ℓ glues each of the sides ℓ_j of the polygon Δ to the side $g_\ell(\ell_j)$. The following claim can be easily verified.

Claim The stabilizer of the vertex $a_i' \in A'$ under the action of the group G_Δ contains r_i elements. The points of E that do not belong to the orbits of the points $a_i' \in A'$ have trivial stabilizers.

Consider a Riemann mapping f of the polygon $\Delta \in \overline{E}$ with signature (A', B', R) onto the upper half-plane. We introduce the following notation:

1. A is the set $f(A') = \{a_k = f(a'_k) : a'_k \in A'\}$.
2. B is the set $f(B') = \{b_j = f(b'_j) : b'_j \in B'\}$.

Theorem C.13 *The mapping $f : \Delta \to \mathbb{C} \cup \{\infty\}$ can be extended to E, and it defines a universal branched covering with signature (A, B, R) over the Riemann sphere. The mapping f realizes the quotient of the space E by the action of the group G_Δ.*

Proof the result follows from the Riemann–Schwarz reflection principle. □

C.2 The Spherical Case

C.2.1 Application of the Riemann–Hurwitz Formula

Suppose that a discrete group of automorphisms G acts on the sphere Z. Then the group G is finite, and the quotient space Z/G is a sphere (since there are no nonconstant analytic mappings of the sphere to a higher-genus Riemann surface). The quotient mapping $Z \to Z/G$ defines (up to composition with an automorphism of the sphere S) a universal covering $\pi : Z \to S$ with elliptic signature (A, R).

Claim The signature (A, R) has an elliptic set R.

Proof Let $\#G = N$. The Riemann–Hurwitz formula implies

$$2 = 2N - \sum_{a_i \in A} \left(N - \frac{N}{r_i} \right) = N(2 - \chi(R)).$$

Hence $\chi(R) < 2$. □

We now give names to the following sets:

1. (k, k) is the set of a k-gon.
2. $(2, 2, k)$ is the set of the dihedron D_k.
3. $(2, 3, 3)$ is the set of a tetrahedron.
4. $(2, 3, 4)$ is the set of a cube or octahedron.
5. $(2, 3, 5)$ is the set of a dodecahedron or icosahedron.

These sets are elliptic.

Claim If the signature (A, R) is elliptic, then the set R is among the five sets mentioned above.

Proof It is enough to find all solutions of the inequality $\chi(R) < 2$ satisfying the restrictions imposed on R for $n \leq 2$. □

C.2.2 Finite Groups of Rotations of the Sphere

Consider the following polyhedra in $\mathbb{R}^3$ having their centers of mass at the origin:

1. A pyramid with a regular k-gon as its base.
2. A dihedron with k vertices, or equivalently, a polyhedron consisting of two pyramids as in the previous item joined along their base face.
3. A regular tetrahedron.
4. A cube or octahedron.
5. A dodecahedron or icosahedron.

The symmetry planes of each of these polyhedra cut a net of great circles on the unit sphere. This net divides the sphere into a union of equal spherical polygons Δ (triangles in cases (2)–(5) and digons in case (1)). The stereographic projections of these nets is shown in Fig. 6.1. One sees that the signatures (A', R) of polygons Δ in cases (1)–(5) have a set R that is equal to the set having the same name (and the same parameter k in cases (1) and (2)).

Each polyhedron Δ defines a group $\tilde{G}_\Delta$ of isometries of the unit sphere generated by reflections in its sides and its index-2 subgroup G_Δ of orientation-preserving isometries from $\tilde{G}_\Delta$.

Definition C.14 The groups of rotations of the sphere described above are called (1) the group of the k-gon, (2) the group of the dihedron D_k, (3) the group of the tetrahedron, (4) the group of cube/octahedron, (5) the group of the icosahedron/dodecahedron.

Claim A spherical polyhedron with signature (A', R) exists if and only if R is one of the elliptic sets described above.

Proof For one direction, it is enough to find all the solutions of the inequality $\chi(R) < 2$ (keeping in mind the restrictions imposed on R when $n \leq 2$). For the other direction, it is enough to exhibit examples of the spherical polygons. All the examples are given by triangles and dihedra that appear when the sphere is divided into equal polygons by the symmetry planes of the polyhedra described above (see Fig. 6.1). □

Theorem C.15 *A finite group of automorphisms of the Riemann sphere with a given signature coincides up to an automorphism of the Riemann sphere with a group of rotations of the sphere with the same name as its signature.*

C.2.3 Coverings with Elliptic Signatures

Every automorphism of the sphere has fixed points, and thus the automorphism group of the sphere has no free normal subgroups. Fix an elliptic signature. The universal covering with this signature is the Riemann sphere Z equipped with the deck transformation group G, the quotient map $Z \to Z/G$, as well as an isomorphism $Z/G \simeq S$.

The coverings with a given elliptic signature are in one-to-one correspondence with conjugacy classes of subgroups of G that do not have nontrivial normal subgroups of the group G. Each such covering has a normalization that is equivalent to the universal covering with the same signature and monodromy group isomorphic to the group G. Thus the monodromy group of a covering with an elliptic signature is determined by its signature.

C.2.4 Equations with an Elliptic Signature

Theorem C.16 *An algebraic function with an elliptic signature and a set of orders not equal to the set $\{2, 3, 5\}$ can be represented in radicals. If the set of orders is equal to $\{2, 3, 5\}$, then it can be represented by radicals and solutions of equations of degree at most 5.*

Example C.17 The inverse of the Chebyshev polynomial of degree n has signature $A = \{1, -1, \infty\}$, $R = (2, 2, n)$ of elliptic type (the case of the dihedron D_n). This explains why the Chebyshev polynomials are invertible in radicals.

Theorem C.18 *A linear differential equation of Fuchsian type with elliptic signature and the set of orders different from the set $\{2, 3, 5\}$ can be solved in radicals. If the set of orders is $\{2, 3, 5\}$, then it can be solved in radicals and the solution of algebraic equations of degree at most 5.*

C.3 The Case of the Plane

C.3.1 Discrete Groups of Affine Transformations

Every automorphism of the complex line is an affine transformation $z \mapsto az + b$ with $a \neq 0$. The group of affine transformations has an abelian normal subgroup $\mathbb{C}$ consisting of translations with an abelian factor group $\mathbb{C}^*$. The group of automorphisms of the line is thus solvable, and hence all its discrete subgroups are solvable as well. The affine transformations with no fixed points are precisely the translations.

The discrete groups G of the group of affine transformations of the complex line can be classified up to an affine change of coordinates as having one of the eight types listed below. The space $\mathbb{C}/G$ for each group G, except the groups in case (4), is a sphere or a sphere minus one or two points. The quotient $\mathbb{C} \to \mathbb{C}/G$ defines in these cases a covering with parabolic signature. For all the groups except the group in case (5), the set $A \cup B$ for these signatures consists of at most three points. Hence in this case, the signature is defined up to an automorphism of the sphere by the set of its orders R.

We use the following notation: $S_k \subset \mathbb{C}^*$ is the multiplicative subgroup of order k, $\Lambda_2 = (1, c)$ is the additive group $\Lambda_2 \subset \mathbb{C}$ generated by the numbers 1 and c, where $c \notin \mathbb{R}$ is defined up to the action of the modular group; the number $\lambda \notin \{0, 1, \infty\}$ denotes a number under the equivalence whereby λ, $1-\lambda$, λ^{-1}, $(1-\lambda)^{-1}$, $\lambda(\lambda-1)^{-1}$, $\lambda^{-1}(\lambda-1)$ are equivalent, where τ_6 is a primitive root of unity of order 6.

The groups G consist of transformations $x \mapsto ax + b$, where a, b, and the signature are of one of the following eight types:

1. $a \in S_k, b = 0;\ \ R = (k, \infty)$.
2. $a = 1, b \in \mathbb{Z};\ \ R = (\infty, \infty)$.
3. $a \in S_2, b \in \mathbb{Z};\ \ R = (2, 2, \infty)$.
4. $a = 1, b \in \Lambda_2 = (1, c);\ \ \mathbb{C}/G$ is a curve of genus one.
5. $a \in S_2, b \in \Lambda_2 = (1, c);$ signature $A = \{0, 1, \infty, \lambda, \}$, $R = (2, 2, 2, 2)$.
6. $a \in S_4, b \in \Lambda_2 = (1, i);\ \ R = (4, 4, 2)$.
7. $a \in S_3, b \in \Lambda_2 = (1, \tau_6);\ \ R = (3, 3, 3)$.
8. $a \in S_6, b \in \Lambda_2 = (1, \tau_6);\ \ R = (6, 3, 2)$.

Theorem C.19 *A discrete group G of affine transformations is, up to an affine change of coordinates, one of the groups from the list above. The signature of the coverings related to the action of the group is defined up to an automorphism of the sphere by the data from the list.*

Below, we sketch a proof of this result. If G does not contain translations and only one point is fixed under transformations $g \in G \setminus \{e\}$, then G is of type (1). If G consists of translations only, then G is of type (2) or (4). If transformations $g_1, g_2 \in G$ have different fixed points, then $g_1 g_2 g_1^{-1} g_2^{-1} \neq e$, and hence G contains a discrete subgroup of translations $\Lambda_G \neq G$ and hence is of type (2) or (4). If $g(z) = az + b$ and $g \in G$, then the multiplication $z \mapsto az$ defines an automorphism of the lattice Λ_G. The group of automorphisms of a lattice is a group S_k having at most two elements linearly independent over $\mathbb{Q}$. Hence the order k of group S_k must be in the set $\{1, 2, 3, 4, 6\}$. This leads to the remaining cases.

A group of type (4) does not belong to our subject, since $\mathbb{C}/\Lambda_2$ is a torus rather than a sphere. A group of type (1) is of interest for our purposes: it uniformizes functions with sets of orders (k, ∞), among which only the functions with sets of orders (k, k) are of interest to us. These functions have already been considered above. All other groups are of interest to us.

These groups (with the exception of the majority of groups of type (5)) can be described geometrically by means of planar polygons.

C.3.2 Affine Groups Generated by Reflections

We name sets of orders as follows:

1. (∞, ∞): the set of a strip.
2. $(2, 2, \infty)$: the set of a half-strip.
3. $(2, 2, 4)$: the set of a half of a square.
4. $(3, 3, 3)$: the set of a regular triangle.
5. $(2, 3, 6)$: the set of a half of a regular triangle.
6. $(2, 2, 2, 2)$: the set of a rectangle.

All these sets are parabolic.

Claim A planar polygon with signature (A', B', R) exists if and only if R is one of the sets mentioned above. The polygon is defined uniquely by R up to affine transformations in all cases but the last one. A rectangle is defined up to such a transformation by the quotient of its side lengths.

Proof For the proof, it is enough to find all the solutions of the equation $\chi = 2$, exhibit examples of the required polygons, and classify those polygons up to affine transformations. Here we consider only examples: in case (1), it is a strip between two parallel lines. In case (2), it is the triangle obtained by cutting the strip from the first example by a line perpendicular to its sides. In other cases, these are the triangles and quadrilaterals appearing in the names of the cases. □

By comparing the lists in Sects. C.3.1 and C.3.2, we see that the groups of types (2), (3), (6), (7), (8) are subgroups of index 2 in the groups generated by reflections in a two- or three-gon with the same set R. For a group of type (5), this is so if $\lambda \in \mathbb{R}$: in this case, the covering is given by the inverse of the elliptic Schwartz–Christoffel integral

$$\int \frac{dz}{\sqrt{p(z)}}$$

with $p(z) = z(z-1)(z-\lambda)$. This integral transforms the upper half-plane into a rectangle.

C.3.3 Coverings with Parabolic Signatures

Let a parabolic signature (A, B, R) be fixed. The universal covering with this signature consists of the line $\mathbb{C}$ equipped with a discrete group of its transformations G, the factorization mapping $\mathbb{C} \to \mathbb{C}/G$, and the isomorphism $\mathbb{C}/G \tilde{\to} S$. If $\# A \cup B \leq 3$, then the position of the points $A \cup B$ has no significance, since any configuration of at most three points on the sphere can be transformed to any other configuration by an automorphism of the sphere. In this case, we know the group G and its geometric description.

Consider the case of signature $A = \{a_1, a_2, a_3, a_4\}$, $R = \{2, 2, 2, 2\}$. If the points of the set A lie on a circle, they can be transformed into points $0, 1, \infty, \lambda$ with real λ. For such points, we have described the universal covering above as the inverse of the elliptic Schwartz–Christoffel integral of the form

$$\int \frac{dz}{p(z)}, \quad p(z) = z(z-1)(z-\lambda).$$

If the points of the set A do not lie on a circle, the universal covering can be described as follows. We can assume that $\infty \notin A$, in which case the universal covering $I^{-1} : \mathbb{C} \to S$ is given by the inverse of the integral

$$I = \int \frac{dz}{\sqrt{p(z)}}$$

with $p(z) = (z - a_1)(z - a_2)(z - a_3)(z - a_4)$. The group of deck transformations of this covering is generated by shifts by the elements of the lattice of periods Λ_2 of the integral I and by multiplication by (-1). The quotient of $\mathbb{C}$ by the group of translations from Λ_2 is a torus, which is a two-sheeted branched covering over the sphere with branch points A.

We now consider the general case of coverings with parabolic signature. The commutator of the group of all automorphisms of the complex line consists of all the translations. The translations are the only transformations that have no fixed points.

To a given parabolic signature, one associates the universal covering with this signature and a group G of automorphisms of the line acting as the group of deck transformations of the covering. Coverings with this signature are in one-to-one correspondence with conjugacy classes of subgroups of G whose intersection H consists only of translations. The monodromy group of this covering $\pi_1 : Y \to S$ is isomorphic to the group G/H and is determined by the signature up to a quotient by a subgroup H in the commutator of the group G.

C.3.4 Equations with Parabolic Signatures

Theorem C.20 *A linear differential equation of Fuchsian type with parabolic signature can be solved by quadratures. Its monodromy group is a factor group of a group G by an abelian normal subgroup, where the group G depends only on the signature.*

Theorem C.21 *An algebraic function with parabolic signature is expressible in radicals. Its monodromy group is a factor group of a group G by an abelian normal subgroup, where the group G depends only on the signature.*

Example C.22 Coverings with the set of orders (∞, ∞) are uniformized by a group of type (1). Equations $y^{(n)} + a_1 y^{(n-1)}x^{-1} + \cdots + a_n y x^{-n} = 0$ of Euler type are of this kind.

Example C.23 Coverings with the set of orders $(2, 2, \infty)$ are uniformized by a group of type (2). Equations of the form

$$\sum_{i=0}^{n} a_i \left((1 - x^2)\frac{d^2}{dx^2} - x\frac{d}{dx}\right)^i y = 0$$

have this signature. By means of a change of variables $x = \cos z$, such an equation can be reduced to an equation with constant coefficients:

$$\sum_{i=0}^{n} a_i \frac{d^{2i} y}{d\, z^{2i}} = 0.$$

Hence the solutions of this equation are of the form

$$y(x) = \sum_j p_j(\arccos x)\cos(\alpha_j \arccos x) + q_j(\arccos x)\sin(\alpha_j \arccos x),$$

where p_j, q_j are polynomials. In particular, all the (multivalued) Chebyshev functions f_α defined by the property

$$f_\alpha\left(\frac{x + x^{-1}}{2}\right) = \frac{x^\alpha + x^{-\alpha}}{2}$$

are solutions of such equations. For integer α, these are Chebyshev polynomials, and for $\alpha = 1/n$ with integer n, these are the inverses of Chebyshev polynomials.

Example C.24 If $p_4(z)$ is a fourth-degree polynomial with roots $z_1, \ldots, z_4$, then the elliptic integral

$$y(z) = \int_{z_0}^{z} \frac{d\, z}{\sqrt{p_4(z)}}$$

has signature $(z_1, \ldots, z_4; 2, 2, 2, 2)$, and it is a solution of the differential equation

$$y'' + \frac{1}{2}\frac{p_4'(z)}{p_4(z)} y' = 0$$

of Fuchsian type with the same signature.

C.4 Functions with Nonhyperbolic Signatures in Other Contexts

Algebraic functions with elliptic signatures are classical objects. For instance, the first part of Klein's book [63] is devoted to them. Algebraic functions with nonhyperbolic signatures play a central role in the works of Ritt on rational mappings of prime degree invertible in radicals (see [19, 55, 84]). The reason for their appearance in these works is as follows.

By a result of Galois, an irreducible equation of prime degree p can be solved in radicals if and only if its Galois group is a subgroup of the metacyclic permutation group containing only permutations g representable in the form

$$g(x) \equiv ax + b \,(\mathrm{mod}\ p), \quad a \not\equiv 0\,(\mathrm{mod}\ p). \tag{C.1}$$

A nonidentity permutation (C.1) splits as a product of $1 + (p-1)/m(g)$ disjoint cycles, where $m(g)$ is the order of the element a in the group $\mathbb{Z}_p$ if a is not the identity element, and $m(g) = \infty$ otherwise.

Let f be the inverse function for a degree-p rational map, and let $A = \{a_1, \ldots, a_n\}$ be the branching locus of f. Assume that f is representable by radicals. Fix loops $\gamma_i \in \pi_1(S \setminus A)$ running around points a_i. Denote by g_i a permutation (C.1) corresponding to γ_i. The branching number m_i at a_i is equal to $m(g_i)$ if $m(g_i) \neq \infty$, and is equal to p otherwise.

The Riemann surface of f is the Riemann sphere Y. According to the Riemann–Hurwitz formula for Y, the formula

$$2 = 2p - \sum_{1 \le i \le n} \left(p - 1 - \frac{p-1}{m(g_i)} \right) \tag{C.2}$$

holds. The relation (C.2) means that

$$\sum \left(1 - \frac{1}{m(g_i)} \right) = 2.$$

So $\sum (1 - 1/m_i) \le 2$, and the signature of f is nonhyberbolic.

In dynamics, Lattès maps are studied as examples of rational mappings with exceptional (usually exceptionally simple) dynamics—these are rational mappings induced by an endomorphism of an elliptic curve (see [70, 79]). These mappings have parabolic signature (but they do not exhaust all the examples of rational mappings with parabolic signatures: to describe all such examples, one has to include all the mappings of a sphere to itself induced by a homomorphism between two different elliptic curves). Lattès maps provided the first examples of rational mappings with Julia set equal to the whole Riemann sphere.

C.5 The Hyperbolic Case

Let R be a signature of an algebraic function. If the universal covering with this signature is the Riemann sphere or the complex line, then the monodromy group of every algebraic function with signature R can be described explicitly: it contains a normal subgroup that is an abelian group with at most two generators, and the quotient by this group is a finite group from a finite list of groups associated with the given signature. In contrast, if the universal covering with signature R is the hyperbolic plane, then the monodromy group of an algebraic function with such signature can be arbitrarily complicated, as the next theorem shows.

Theorem C.25 *Let R be a signature of an algebraic function, and let the universal covering with signature R have the hyperbolic plane as its total space. Let G be an arbitrary finite group. Then there exist a covering with signature R and monodromy group H containing a subgroup H_1 that has a normal subgroup such that the quotient of H_1 by that subgroup is isomorphic to G (i.e., the monodromy group H has a subquotient isomorphic to G).*

Proof If $\pi : Y \to S$ is the normalization of the covering associated to an algebraic function with signature R, then the universal covering $Z \to S$ with signature R can be obtained as the composition of π and the universal (unbranched) covering $Z \to Y$. In particular, if Z is the hyperbolic plane, then Y is topologically a sphere with at least two handles.

Fix a representation of the group G as a factor group of a free group on k generators. Replace the covering $\pi : Y \to S$ by a covering $\pi_1 : Y_1 \to S$, where Y_1 is an unbranched covering of Y with the topological type of a sphere with at least k handles. The fundamental group of the surface Y_1 admits a homomorphism onto the free group with k generators, and hence onto G. Let $\pi_1 : Y_2 \to Y$ be the unbranched covering associated with the kernel of this homomorphism. Then the composition $\pi \circ \pi_1 : Y_2 \to S$ is a covering with signature R whose monodromy group contains a subgroup admitting a mapping onto G (more precisely, it is the subgroup of permutations of the fiber that correspond to loops in the base space that can be lifted to loops in Y_1). □

Corollary C.26 *Let R be a signature of an algebraic function. If the signature is elliptic or parabolic and different from $(2, 3, 5)$, then every algebraic function having this signature can be represented by radicals. If the signature is $(2, 3, 5)$, then every such function can be expressed by radicals and solutions of algebraic equations of degree at most 5. Finally, if the signature is hyperbolic, then for a given integer k, there exists an algebraic function having this signature that cannot be represented by k-radicals, i.e., radicals and solutions of algebraic equations of degree at most k, or k-quadratures, i.e., quadratures and solutions of algebraic equations of degree at most k.*

Corollary C.27 *Let R be a signature of an algebraic function. If the signature is elliptic and different from $(2, 3, 5)$, then every linear differential equation of*

Fuchsian type with this signature can be solved in radicals. If the signature is $(2,3,5)$*, then every such equation can be solved in radicals and solutions of algebraic equations of degree at most 5. If the signature is parabolic, then every such equation can be solved by quadratures. Finally, if the signature is hyperbolic, then for a given integer* k*, there exists a linear differential equation of Fuchsian type with this signature that is not solvable in* k*-quadratures, i.e., quadratures and solutions of algebraic equations of degree at most* k*.*

Proof The result follows from Theorems C.16, C.18, C.20, C.21, C.25 as well as Theorems 2.57 and 6.4 (see also Theorem 6.16). □

Appendix D
On an Algebraic Version of Hilbert's 13th Problem

Yuri Burda and Askold Khovanskii

D.1 Versions of Hilbert's 13th Problem

D.1.1 Simplification of Equations of High Degree

It is known that the general equation of degree 5 and higher cannot be solved in radicals. It is natural then to ask to what extent one can simplify an equation of high degree by solving auxiliary simpler equations, making algebraic changes of variables, etc. For instance, Hermite proved the following theorem:

Theorem D.1 *The general equation of fifth degree*

$$x^5 + a_1x^4 + \cdots + a_5 = 0 \tag{D.1}$$

can be reduced to the form

$$y^5 + ay^3 + by + b = 0,$$

where a, b are rational functions of the parameters $a_1, \ldots, a_5$, and x is a rational function of y and the parameters $a_1, \ldots, a_5$.

Bring proved the following result.

Theorem D.2 *Equation* (D.1) *can be reduced to the equation*

$$z^5 + z + c = 0, \tag{D.2}$$

where c can be expressed by means of the parameters $a_1, \ldots, a_5$, arithmetic operations, and extracting radicals, and x can be obtained from y and the coefficients $a_1, \ldots, a_5$ by arithmetic operations and extracting radicals.

A. Khovanskii, *Topological Galois Theory*, Springer Monographs in Mathematics,
DOI 10.1007/978-3-642-38871-2

Thus even though the general equation of fifth degree cannot be solved in radicals, it can be solved in radicals and special algebraic functions of one variable (more precisely, the function (D.2)). The general equation $x^n+a_1x^{n-1}+\cdots+a_n=0$ of degree n can be reduced by means of a change of variables $y = x + a_1/n$ to an equation whose coefficient at y^{n-1} vanishes. By means of a change of variables that uses arithmetic operations and extracting radicals, one can make the coefficients at y^{n-1}, y^{n-2}, y^{n-3} vanish, and make the constant equal to 1. In particular, the solution of the general equation of degree 7 can be expressed in terms of arithmetic operations, radicals, and an algebraic function of three variables defined by the equation

$$z^7 + az^3 + bz^2 + cz + 1 = 0.$$

Examples of this kind gave rise to the following classical problems.

Problem D.3 (Klein's resolvent problem) Can the general algebraic function of degree n be represented as a composition of rational functions and one algebraic function of a small number of variables?

Problem D.4 (Algebraic version of Hilbert's problem) Is the solution of the equation

$$z^7 + az^3 + bz^2 + cz + 1 = 0 \tag{D.3}$$

a branch of composition of algebraic functions of two variables?

There exist many other versions of these problems. For instance, in Klein's problem, one can allow the use of the square root of the discriminant in the composition. One can also allow the use of arbitrary radicals. In Hilbert's problem, one can ask whether a germ of the solution of (D.3) is a composition of continuous functions of two variables (as Hilbert did in his famous list of 23 problems), smooth functions from a given differentiability class, or analytic functions. All these questions can also be asked about any algebraic, analytic, smooth, or continuous functions of several variables. These classical problems have been the subject of many wonderful papers, some of which are mentioned below.

D.1.2 *Versions of the Problem for Different Classes of Functions*

Hilbert was certain that the algebraic function (D.3) could not be represented as a composition of functions of two variables from any reasonable class of functions. For this reason, he formulated his problem for the class of continuous functions.

The next theorem is apparently due to Hilbert (see [101]):

Theorem D.5 *There exists a germ of an analytic function of n variables that is not representable as a composition of analytic functions of fewer than n variables.*

One can also show that the functions representable as compositions of analytic functions of fewer than n variables form a small set in the space of analytic functions of n variables.

Consider the space $C^p(I^n)$ of functions on the unit cube of dimension n of differentiability class C^p. The complexity of this space is defined to be the number n/p.

Theorem D.6 (Vitushkin) *There exists a germ of a function of n variables of smoothness p that is not representable as a composition of functions from the spaces $C^r(I^m)$ of complexity less than n/p.*

One can also show that the functions representable as compositions of functions from spaces of complexity smaller than n/p form a meager set in the space $C^p(I^n)$.

Vitushkin's theorem has further convinced everyone that the answer to Hilbert's original question is negative. However, Kolmogorov and Arnold proved the following result:

Theorem D.7 *Every continuous function of n variables can be represented as a composition of continuous functions of one variable and the operation of addition.*

These amazing theorems have resonated with the mathematical community. There are several very good expositions of these theorems, so we shall not dwell on them further (see [100] for a recent survey). Another reason for us not to discuss these theorems is that they shed no light on the algebraic versions of Klein's and Hilbert's problems.

D.2 Arnold's Theorem

D.2.1 Formulation of the Theorem

Arnold noticed that the classical formula for the solution of the equation of degree four in radicals defines a 72-valued algebraic function, while the actual solution is only one branch of that function. So in fact, the original algebraic function and the function defined by the formula for the solution of the equation of fourth degree are different functions. Arnold then asked the following natural question: can a given algebraic function be represented as an exact composition of algebraic functions of a small number of variables, i.e., represented in such a way that the original algebraic function and the function defined by the composition are in fact the same function?

In this context, Arnold proved that the question of representability of an algebraic function as an exact composition of functions of a smaller number of variables is in fact equivalent to Klein's problem:

Theorem D.8 *The covering* $\pi : R \to U$ *given by the Riemann surface* R *of the general algebraic function* $z^n + a_1 z^{n-1} + \cdots + a_n = 0$ *over a Zariski open subset* U *of the complement to its discriminant is not a composition of coverings* $R \to R_1 \to \cdots \to R_k \to U$ *of degree at least* 2 *for* $k > 0$.

Proof The monodromy group of a composition of coverings acts imprimitively on a fiber of the map π if more than one of the coverings in the composition is nontrivial. The monodromy group of the restriction of the general algebraic function of degree n to a Zariski open subset of the complement of the discriminant acts primitively on the fiber. □

According to Galois theory, if an entire algebraic function is representable in radicals, then there exists in addition a representation that uses only the operations of addition, multiplication, extracting radicals, and multiplication by complex numbers, but does not use the operation of division. This observation naturally leads to consideration of the following version of Hilbert's problem: can a given entire algebraic function of n variables be represented as a composition of polynomial functions and entire algebraic functions of k variables for a given $k < n$?

Arnold proved the following theorem [3]:

Theorem D.9 *The general algebraic function of degree n cannot be represented as an exact composition of polynomials and algebraic functions of* $\phi(n)$ *variables, where* $\phi(n) = n - d(n)$, *with* $d(n)$ *the number of occurrences of* 1 *in the binary expansion of the number n. For n of the form* $n = 2^k$, *we have* $\phi(n) = n - 1$.

According to Theorem D.8, Theorem D.9 actually deals with the polynomial version of Klein's problem, i.e., with the question of representability of an algebraic function as a composition of polynomials and one algebraic function of k variables. Even though stronger results have been obtained in the polynomial version of Klein's problem (see Sect. D.2.4 below), the proof of Arnold's theorem remains a very beautiful piece of mathematics that has inspired many other investigations.

D.2.2 Results Related to Arnold's Theorem

The proof of Arnold's theorem is based on consideration of the topology of the complement of the discriminant set of the general algebraic function of degree n. In this context, Arnold made an observation of very general character. In many problems, we are interested in elements in general position in a space of functions, for instance polynomials without multiple roots. The degenerate elements, i.e., those not in general position, form a subset called the *discriminant* in the given space of functions. We are usually interested in the topology of the space of nondegenerate elements, i.e., in the topology of the complement of the discriminant. This topology is related by means of duality to the topology of the discriminant itself. The discriminant usually comes with a natural geometric structure that is useful for its study. More precisely, the discriminant comes with a stratification according to degeneracy types of the objects that it parameterizes. This stratification gives nontrivial information about the topology of the discriminant. Arnold proposed the investigation of the topology of complements of discriminants in this general setting. Research into this question has turned out to be extremely fruitful: for instance, it led Vassiliev to the discovery of knot invariants of finite type [97, 98].

The complement of the discriminant of the general algebraic function of degree n can be considered the space of unordered sets of n distinct complex numbers (instead of a point in the complement of the discriminant in the space of parameters of the algebraic function, one can consider the set of its n values at that point). This space admits a natural covering by the space of ordered n-tuples of distinct complex numbers (the covering map being the map that forgets the ordering). This space can be described as the complement of the hyperplanes $x_i = x_j$. Thus there is an $n!$-sheeted covering

$$\pi : \mathbb{C}^n \setminus \cup L_{ij} \to \mathbb{C}^n \setminus D,$$

where $L_{ij} = \{(x_1, \ldots, x_k) \mid x_i = x_j\}$ and D is the discriminant. Arnold obtained an explicit description of the cohomology ring of this space:

Theorem D.10 (Arnold) *The cohomology ring with integer coefficients of the space $\mathbb{C}^n \setminus \cup L_{ij}$ is generated by the cohomology classes of forms*

$$w_{ij} = \frac{1}{2\pi i} d \log(x_i - x_j)$$

subject to the relations

$$w_{ij} = w_{ji} \quad \textit{and} \quad w_{ij} \wedge w_{jk} + w_{jk} \wedge w_{ki} + w_{ki} \wedge w_{ij} = 0.$$

This theorem served as the beginning of the modern theory of hyperplane arrangements; see, for example, [81].

The complement of the discriminant of the general algebraic function of degree n is a $K(\pi, 1)$ space for Artin's braid group on n generators. Arnold initiated the study of the cohomology of this group, which led to its complete description in [95].

It is evident from these remarks that Arnold's work has stimulated many important developments in mathematics.

D.2.3 The Proof of the Theorem

An algebraic function defines a covering over the complement of its discriminant. Arnold considered a characteristic class for coverings that assigns to a covering a $\mathbb{Z}_2$ cohomology class of its base space. If this class does not vanish in dimension k, then the algebraic function to which it is assigned cannot be induced from an algebraic function on a space of dimension less than k.

Theorem D.11 *For the general algebraic function of degree n, the largest dimension in which a characteristic class with $\mathbb{Z}_2$ coefficients does not vanish is equal to $\phi(n) = n - d(n)$, where $d(n)$ is the number of 1's in the binary expansion of the number n.*

Below, we present another proof of Arnold's bound $k \geq \phi(n)$. It serves as a demonstration of his thesis that information about the topology of the complement of the discriminant can be obtained by studying the stratification of the discriminant itself (this proof was found shortly after Arnold's theorem appeared by Khovanskii, who was Arnold's student at that time).

Define the notion of a chain of branching sets for an algebraic function. The first element of a chain is an irreducible component of the set of points for which the local monodromy group is nontrivial. Each succeeding set in the chain is obtained by application of the same construction to the restriction of the algebraic function to the previous set in the chain. The restriction of the algebraic function to one of the sets in the chain has roots of different multiplicities. Observe that roots with different multiplicities cannot be interchanged by going around loops inside this set that avoid sets with more complicated degeneracies.

Thus we get a chain of algebraic subsets of decreasing dimension. This chain is characteristic in the following sense: let an algebraic function z be induced from the function w by means of a continuous mapping. Then the chains of branching sets of the function z get mapped to the chains of branching sets of the function w. Thus if the function z has a chain of branching sets of length k, then it is impossible to induce it from an algebraic function defined on a space of dimension less than k by means of a continuous mapping.

For the general algebraic function, one of the chains of branching sets can be described as follows: the ith set S_i from the first $[n/2]$ sets of the chain consists of points where the equation

$$z^n + a_1 z^{n-1} + \cdots + z_n = 0 \tag{D.4}$$

has i double roots (in a neighborhood of such points, the local monodromy group is isomorphic to the group $\mathbb{Z}/2\mathbb{Z}^i$). The next $[n/4]$ sets in the chain are the sets $S_{[n/2]+j}$, where the double roots of the restriction of equation (D.4) to the set $S_{[n/2]}$ coincide in pairs, forming j roots of multiplicity 4. The next $[n/8]$ sets are sets in which pairs of roots of multiplicity 4 coincide, and so on. In this way, we get a chain of length $[n/2] + [n/4] + \cdots$. This length is equal to the dimension $\phi(n)$ of the nonvanishing class in cohomology with coefficients in $\mathbb{Z}/2\mathbb{Z}$ of the complement of the discriminant of the general algebraic function of degree n that appeared in Arnold's proof. For the general algebraic function of degree n, the characteristic class of dimension $\phi(n)$ used by Arnold should be considered to be in some sense dual to this chain of branching sets: in a neighborhood of this chain of branching sets, there exists a cycle on which the characteristic class from Arnold's proof does not vanish.

D.2.4 Polynomial Versions of Klein's and Hilbert's Problems

The following theorem is a known result about the polynomial version of Hilbert's problem.

Theorem D.12 *If an entire algebraic function can be represented as a composition of polynomials and entire algebraic functions of one variable, then its local monodromy group at each point is solvable.*

The proof is based on the fact that the local monodromy group of an algebraic function of one variable is cyclic and on the fact that the permissible operations exclude the operation of division, which destroys locality. A completely analogous argument can be found in Sect. 5.2.3. This theorem was proved shortly after Arnold's theorem appeared by Khovanskii, who was Arnold's student at that time [46].

Unfortunately, in the polynomial version of Hilbert's problem, no one has succeeded in proving that some particular algebraic function cannot be represented as a composition of algebraic functions of two variables.

Progress has been much better in the polynomial version of Klein's problem. The next theorem, along with a series of similar results, was inspired by Arnold's theorem and proved by Lin [72–74]. It provides a complete answer to the polynomial version of Klein's problem for the general algebraic function of n variables.

Theorem D.13 (Lin) *The general algebraic function of degree $n > 2$ cannot be represented as a composition of polynomials and an algebraic function of fewer than $n - 1$ variables.*

In contrast to the results of Arnold, which use only the topology of the complement of the discriminant, Lin's results are based on the analytic structure of this set. For instance, one of Lin's proofs is based on the following simple and beautiful fact.

Theorem D.14 *A holomorphic function on $\{(x_1, \dots, x_k) \in \mathbb{C}^n \mid x_i \neq x_k$ for $i \neq k\}$ that is never equal to 0 or 1 is a simple or double ratio of the coordinates.*

D.3 Klein's Problem

D.3.1 Birational Automorphisms and Klein's Problem

The solution of the equation of degree 5 was the subject of Klein's wonderful book [63], which contains, in particular, the first negative result on Klein's problem, due to Kronecker:

Theorem D.15 (Kronecker) *The equation $z^5 + a_1 z^4 + \cdots + a_5 = 0$ cannot be reduced to an equation depending on one parameter by a change of variables that* *uses the coefficients of the equation, arithmetic operations, and the square root of the discriminant.*

To show that this equation cannot be reduced to an equation depending on one parameter, Kronecker used the classification of one-dimensional rational algebraic varieties admitting a generically free action of the group A_5.

We indicate the spirit of Kronecker's arguments by proving that the algebraic function defined by the equation

$$z^5 + a_1 z^4 + \cdots + a_5 = 0 \tag{D.5}$$

cannot be represented as a composition of a rational function of one parameter and an algebraic function y depending on one parameter: $y = y(c)$.

Indeed, if such a representation were possible, there would exist a formula of the form $z = R(y(c(a)), a)$, where $a = (a_1, \ldots, a_5)$ and $c(a)$, $R(y, a)$ are rational functions. Since the coefficients of an equation are symmetric functions of its roots, we would obtain in this way a rational mapping $r : \mathbb{C}^5 \to C \subset \mathbb{C}^5$ that sends 5-tuples of the roots $z_1, \ldots, z_5$ of the equation (D.5) to the 5-tuples of values $y_1, \ldots, y_5$ of the function y. The image of this mapping is a one-dimensional curve C, and this mapping is S_5-equivariant: if $r(z_1, \ldots, z_5) = (y_1, \ldots, y_5)$, then for every permutation $\sigma \in S_5$, one has $r(z_{\sigma(1)}, \ldots, z_{\sigma(5)}) = (y_{\sigma(1)}, \ldots, y_{\sigma(5)})$.

The domain of the rational mapping r contains a line that is not mapped to a point by means of r. This means that the restriction of r to this line is a rational mapping from the Riemann sphere to the curve C whose image is an open subset of C. This means that C is a curve of genus zero. However, every algebraic action of the group S_5 on a genus-zero curve has a nontrivial kernel, which contradicts the fact that z is a rational function of y and symmetric functions of $z_1, \ldots, z_5$.

Using the Manin–Iskovskih classification of minimal rational surfaces with a group action [41, 78], Serre solved an analogous problem for the equation of degree 6:

Theorem D.16 (Serre) *The general algebraic function of degree 6 cannot be represented as a composition of algebraic functions of two variables and rational functions of the coefficients of the equation and the square root of the discriminant.*

Very recently, Duncan proved the following theorem.

Theorem D.17 (Duncan) *The general algebraic function of degree 7 cannot be represented as a composition of algebraic functions of three variables and rational functions of the coefficients of the equation and the square root of the discriminant.*

The proofs of Duncan and Serre are similar in spirit to Kronecker's proof; they are, however, much more complicated, because the theory of surfaces and algebraic threefolds is much more complicated than the theory of algebraic curves.

Their results can be found in [28, 90] and are formulated in terms of the notion of essential dimension, which we now discuss.

D.3.2 Essential Dimension of Groups

Buhler and Reichstein proved in [17] the following result about the general algebraic function of degree n.

Theorem D.18 (Buhler and Reichstein) *The general algebraic function of degree n cannot be represented as a composition of rational functions and an algebraic function of fewer than* $\lfloor n/2 \rfloor$ *variables.*

The general algebraic function of degree n cannot be represented as a composition of rational functions of its parameters and the square root of the discriminant and an algebraic function of fewer than $2\lfloor n/4 \rfloor$ *variables.*

For the proof of this theorem, they introduced the notion of essential dimension of a finite group [17]:

Definition D.19 The essential dimension of an algebraic action of a finite group G on an algebraic variety X is the smallest dimension of an algebraic variety Y with a generically free action of the group G for which there exists a G-equivariant dominant rational morphism $X \dashrightarrow Y$.

The essential dimension of a finite group is the essential dimension of any one of its faithful linear representations.

The definition of essential dimension relies on the following theorem:

Theorem D.20 *Let G be a finite group,* $G \to \mathrm{GL}(V)$ *and* $G \to \mathrm{GL}(W)$ *faithful representations of G. Then the essential dimensions of the G-varieties V and W are the same.*

In these terms, Klein's problem is equivalent to the computation of the essential dimension of the group S_n, while its version in which the square root of the discriminant is adjoined to the domain of rationality is equivalent to the computation of the essential dimension of the group A_n. To see this, it is enough to reformulate Klein's problem, which was previously formulated in terms of mappings between spaces of parameters of algebraic equations, in terms of equivariant mappings between the spaces of their roots, as we did above in Sect. D.3.1 in demonstrating Kronecker's arguments.

To prove Theorem D.18, Buhler and Reichstein used the following corollary of Theorem D.20:

Theorem D.21 *The essential dimension of a finite group is greater than or equal to the essential dimension of each of its subgroups.*

Definition D.22 The rank $r(G)$ of a finitely generated abelian group G is the smallest number of its generators.

The group S_n contains a subgroup isomorphic to $\mathbb{Z}_2^{\lfloor n/2 \rfloor}$, namely the group $\langle (1,2), (3,4), \ldots, (2\lfloor n/2 \rfloor - 1, 2\lfloor n/2 \rfloor) \rangle$. The bound $\lfloor n/2 \rfloor$ on the essential dimension of the group S_n follows then from Theorem D.21 and the following result.

Theorem D.23 *The essential dimension of a finite abelian group is equal to its rank.*

In the same way, the bound $2\lfloor n/4 \rfloor$ on the essential dimension of the group A_n can be deduced from the fact that the group A_n contains the subgroup

$$\langle (1,2)(3,4), (1,3)(2,4), (5,6)(7,8), (5,7)(6,8), \ldots \rangle,$$

which is isomorphic to the group $Z_2^{2\lfloor n/4 \rfloor}$ of rank $2\lfloor n/4 \rfloor$.

The branch of mathematics dealing with the computation of essential dimensions of finite groups and other algebraic objects has been developing rapidly in recent years. For instance, Karpenko and Merkurjev recently computed the essential dimension of p-groups [44]:

Theorem D.24 *The essential dimension of a p-group G is equal to the smallest dimension of a faithful linear representation of G.*

Theorem D.18, of Buhler and Reichstein, was later proved again by Serre in [18]. His proof uses an algebraic version of characteristic Stiefel–Whitney classes introduced by Delzant. These characteristic classes can be applied to finite extensions of fields, and their values lie in Galois cohomology. Serre showed that for the extension related to the general algebraic function of degree n, the corresponding Stiefel–Whitney class does not vanish in dimension $\lfloor n/2 \rfloor$.

D.3.3 A Topological Approach to Klein's Problem

Let $\pi : Y \to X$ be a covering (possibly not connected) over a topological space X. Its topological essential dimension k is the minimal dimension of a CW-complex over which there exists a covering such that the covering $\pi : Y \to X$ can be induced from it by means of a continuous mapping.

Let T^n denote a real n-dimensional torus and let $\pi : R \to T^n$ be a covering over it (possibly not connected). Its monodromy group G is a finitely generated abelian group, since the fundamental group of the torus T^n is a free abelian group on n generators.

Theorem D.25 *The topological essential dimension of the covering $\pi : R \to T^n$ is equal to the rank $r(G)$ of its monodromy group.*

In one direction, this theorem is proved by an explicit construction: the covering π can be induced from a restriction of π to a subtorus of dimension $r(G)$.

The proof of the other direction uses characteristic classes for covering with a fixed abelian monodromy group G. More precisely, one can prove that if the prime number p is such that $r(G \otimes_Z Z_p) = r(G)$, then there exists a characteristic class with coefficients $\mathbb{Z}_p$ of dimension $r(G)$ that does not vanish for the covering π (see [20]).

This result allows us to prove Theorems D.18 and D.23 using clear topological arguments, and also to prove the following result.

Theorem D.26 *A generic m-valued algebraic function of n variables with $m \geq 2n$ cannot be reduced to an algebraic function of fewer than n variables by means of a rational change of variables.*

The proofs of Theorems D.18, D.23, and D.26 based on Theorem D.25 follow the same line of argument. One exhibits a family of tori in the complement of the discriminant set of the algebraic function having the property that the algebraic function defines a covering of a high enough topological essential dimension over each of them (for instance, for the general algebraic function, this is a family of tori over which the algebraic function defines a covering of topological essential dimension $\lfloor n/2 \rfloor$). This family is chosen so that it has the following property: for every algebraic subvariety Σ in the space of coefficients, there exists a torus from the family lying in the complement of Σ.

These arguments allow one to prove again the bounds in Theorems D.18 and D.23 on the essential dimension of groups S_n, A_n, and finite abelian groups, but they do not allow us to prove the results of Theorems D.16, D.17, and D.24. The structure of these arguments, however, is reminiscent of Arnold's proof of the bound $\phi(n)$ in the polynomial version of Klein's problem.

D.4 Arnold's Proof and Further Developments in Klein's Problem

As we saw above, Arnold's proof of his theorem has influenced the development of many areas of mathematics. However, the original result on Klein's problem was left in the shadows. The reason for this lies in the fact that much stronger results have been obtained in the polynomial version of Klein's problem [73], while it seems that Arnold's results are inapplicable to the rational version of Klein's problem. Indeed, the topology of the complement to the discriminant set changes drastically when one adds to the discriminant the set on which some rational mappings are not defined. Many cohomology classes that Arnold used in his paper do not survive such changes. However, as Burda has shown, one can find families of cycles in the complement of the discriminant that do survive such changes [20]. Geometrically, these cycles are small tori in a small neighborhood of chains of strata of the discriminant. For the general algebraic function of degree n, one can take the tori that correspond to the first $\lfloor n/2 \rfloor$ elements of the chain of branching sets from Sect. D.2.3, i.e., the chain of strata of the discriminant that correspond to loci at which several pairs of roots coincide. However, if this chain is continued further, then the cycle that corresponds to it in the sense of Sect. D.2.3 does not survive the removal of some hypersurfaces from the space of parameters.

References

1. V.B. Alekseev, *Abel's Theorem in Problems and Solutions, Based on the Lectures of Professor V.I. Arnold* (Kluwer Academic, Dordrecht, 2004)
2. V.I. Arnold, Algebraic unsolvability of the problem of Lyapunov stability and the problem of the topological classification of the singular points of an analytic system of differential equations. Funct. Anal. Appl. **4**, 173–180 (1970)
3. V.I. Arnold, Topological invariants of algebraic functions. II. Funct. Anal. Appl. **4**, 91–98 (1970)
4. V.I. Arnold, Superpositions, in *Collected Works, Mathematics and Mechanics*, ed. by A.N. Kolmogorov (Nauka, Moscow, 1985), pp. 444–451
5. V.I. Arnold, Topological proof of the transcendence of the abelian integrals in Newton's *Principia*. Istoriko-matematicheskie issledovaniya **31**, 7–17 (1989)
6. V.I. Arnold, Problèmes résolubles et problèmes irrésolubles analytiques et géométriques, in *Passion des Formes. Dynamique Qualitative Sémiophysique et Intelligibilité*, Dédié à R. Thom. Fontenay-St Cloud: ENS Éditions, 1994, pp. 411–417
7. V.I. Arnold, On some problems in the theory of dynamical systems. Topol. Methods Nonlinear Anal. **4**, 209–225 (1994)
8. V.I. Arnold, O.A. Oleinik, Topology of real algebraic varieties. Vestnik MGU, Ser. 1, Matem, Mekhan. **6**, 7–17 (1979)
9. V.I. Arnold, I.G. Petrovskii, Hilbert's topological problems, and modern mathematics. Russ. Math. Surv. **57**, 833–845 (2002)
10. V.I. Arnold, V.A. Vassiliev, Newton's *Principia* read 300 years later. Not. Am. Math. Soc. **36**, 1148–1154 (1989). Addendum: Not. Am. Math. Soc. **37**, 144 (1990)
11. F. Baldassarri, B. Dwork, On second order linear differential equations with algebraic solutions. Am. J. Math. **101**, 42–76 (1979)
12. M. Berger, *Geometry*, translated from French by M. Cole and S. Levy. Universitext (Springer, Berlin, 1987)
13. A.A. Bolibruch, *Fuchsian Differential Equations and Holomorphic Vector Bundles* (MCCME, Moscow, 2000)
14. A.A. Bolibruch, Inverse monodromy problems in analytic theory of differential equations, in *Matematicheskie sobytiya XX veka* (Fazis, Moscow, 2004), pp. 53–79
15. M. Bronstein, Integration of elementary functions. J. Symbolic Comput. **9**, 117–173 (1990)
16. M. Bronstein, *Symbolic Integration I: Transcendental Functions*. Algorithms and Computation in Mathematics, vol. 1, 2nd edn. (Springer, Berlin/New York, 2010)
17. J. Buhler, Z. Reichstein, On the essential dimension of a finite group. Compos. Math. **106**, 159–179 (1997)

A. Khovanskii, *Topological Galois Theory*, Springer Monographs in Mathematics,
DOI 10.1007/978-3-642-38871-2

18. J. Buhler, Z. Reichstein, On Tschirnhaus transformations, in *Topics in Number Theory*, ed. by S.D. Ahlgren et al. (Kluwer Academic, Dordrecht/Bosto 1999), pp. 127–142
19. Y. Burda, Around rational functions invertible in radicals (2010). arXiv:1005:4101
20. Y. Burda, Coverings over tori and topological approach to Klein's resolvent problem. Transform. Groups **17**, 921–951 (2012)
21. Y. Burda, L. Kadets, Construction of the Heptadecagon and quadratic reciprocity. Math. Rep. Acad. Sci. (R. Soc. Can.) **35**(1), 16–21 (2013)
22. Y. Burda, A. Khovanskii, Polynomials invertible in k-radicals (2012). arXiv:1209.5137
23. Y. Burda, A. Khovanskii, Signatures of branched coverings and solvability in quadratures. Mosc. Math. J. **14**, 225–237 (2014)
24. N.G. Chebotarev, *Basic Galois Theory 1* (URSS, Moskva, 1934)
25. N.G. Chebotarev, *Basic Galois Theory 2* (URSS, Moskva, 1937)
26. N.G. Chebotarev, *Theory of Algebraic Functions* (URSS, Moscow, 2007)
27. J.H. Davenport, *On the Integration of Algebraic Functions* (Springer, Berlin/New York, 1981)
28. A. Duncan, Essential dimensions of A_7 and S_7. Math. Res. Lett. **17**, 263–266 (2010)
29. A.È. Eremenko, Some functional equations connected with iteration of rational functions. Algebra Anal. **1**, 102–116 (1989) (Russian); English translation: Leningrad Math. J. **1**, 905–919 (1990)
30. P. Fatou, Sur l'itértation analytique et les substitutions permutables. J. Math. Pure Appl. **23**, 1–49 (1924)
31. W. Feit, Some consequences of the classification of finite simple groups, in *The Santa Cruz Conference on Finite Groups (University of California, Santa Cruz, 1979)*. Proceedings of Symposia in Pure Mathematics, vol. 37 (American Mathematical Society, Providence, 1980), pp. 175–181
32. M. Fried, On a conjecture of Schur. Michgan Math. J. **17**, 41–55 (1970)
33. B.A. Fuchs, *Introduction to the Theory of Analytic Functions of Several Complex Variables* (Fizmatgiz, Moscow, 1962)
34. D.B. Fuchs, A.T. Fomenko, *A Course in Homotopic Topology* (Nauka, Moscow, 1989); English translation of a preliminary edition: A.T. Fomenko, D.B. Fuchs, V.L. Gutenmakher, *Homotopic Topology* (Akademiai Kiado, Budapest, 1986)
35. M. Goreesky, R. MacPherson, *Stratified Morse Theory* (Springer, Berlin/New York, 1988)
36. A. Hurwitz, R. Courant, *Vorlesungen über allgemeine Funktionentheorie und elliptische Funktionen* (Springer, Berlin, 1922)
37. A. Hurwitz, R. Courant, *Funktionentheorie*, 4th edn. (Springer, Berlin/Heidelberg/New York, 1964)
38. V.V. Golubev, *Lectures on the Analytic Theory of Differential Equations*, 2nd edn. (Izd. tekhniko-teor. lit., Moscow–Leningrad, 1950)
39. Yu.S. Ilyashenko, A.G. Khovanskii, *Galois Theory of Systems of Fuchs-Type Differential Equations with Small Coefficients*. Preprint IPM AS USSR, Moscow, 117, 1974
40. E.L. Ince, *Ordinary Differential Equations* (Dover, New York, 1944)
41. V.A. Iskovskih, Minimal models of rational surfaces over arbitrary fields. Math. USSR-Izv. **14**, 17–39 (1980)
42. G. Julia, Mémoire sur la permutabilité des fractions rationelles. Ann. Sci. Ec. Super. **39**, 131–152 (1922)
43. I. Kaplansky, *An Introduction to Differential Algebra* (Hermann, Paris, 1976)
44. N.A. Karpenko, A.S. Merkurjev, The essential dimension of finite p-groups. Invent. Math. **172**, 491–508 (2008)
45. A.G. Khovanskii, The representability of algebroidal functions by superpositions of analytic functions and of algebroidal functions of one variable. Funkts. Analiz **4**(190), 74–79; translation in Funct. Anal. Appl. **4**, 152–156 (1970)
46. A.G. Khovanskii, Superpositions of holomorphic functions with radicals. Uspehi Matem. Nauk **26**, 213–214 (1971). (Russ. Math. Surv.)
47. A.G. Khovanskii, On representability of functions in quadratures. Uspehi Matem. Nauk **26**, 251–252 (1971). (Russ. Math. Surv.)

48. A.G. Khovanskii, On representability of functions by quadratures. Ph.D. thesis, Steklov Mathematical Institute, Moscow, 1973
49. A.G. Khovanskii, *Fewnomials*. Translations of Mathematical Monographs, vol. 88 (American Mathematical Society, Providence, 1991)
50. A.G. Khovanskii, Topological obstructions for representability of functions by quadratures. J. Dyn. Control Syst. **1**, 99–132 (1995)
51. A.G. Khovanskii, On the continuability of multivalued analytic functions to an analytic subset. Funkts. Analiz **35**, 62–73 (2001); translation in Funct. Anal. Appl. **35**, 51–59 (2001)
52. A.G. Khovanskii, On the monodromy of a multivalued function on its ramification set. Funkts. Analiz **37**, 65–74 (2003); translation in Funct. Anal. Appl. **37**, 134–141 (2003)
53. A.G. Khovanskii, Multidimensional results on the nonrepresentability of functions by quadratures. Funkts. Analiz **37**, 74–85 (2003); translation in Funct. Anal. Appl. **37**, 302–310 (2003)
54. A.G. Khovanskii, On solvability and unsolvability of equations in explicit form. Uspehi Matem. Nauk **59**, 69–146 (2004); translation in Russ. Math. Surv. **59**, 661–736 (2004)
55. A.G. Khovanskii, Variations on solvability by radicals. Trudy Mat. Inst. Steklova **259**, 86–105 (2007) (in Russian); English translation: Proc. Steklov Inst. Math. **259**, 82–100 (2007)
56. A.G. Khovanskii, *Galois Theory, Coverings and Riemann Surfaces* (MCCME, Moscow, 2007); English translation (Springer, 2013)
57. A. Khovanskii, Ruler and compass constructions. Matematicheskoe Prosveshenie Ser. 3 **17**, 42–60 (2013) (in Russian)
58. A. Khovanskii, Chebyshev polynomials and their inverses. Matematicheskoe Prosveshenie Ser. 3 **17**, 93–106 (2013) (in Russian)
59. A.G. Khovanskii, On algebraic functions integrable in finite terms. Funct. Anal. Appl. **49**(1) (2015, to appear)
60. A.G. Khovanskii, *Interpolation Polynomials and Their Applications in Pure Mathematics*. To be publihed by MCCME, Moscow
61. A.G. Khovanskii, S.P. Chulkov, *Geometry of the Semigroup* $\mathbb{Z}^n_{\geq 0}$. Applications to Combinatorics, Algebra and Differential Equations (MCCME, Moscow, 2006)
62. A.G. Khovanskii, O.A. Gelfond, Real Liouville functions. Funkts. Analiz **14**, 52–53 (1980); translation in Funct. Anal. Appl. **14**, 122–123 (1980)
63. F. Klein, *Lectures on the Icosahedron and the Solution of Equations of the Fifth Degree* (Dover, New York, 1956)
64. E.R. Kolchin, Algebraic matrix groups and the Picard–Vessiot theory of homogeneous linear ordinary differential equations. Ann. Math. **49**, 1–42 (1948)
65. E.R. Kolchin, Galois theory of differential fields. Am. J. Math. **75**, 753–824 (1953)
66. E.R. Kolchin, *Differential Algebra and Algebraic Groups*. Pure and Applied Mathematics, vol. 54 (Academic, New York, 1973)
67. E.R. Kolchin, in *Selected Works of Ellis Kolchin with Commentary (Collected Works)*, ed. by H. Bass, A. Buium, P.J. Cassidy (American Mathematical Society, Providence, 1999)
68. A.G. Kurosh, *Lectures on Abstract Algebra* (FM, Moscow, 1962)
69. I.A. Lappo-Danilevsky, *Applications of Matrix Functions to the Theory of Linear Systems of Ordinary Differential Equations* (GITTL, Moscow, 1957)
70. S. Lattès, Sur l'itération des substitutions rationelles et les fonctions de Poincaré. C. R. Acad. Sci. Paris **166**, 26–28 (1918)
71. V.P. Leksin, Riemann–Hilbert problem for analytic families of representations. Math. Notes **50**, 832–838 (1991)
72. V.Ya. Lin, On Superpositions of algebraic functions. Funct. Anal. Appl. **6**, 240–241 (1972)
73. V.Ya. Lin, Superpositions of algebraic functions. Funct. Anal. Appl. **10**, 32–38 (1976)
74. V.Ya. Lin, Algebraic functions, configuration spaces, Teichmüller spaces, and new holomorphically combinatorial invariants. Funct. Anal. Appl. **45**, 204–224 (2011)
75. J. Liouville, Sur la détermination des intégrales dont la valeur est algébrique. J. de l'École Poly. **14**, 124–148 (1833)

76. J. Liouville, Mémoire sur l'intégration d'une classe de fonctions transcendentes. J. Reine und Angewandte math. **13**, 93–118 (1835)
77. J. Liouville, Mémoire sur l'intégration d'une classe d'équations différentielles du second ordre en quantités finies explicites. Journal de mathématiques, pures et appliquees **IV**, 423–456 (1839)
78. Yu.I. Manin, Rational surfaces over perfect fields (in Russian). Inst. Hautes Etudes Sci. Publ. Math. **30**, 55–113 (1966)
79. J. Milnor, On Lattès Maps. Stony Brook IMS Preprint # 2004/01 (2004)
80. P. Muller, Primitive monodromy groups of polynomials, in *Recent Developments in the Inverse Galois Problem (Seattle, WA, 1993)*. Contemporary Mathematics, vol. 186 (American Mathematical Society, Providence, 1995), pp. 385–401
81. P. Orlik, H. Terao, *Arrangements of Hyperplanes*. Fundamental Principles of Mathematical Sciences, vol. 300 (Springer, Berlin, 1992)
82. V.V. Prasolov, Non-elemetarity of integrals of some elementary functions. Mat. prosveschenie, Ser. III 126–135 (2003)
83. R.H. Risch, The solution of the problem of integration in finite terms. Bull. Am. Math. Soc. **76**, 605–608 (1970)
84. J.F. Ritt, On algebraic functions which can be expressed in terms of radicals. Trans. Am. Math. Soc. **24**, 21–30 (1922)
85. J.F. Ritt, Permutable rational functions. Trans. Am. Math. Soc. **25**, 1–49 (1923)
86. J. Ritt, *Integration in Finite Terms. Liouville's Theory of Elementary Methods* (Columbia University Press, New York, 1948)
87. M. Rosenlicht, Liouville's theorem on functions with elementary integrals. Pac. J. Math. **24**, 153–161 (1968)
88. M. Rosenlicht, Integration in finite terms. Am. Math. Mon. **79**, 963–972 (1972)
89. M. Rosenlicht, On Liouville's theory elementary of functions. Pac. J. Math. **65**, 485–492 (1976)
90. J.-P. Serre, *Le groupe de Cremona et ses sous-groupes finis* (Séminaire Bourbaki, Paris, 2008)
91. I. Shur, Über den Zusammenhang zwischen einem Problem der Zahlentheorie und einem Satz über algebraische Funktionenpolynomials. Acta Arith. **12**, 289–299 (1966/1967)
92. M.F. Singer, Liouvillian solutions of nth order homogeneous linear differential equations. Am. J. Math. **103**, 661–682 (1981)
93. M.F. Singer, Formal solutions of differential equations. J. Symbolic Comput. **10**, 59–94 (1990)
94. M.F. Singer, Direct and inverse problems in differential Galois theory, in *Selected Works of Ellis Kolchin with Commentary (Collected Works)*, ed. by H. Bass, A. Buium, P.J. Cassidy (American Mathematical Society, Providence, 1999), pp. 527–554
95. F.V. Vainstein, The cohomology of braid groups. Funct. Anal. Appl. **12**, 135–137 (1978)
96. M. van den Put, M.F. Singer, *Galois Theory of Linear Differential Equations* (Springer, Berlin/New York, 2003)
97. V.A. Vassiliev, Cohomology of knot spaces, in *Theory of Singularities and Its Applications*. Advances in Soviet Mathematics, vol. 1 (American Mathematical Society, Providence, 1990), pp. 23–69
98. V.A. Vassiliev, Topology of discriminants and their complements, in *Proceedings of the International Congress of Mathematicians*, Zurich, 1994, vol. 1 (Birkhäuser, 1995), pp. 209–226
99. A.P. Veselov, Integrable mappings. UMN, T. 45 vyp. 5 (281), 3–45 (1991) (in Russian)
100. A.G. Vitushkin, Hilbert's thirteenth problem and related questions. Russ. Math. Surv. **59**, 11–25 (2004)
101. A.G. Vitushkin, G.M. Henkin, Linear superpositions of functions. Russ. Math. Surv. **22**, 77–125 (1967)

102. I.V. Vyugin, R.R. Gontsov, On the question of solubility of Fuchsian systems by quadratures. Uspekhi Mat. Nauk **67**, 183–184 (2012); English translation in Russ. Math. Surv. **67**, 585–587 (2012)
103. M. Zieve, P. Muller, On Ritt's polynomial decomposition theorems. arXiv:0807.3578v1 [math.AG], Jul7-2008

Index

A. Khovanskii, *Topological Galois Theory*, Springer Monographs in Mathematics,
DOI 10.1007/978-3-642-38871-2

MIX
Papier aus verantwortungsvollen Quellen
Paper from responsible sources
FSC® C105338

If you have any concerns about our products,
you can contact us on
ProductSafety@springernature.com

In case Publisher is established outside the EU,
the EU authorized representative is:
Springer Nature Customer Service Center GmbH
Europaplatz 3, 69115 Heidelberg, Germany

Printed by Libri Plureos GmbH
in Hamburg, Germany